AF758179

Constructions of Time and History in the Pre-Columbian Andes

Constructions of Time and History in the Pre-Columbian Andes

Edited by Edward Swenson and Andrew P. Roddick

UNIVERSITY PRESS OF COLORADO
Boulder

© 2018 by University Press of Colorado

Published by University Press of Colorado
245 Century Circle, Suite 202
Louisville, Colorado 80027

All rights reserved
Printed in the United States of America

The University Press of Colorado is a proud member
of the Association of University Presses.

The University Press of Colorado is a cooperative publishing enterprise supported, in part, by Adams State University, Colorado State University, Fort Lewis College, Metropolitan State University of Denver, Regis University, University of Colorado, University of Northern Colorado, Utah State University, and Western State Colorado University.

∞ This paper meets the requirements of the ANSI/NISO Z39.48-1992 (Permanence of Paper).

ISBN: 978-1-60732-641-0 (cloth)
ISBN: 978-1-60732-642-7 (ebook)
DOI: https://doi.org/10.5876/9781607326427

Library of Congress Cataloging-in-Publication Data

Names: Swenson, Edward, editor. | Roddick, Andrew P., editor.
Title: Constructions of time and history in the pre-Columbian Andes / edited by Edward Swenson and Andrew Roddick.
Description: Boulder : University Press of Colorado, [2017] | Includes bibliographical references and index.
Identifiers: LCCN 2017030888 | ISBN 9781607326410 (cloth) | ISBN 9781607326427 (ebook)
Subjects: LCSH: Indians of South America—Andes Region—Antiquities. | Time—Social aspects—Andes Region—History. | Memory—Social aspects—Andes Region—History. | Archaeology and history—Andes Region.
Classification: LCC F2229 .C664 2017 | DDC 980/.01—dc23
LC record available at https://lccn.loc.gov/2017030888

Cover illustration: "El sistema de seques del Cusco" © 2005 by Miguel Aráoz Cartagena

Contents

Figures

Tables

Acknowledgments

Archaeologists have long held that they secure a privileged position in explaining long-term historical change. Despite this overarching concern with chronology and process, archaeologists have only recently considered indigenous temporalities and modes of historical production. This book springs from a Theoretical Archaeological Group symposium (Chicago, 2013) organized to explore how time, the past, and memory were conceptualized by the ancient people of the Americas. In turn we sought to compare indigenous philosophies of time with contemporary archaeological approaches to theorizing and writing history. Entitled Envisioning Time and Imagining Place in pre-Columbian Landscapes, the symposium involved the participation of number of scholars working throughout the Americas. Although this book focuses exclusively on the Andes, the volume benefited from the exchange of ideas and compelling arguments made by conference participants. We thus extend our sincere thanks to Jonah Augustine, Katrina Burch Joosten, John Creese, Nicole Couture, Dave Haskell, Christopher Stawski, Alexi Vranich, and Mary Weismantel. Although they did not originally present papers at the TAG conference, we are also grateful that

Zach Chase, María José Culquichicón-Venegas, Francisco Seoane, and Matt Sayre accepted our invitation to submit chapters to the volume. Furthermore, we express our gratitude to Tamara Bray for agreeing to write the concluding chapter in a short period of time. We also extend our deepest thanks to the acquisitions editor at the University Press of Colorado, Jessica d'Arbonne, for all her hard work in ensuring this volume would come to print. Despite many delays and setbacks, Jessica was always patient, encouraging, and extremely helpful. Of course, we appreciate all the support of the University Press of Colorado staff. Finally, we thank the anonymous reviewers of the chapters for their insights and criticisms, which helped improve the final product.

Constructions of Time and History in the Pre-Columbian Andes

ONE

Introduction

Rethinking Temporality and Historicity from the Perspective of Andean Archaeology

EDWARD SWENSON AND ANDREW P. RODDICK

For many Andean archaeologists questions of time are limited to chronological questions, rather than indigenous temporalities or conceptions of time (but see Dillehay 2004; Hocquenghem 2008; Roddick 2013; Weismantel 2004). Scholars have interpreted social change, as expressed in shifts in material styles or settlement patterns, as a strictly etic problem, separate from how past communities experienced time's passage, understood historical process, or ritually constructed social memory. However, sociopolitical transformation is often directly related to changes in temporal cycles and the ideological regulation of time itself, and the intersection of chronology and temporality demands consideration in Andean archaeology. If the tempo of culturally specific practices leaves distinct material signatures, then the formulation of both relative and absolute chronologies should be sensitive to lived temporal rhythms. In other words, stratigraphic and stylistic analyses should be designed not simply to demarcate chronological phases and social boundaries; they should be geared to exploring how past subjects actively created and managed time itself (Alcock 2002; Bailey 2007; Bradley 2002; Gosden 1994; Lucas 2005; Murray 1999; Olivier 2004).

DOI: 10.5876/9781607326427.c001

In this chapter, we outline why archaeologists have become increasingly critical of chronologically based models of historical change. We then briefly review influential theories on temporality and relate these perspectives to interpretations of time, historical consciousness, and memory in the Andes. Ultimately, the chapter evaluates innovative anthropological approaches to the complexity of time and its inextricable relationship to the construction and experience of Andean political landscapes. This discussion includes a critical reappraisal of the horizon chronological schema that has long been employed to both order and explain the culture and political history of the pre-Columbian Andes. The second half of the chapter explores some of the overarching commonalities in Andean historical consciousness and conceptions of time. Finally, we review the contributions of the volume, highlighting the diversity of Andean constructions of history, memory, and temporality.

ARCHAEOLOGIES OF TIME AND HISTORY

The Limitations of Chronology

Recently archaeologists have endeavored to uncover subjective experiences of time and "period making" as a complement to the traditional construction of regional and site-based chronologies (Lucas 2005). Of course, an investigation of past temporalities cannot simply replace traditional (often ceramic-based) relative chronologies, and the latter constitutes the inevitable starting point for analysis. However, archaeologists are devising new interpretive frameworks to move "beyond chronology" (see Lucas 2005: 1–32), and it has long been recognized that chronologies alone provide a poor means of explaining historical process (Adams 1979; Kubler 1970). As Lucas notes: the "restrictive conception of time in chronology" fails to disentangle time and historical transformation and has led to "impoverished interpretations of cultural change" in archaeology (Lucas 2005: 2) (see also Sayre [chapter 2], Roddick [chapter 3], this volume). As has been criticized by archaeologists and philosophers alike, standard chronological schemes reduce time to a uniform, linear phenomenon and by extension history is construed in terms of the homogenized passage of artificially bounded events (Gell 1992; McGlade 1999; Munn 1992; Murray 1999; Olivier 2004: 208). Indeed, chronologies in use in South American and elsewhere often resonate with social evolutionary models; time moves linearly and in directed fashion from one circumscribed stage to another. To quote Lucas once again (Lucas 2005: 27, 52): "Cultural evolution, as articulated in terms of a social typology and stages of historical change, reproduces the same basic temporal structure as chronology. . . . Most relative chronologies reflect temporal narratives of progress" (see also Olivier 2011; Roddick this volume [chapter 3]).

The major problem with stand-alone chronologies is that successive events are plotted in linear sequence with little attention to their duration or the retentions of such actions on later social practices (Husserl 1966). Of course, chronology, as an exemplar of John McTaggart's B series of time, is an indispensable tool for representing the abstraction of time's phenomenal passage (McTaggart's so-called A series) (see Gell 1992; Gosden 1994); however, when employed as historical explanation, it inevitably effaces the culturally specific experience and sensual flux of time itself (Munn 1992). In other words, chronology must be expanded to encompass different modalities and durations of time if archaeologists wish to reconstruct both unique histories and temporalities (Bailey 2007; Lucas 2005). In a similar light, anthropologists increasingly recognize that objects and landscapes (and archaeological remains in general) defy the successive event-based timescales of conventional historiography. The physical persistence of the past in the present and the efficacy of buildings and accumulated artifacts to direct future action reveal that "archaeological time" is dynamic and "pluritemporal," one of diverse retentions, protentions, and discontinuities (Swenson 2017; see also Dawdy 2010; González-Ruibal 2014; Gosden 1994; Olivier 2011).

Andean archaeologists have recently recognized such problems when dealing with standard chronologies, in particular the uncritical conflation of ceramic seriation with historical process. For instance, Rafael Larco's sequence of five phases is still commonly used as a general benchmark on the North Coast of Peru. Yet the considerable diversity that characterized ceramic industries in different valleys of the North Coast suggests it only has limited applicability in the larger region (Bawden 1996: 193–95; Quilter and Castillo 2010; Kaulicke 1991; Koons and Alex 2014). In some cases, stylistic types were likely related to functional differences (such as vessels used in funerary contexts as opposed to other ritual or political events) rather than to separate temporal phases (Bawden 1996: 194–95; and see Janusek and Alconini 1994 for a similar argument for Tiwanaku phases in the Titicaca Basin). Bawden notes that the Larco sequence, which highlights systematic difference between the various phases, tacitly supports interpretations of Moche sociopolitical dynamics as having been shaped by abrupt and pervasive change (where history is viewed as discontinuous; see Stone-Miller 1993: 22–25). That is to say, the sequence implicitly reifies each phase as a static sociopolitical phenomenon that suddenly and systematically transforms into the next, relatively static social formation of the following period. The typology leaves little room for interpreting the actual meaning of stylistic differences or how they might relate to both historical transformations and possible shifts in the spatiotemporal modalities of social practice.

Moreover, it is now evident that styles fashionable in an earlier phase in a particular valley gained popularity in other regions during later periods. In other words, the various "phases" existed at the same time in different areas (Bawden 1996: 196; Chapdelaine 2011; Donnan 2009; Koons and Alex 2014; Millaire 2009), with various incongruities between supposed temporal phases and artifact styles.

For instance, the Moche I molded ceramics from the tombs at Dos Cabezas appear to date to the Middle Moche Period (AD 400–600) (Donnan 2001: 59; 2007: 197–98). More recent radiometric analysis suggests that the Moche III assemblage was the first to emerge and predated the Moche I and II styles, a tradition that now seems to have been largely confined to the Chicama and Jequetepeque regions (Koons and Alex 2014). Ceramics diagnostic of the Moche IV style are also conspicuously absent in Jequetepeque and other valleys north of Chicama. Chronological work with Tiwanaku ceramics has presented similar kinds of problems. The pioneering work of C. Ponce Sanginés (1981) defined a sequence of five cultural phases, which have served to frame conventional temporal narratives in the Lake Titicaca Basin (Burkholder 1997; Isbell 2013: 169–70; Janusek 2003; Marsh 2012). The Tiwanaku I, II, and III phases (today commonly referred to as Late Formative I and II), however, remain problematic. While settlement surveys have been conducted for the region based on Ponce's Tiwanaku-centered timeline, it is now clear that there were many more localized elements that do not fit into this scheme (Roddick 2009, this volume [chapter 3]). Furthermore, key phases in Ponce's scheme, such as Tiwanaku III, now appear to be limited to the urban center itself.

As archaeologists produce more radiometric dates and fine-grained stylistic analyses, they continue to encounter such confusing spatial and temporal overlaps. For instance, Owen and Goldstein's work in the middle Osmore drainage, near Moquegua, has explored the relationship between the Chen Chen and Omo styles of ceramics (Goldstein and Owen 2001; Owen 2001, 2005). Once thought to be a chronological sequence demonstrating Tiwanaku colonization in the region, their stylistic analyses, radiocarbon dates, and stratigraphic analyses suggest that rather than the Omo style marking an earlier time, Omo style sites and Chen Chen sites were actually occupied at the same time. As Owen stresses, such chronological refinements can change "the whole nature of the Tiwanaku occupation of the Osmore, and raises new possibilities for understanding Tiwanaku itself" (Owen 2001: 2). Similarly, Claude Chapdelaine's (2000, 2001, 2011: 196) research in the urban sector of Cerro Blanco on the North Coast has found that the Moche V pottery styles of Galindo emerged concurrently with later manifestations of the Moche IV style at Huacas de Moche (see also Lockard 2009). In response to these finding, some have argued for the formulation of valley-specific ceramic sequences for proper understanding of local manifestations of Moche history (Castillo and Donnan 1994; Chapdelaine 2011: 195–97).

The differential retentions of specific styles and other corporate traditions demonstrate the urgent need to consider Andean temporal practices and conceptions of history if we wish to improve our understanding of the diverse and changing political landscape across the Andes. As Chapdelaine (2009) notes, the continuation of Moche IV ceramic styles at Huaca de La Luna, well into the

Middle Horizon, suggests that Huacas de Moche became defined by an uncompromising "conservativism" that may have translated to the maintenance of traditional political forms (but see Uceda 2010). This emphasis on tradition and continuity no doubt speaks to a specific ideology of history and memory (González-Ruibal 2014). Christopher Donnan (2009) has raised a similar point in invoking the "Gallinazo illusion" and the failure of archaeologists in general to recognize that the long-term retention of utilitarian wares should not necessarily be confused with cultural continuity or the ascription of essentialized ethnic labels to past archaeological cultures (see also Millaire 2009).

Of course, the mobilization of ceramic-based chronologies to explain historical process has been the subject of sustained critiqued in the Andes and beyond. George Kubler (1970: 133) long ago questioned whether ceramics are effective indexes of periods and meaningful historical transformation. He showed that major shifts in fine pottery in Classical Greece had nothing to do with political, religious, or cultural change (see also Adams 1979; and Stone-Miller 1993: 18, 26). In contrast, the establishment of the Choson Dynasty in Korea (1392–1911) resulted in new sumptuary laws, and new perspectives on fine celadon wares, which ultimately led to the abrupt cessation of the production of these vessels, which were the hallmark of the preceding Koryo Dynasty (AD 918–1392) (Finlay 2010: 181–83). In a more bottom-up perspective, Olivier Gosselain (2015) traces a recent male potting tradition in Niger, drawing out the wide range of social, economic, and political processes that contribute to what archaeologically emerges as a recognizable tradition. Andrew Roddick and Christine Hastorf have similarly argued that we explore the historic processes and memory work of the Formative Period in the Lake Titicaca Basin, calling attention to the historical dynamics and active strategies of maintenance behind traditions that are too often simply explained away (Roddick and Hastorf 2010; see also Lightfoot 2001 and Bray, this volume [chapter 9]).

Time beyond the Horizon

The same problems undermining the standard Moche and Tiwanaku chronologies have beset traditional applications of the horizon model that has long ordered cultures in space and time in Andean archaeology (Isbell and Silverman 2006; Menzel 1964; Rice 1993; Rowe 1956; Stone-Miller 1993; Willey 1951). A creation of the "cultural-historical goals and efforts" of the early twentieth century, the Horizon/Intermediate Period schema is nonetheless still commonly employed to explain global historical processes in western South America (Rice 1993: 3).[1] Thus, the Early Horizon is correlated not simply with the widespread distribution of a singular material culture tradition ("a spatial continuum . . . of a recognizable art style" (Willey and Phillips 1958: 625), but it is often implicitly equated with a fixed block of time associated with the religious and political influence of the Chavín Cult (see Silverman 2004: 11–14; see also Sayre [chapter 2] this volume). In a similar manner, the imperial conquests of Wari or the religious-commercial allure of Tiwanaku are thought to explain the general homogeneity of iconography, political landscapes, and material culture defining the Middle Horizon (McEwan 2012). The horizons are distinguished from the intensified regional styles of the intermediate periods, interpreted as signaling political balkanization and cultural isolation.

Recently, David Beresford-Jones and Paul Heggarty have argued that the Wari Horizon corresponds to the rapid spread and imposition of Quechua in the south-central Andes (Beresford-Jones and Heggarty 2012). Based on tentative evidence (toponyms and the geographic distribution of linguistic groups), they suggest that the dissemination of Aymara may possibly correlate with the spread of the Chavín Cult during the Early Horizon.[2] Drawing parallels from the Roman Empire, Beresford-Jones and Heggarty contend that conquest and political incorporation of subject peoples better accounts for the adoption of Quechua by nonelites communities than would religious proselytization or long-distance trade. They also suggest that the horizon styles expanded relatively rapidly over a delimited period of time, thus validating the "core-periphery" spatial dynamic "implicit in the construction of horizons" (Rice 1993: 2). However, how and why political administration may have led to pervasive and long-lasting linguistic replacement and related shifts in ideology and material culture is never adequately explained. John Hyslop (1993) and Rebecca Stone-Miller (1993: 34) also caution that the Inca conquest did not lead to "monolithic stylistic takeover" (for a more culturally sensitive interpretation of the possible relationship between horizon, language, and material culture, see Urton 2012).

Problems notwithstanding, Beresford-Jones and Heggarty's linguistic and archaeological study, along with the research of many other archaeologists, proves the potential heuristic value of the horizon in making sense of Chavín, Wari, or Tiwanaku and interpreting their changing political relationship with other Andean polities (Isbell and Silverman 2006). For instance, Michelle Koons and Bridget Alex's systematic radiometric analysis strongly indicates that around AD 600 to 650 much of the North Coast witnessed major transformations in settlement patterns, temple construction, mural art, and the distribution of corporate wares, evidence that points to significant sociopolitical reconstitution at this time (Koons and Alex's 2014). Similar explosive changes appear to mark the Middle Horizon in the south-central Andes, though dates continue to be debated (Isbell and Knobloch 2008; Janusek 2003; Marsh 2012). Evidently, "abrupt" sociopolitical and religious realignments seem to define the first half of the seventh century and the early years of the Middle Horizon.

However, the horizon model fails as a stand-alone explanation of historical developments in the Andes. As Elizabeth Boone (1993: vii] noted more than twenty years ago: "Because of its simplicity and relative neutrality [the horizon concept] is a strong model for structuring the past; but like all universalizing structures, it hides cultural variability and the nuances of the archaeological record, and it does carry its own message." She continues that "the horizon concept is too broad and simple for the scholar, but useful for the student." Indeed, the Early and Middle Horizons defy reduction to unilinear chronological stages, thus countering Gordon Willey and Philip Phillips's assertion (Willey and Phillips 1955: 723) that "the horizon is characterized by its relatively limited time dimension and its significant geographic spread" (see also Rice 1993: 1). Of course, Willey (1945: 55) acknowledged that "horizon styles were not absolutely coeval in all parts of Peru" (cited in Rice 1993: 5), and the spans of time in which the material corpora indexed a horizon clearly differed from region to region in the Andes (see also Rowe 1962: 48). This temporal ambiguity is apparent in the competing timeframes proposed to delineate the Middle Horizon (originally 650–850 and recently expanded to 550–1000), evidence that points to the varied durations and effects of horizon-like material regimes in different regions of the Andes (Cook 2004: 158). In fact, research questions could productively explore how the spread and adoption of such material and spatial regimes led to possible shifts in historical consciousness and the temporalities of everyday practice. Therefore, an analysis of the varied persistence or transformation of definable "horizon markers" should provide a more nuanced appreciation of the scale, intensity, and meaning of interregional political, economic, and religious exchanges. For instance, it has been argued that maize production, feasting, and intensive agriculture expanded significantly during the Middle Horizon, and such developments may well have led to shifts in the experience and construction of temporal rhythms (see Swenson this volume [chapter 6], Roddick 2013). The latter could be conceivably interpreted in changes to household configurations, waste-management practices, irrigation agriculture, and the scheduling of festivals. Thus the presence of Middle Horizon ceramics and architectural forms at a particular settlement are meaningless unless properly contextualized vis-à-vis the structure of quotidian regimes of practice (e.g., continuity or change in the tempo of farming, funerary rites, household renovation). In turn, such an exercise would permit improved interpretations of the scope and nature of Wari influences outside of Ayacucho, and Tiwanaku outside of the Titicaca Basin.

In the end, archaeologists "need to add greater temporal complexity" to the investigation of archaeological deposits so to avoid reducing history to "chronological charts" or to phases and sequences (including horizons and intermediate periods) that provide little recourse to explanation (Lucas 2005: 38; see also Stone-Miller 1993, and Roddick this volume [chapter 3]). In this vein, archaeologists have recently sought inspiration from the Annales School of history to better decipher the complex interrelationship of social change, continuity, and long-term historical developments (Bintliff 1991; Bradley 1991; Gosden 1994; Knapp 1992; Swenson 2015). Proponents of Annales history recognize that social practices and the structures they reproduce occur at different rates of time. Following Fernand Braudel's (1970) classic model, Annales-inspired archaeologists commonly recognize at least three principal timescales in their analysis of historical change, a schema that transcends reductive linear models. They include the *longue durée*; the *moyenne durée* (conjunctures); and the event, or *événement*. In the Andean North Coast, structures of the longue durée (300–1,000 years) were most likely defined by culturally sanctioned labor relations, the technical imperatives of irrigation technology, belief in gender complementary, sociocosmic dualisms, sacrificial political theologies, the millennial trade of spondylus, and so forth. The moyenne durée (50–200 year periods) possibly encompassed generational conflicts between coastal and sierra polities, protracted environmental perturbations such as the sixth-century drought, poorly understood demographic cycles, competition between Moche polities, and the rise and fall of coastal dynasties, including Dos Cabezas, Huacas de Moche, and Pampa Grande. Short-term events—embedded in and shaped by varied middle- and long-term temporal structures—would include the burial of a priestess at San José de Moro, revitalization movements, the violent destruction of Pampa Grande, or a phase of architectural renovation at Huacas de Moche.

John Barrett (2004: 14) admonishes that "the distinction between process as cause (structuring) and event as consequence" is inherently problematic, "for processes must be generated through the working of events." Nevertheless, the Annales timeframes at least serves as a reminder of the complexity of past sociopolitical transformations as having been conditioned by temporally varied structures, landscapes, and social-ecological fluctuations. Although critical of Annales approaches, proponents of "Time Perspectivism" also recognize the vastly different temporalities (in terms of duration and effects) of distinct ecological and social processes (Bailey 2007; but see Harding 2005). Therefore, in order to understand historical change, say in the Moche context, we must grapple with the varied and polyrhythmic tempos of social practice on the North Coast of Peru. Ultimately, time and event are not separable phenomena; rather, collective tasks (that in reality defy reduction to bounded "events") are shaped by specific practices of "temporalization" (see Munn 1992).

In acknowledging the complexity of historical process as underwritten by different rates and scales of time, scholars often make a distinction between temporality and historicity. The temporal refers to duration, continuity, and cyclicality, while historicity pertains to change and conscious time-reckoning. In fact, a number of dichotomous interpretive frameworks have been devised that discuss time as either explicitly marked (history, memory, myth) or as habitually internalized in social practice. Variations on this dichotomy include

Michael Herzfeld's monumental and social time, Christopher Gosden's public and habitual time, Luicen Febvre's *temps mesuré* and *temps vécu*, Paul Connerton's notion of inscription and incorporation, and Alfred Gell's employment of John McTaggart's A and B series (see Febvre 1947: 471; Gell 1992; González-Ruibal 2014: 27; Gosden 1994: 124–26; Heidegger 1962: 374; Herzfeld 1991: 6–10; Ingold 2000; Joyce 2000; Last 1995; Lucas 2005; Munn 1992: 98; Rice 2008: 276; Roddick 2013). Archaeological research has demonstrated that habitual, iterative practices can be read from archaeological deposits, ultimately retracing the "taskscapes" of the past (Ingold 2000; Lucas 2005: Lucero 2003; Olsen 2010; Mills and Walker 2008; Roddick 2015). As a corollary, the varying media in which society memorialized past events and explicitly recorded their succession should be equally informative of specific temporal ideologies (see Wilkinson and D'Altroy this volume [chapter 4]).

This particular dichotomy resonates with theories that postulate the often unique temporal framework of ritual performance as distinct from everyday practice. For instance, Richard Bradley (1991: 212) contrasts mundane with ritual time arguing that "by observing the interplay of ritual and mundane time, we can practice a form of contextual archaeology, but one which makes a proper use of sequence" (see also Dillehay 2004). In a similar manner, Lisa Lucero (2003) contends that archaeological investigations of past historical change should focus on "purposeful deposits"—created by repetitive, formalized actions (see also Richards and Thomas 1984): "rather than just evaluating strata in terms of chronology, we can view them as reflecting sequences of (ritual) behaviours—more specifically, ritual replication, in which similar formal ritual activities took place in a variety of architectural contexts, from houses to palaces to temples" (Lucero 2003: 526). Social scientists have shown that ritual performance is integral to the making of time (Leach 1961; Rappaport 1992). Although ritual can create a place and time "out of time" (i.e., a liminal suspension of temporal routines), its chronometric faculty is evident in its capacity to delineate stages, phases, and series (Rappaport 1992: 6–10; Swenson 2017). In his influential work on social memory, Connerton (1989) makes the distinction between commemorative rites of history making ("inscription") and the more unconscious rhythms of what he calls incorporative practices of a routinized nature. Commemorative rituals conjure up and literally materialize the past as a conscious construct, and rites of this kind often performatively reenact past historical and mythic events, including, most notably cosmogonic narratives. It is in precisely through such ceremonies that history, identity, and community are ideologically constructed.

Nevertheless, temporalized practices often entail both unconscious and conscious behavior, and the temporal cannot be relegated simply to the realm of everyday behavior and the historical or eventful to the domain of ritual, myth, and ideology (Bradley 1991; Connerton 1989; Olsen 2010: 124–25). The periodicities of quotidian practices were often shaped by ritualized spectacles of history and place making, and they no doubt inculcated culturally specific understandings of time's passage (Dillehay 2004; Gosden 1994: 125; see Swenson 2017). Thus the evanescent, peak periods of public ceremonies could create a powerful sense of anticipation that motivated more prosaic tasks defining protracted intervals of "mundane" life (i.e., the production of *chicha* [corn beer] or the construction of a temple). The creation of such anticipation no doubt played a critical role in the construction of past political subjects. Furthermore, routinized practices—irrespective of their degree of synchronization with "public time"—can slowly sow the seeds of environmental, economic, and ideological change (Munn 1992: 102; see also Roddick, Bruno, and Hastorf 2014).

In light of our critique of the temporality/history dichotomy, archaeologists cannot simply presume that corporate fine-wares changed more rapidly than domestic pottery and can thus always be interpreted as transparent markers of "history," political culture, or uniform horizons (Donnan 2009; Millaire 2009; Roddick 2016; see Roddick this volume [chapter 3]). In other words, there is a tendency in archaeology to treat utilitarian ceramics as materialization of the unconscious or doxic mode, Heidegger's "ready at hand," while corporate ceramics are to be analyzed as indexes of the consciously ideological and historical, Heidegger's "present at hand." Alfredo González-Ruibal (2014) has suggested that the anthropological understanding of history *as transformation* betrays Neoliberal values, and he argues that the deep temporalities of certain things speak to the resiliency of particular structures of practice (see also McGlade 1999; Turner 1988). These practices were often the prerogative of women and subalterns, and González-Ruibal remarks that they should be interpreted as strategies of "timework" as equally significant as political revolutions or messianic movements (Flaherty 2011; see also Picazo 1997). In other words, there may be situations when domestic pottery was as actively politicized as corporate wares, and in some instances utilitarian ceramics underwent replacement as rapidly as elite styles, something Swenson has identified with certain cooking pots that disappeared at the end of the Late Moche Period in Jequetepeque (Swenson 2017).

Michel Trouillot's distinction between "historicity I" and "historicity II" provides an alternative to the temporality/history dichotomy. The former signifies the "materiality of the socio-historical process," while the second term refers to the "sociopolitical management" of this material process entailing its ordering, sequencing, and narration (Trouillot 1995: 29, 106; see also Chase [chapter 5] and Sayre [chapter 2], this volume). Historicity I—the actors, events, and accumulated practices that leave material traces—certainly limits the narratives (historicity II) that can be told about these happenings. However, the latter, as either an explicit ideology or as tailored by unconscious political dispositions, can distort, mystify, and inevitably "silence" certain aspects of past events and processes (hence, the title of Trouillot's famed book). The concept of Historicity II recognizes that the creation of narratives forms part of the sociopolitical process and is itself historically mediated (thus making "human beings doubly [fully] historical"; Trouillot 1995; 22–24). The heuristic thus acknowledges considerable cultural diversity in the selective making of history.

It deserves mention that the notion of historicity has a complex genealogy. Among different philosophers, ranging from Marxists to phenomenologists, it originally served as a critique of universalist theories and signified that all concepts, innovations, worldviews, and so forth were the product of distinct historical processes. In other words, no aspects of reality can be explained independently of a particular historical context (for a recent and thorough discussion of the concept, see Hertog 2015). Among anthropologists, the term is often conflated with historiography, epistemology, commemoration, and history itself (see Whitehead 2003). For instance, Neil Whitehead (2003: xi) argues: "Historicity . . . encompasses historiography, which is the culturally particular methodology of how the past may be written or otherwise expressed." This definition parallels Trouillot's emphasis on historicity II as subsuming "the means of historical production," which entails "four crucial moments": the making of sources, the assembly of facts in archives, the ordering of facts into narratives, and the "moment of retrospective significance . . . the making of history in the final instance" (Trouillot 1995: 26, 140). For Whitehead (2003: xi), "histories" refer to the actual texts, ritual performances, oral accounts, monuments, and so forth produced by historiographic agents. At the same time, he expands the definition of historicity to include "the cultural proclivities that lead to certain kinds of historical consciousness within which such histories are meaningful" (Whitehead 2003: xi). Thus, deeply ingrained ontological and epistemological orders of time—whether construed as cyclical, fatalistic, teleological, presentist, and so forth—will partly predetermine political constructions of history and historical knowing. Trouillot shows that early nineteenth-century Europeans were incapable of perceiving the Haitian Revolution as a true "revolution" for their "ready-made categories" on race, colonialism, human nature, and historical progression were "incompatible with the idea of a slave revolution" (Trouillot 1995: 73–74). This particular "cultural proclivity" of the early modern period can partly explain why professional historians and more influential purveyors of history (the press, heritage groups, politicians) continue to downplay the significance of the Haitian Revolution.

In this volume, we avoid ascribing absolute meanings to the above terms and recognize their heuristic utility in capturing the complexity of time and history as social and political creations.[3] Indeed, Trouillot formulated historicity I and II to move beyond reductive theories that history can only be understood as either factual events and processes (positivist approaches) or as invented stories about the past (constructivist perspectives). He writes: "My trust here is that too many conceptualizations of history tend to privilege one side of historicity over the other . . . this one-sidedness itself is possible because most theories of history are built without much attention to the process of production of specific historical narratives" (Trouillot 1995: 157). Trouillot recognized that this production is an exercise of power and is not something restricted to chronicles produced by the "guild" of professional historians. Instead, it commonly entails rituals, calendar rounds, performances, and the construction of meaningful places (Trouillot 1995: 58–66; 116–18). Hence, temporalities—routinized social practices and related phenological processes—cannot simply be relegated to historicity I (the materiality of the sociohistorical process), for they could at once influence or be the partial product of the epistemological ordering, ideological management, or silencing of the significance of past occurrences. Indeed, the production of history—whether in the form of monumental architecture, public commemorative rites, or the remaking of landscapes—can have direct consequences for how time is experienced and conceptualized.

In the end, Trouillot's dual concept of historicity can serve as a reminder that the naturalization of social rhythms could be the direct outcome of historiographic projects. In a similar manner, Henri Lefebvre develops the concept and "method" of "rhythmanalysis" to capture the deep-seated effects of capitalism on the minutiae of social life. He argues that "biological rhythms of sleep, hunger, thirst, walking, excretion and so on" have been largely determined by the commodification of time underwriting the scheduling of the workweek in capitalist societies (Lefebvre 2004: 6). He further contends that the temporalities of the Mediterranean World differed from North Atlantic Europe given variation in ecology, level of industrialization, and religion. Lefebvre argues that Mediterranean cultures better resisted the sublimation of their temporal *habitus* and circadian rhythms to the mechanized time regimes of capitalism. Much has been written on how the mechanical clock in the early modern period led to increased bodily discipline and to the tighter regulation of embodied routines (Thompson 1967). However, this was a process that met fierce resistance as indicated by the tenacious maintenance of irregular feast days and the popularity of "Saint Mondays" among the British artisanal class (Thompson 1967).

In light of the above discussion, the temporal and historical defy rigid dichotomization (Assmann 2006: 8; McGlade 1999); instead, archaeologists should be attentive to the degree in which they may have coincided in past cultures (see Swenson, this volume [chapter 6]). As mentioned above, an investigation of the intersection of distinct regimes of temporalized practices with culturally specific ideologies of history could provide a more a fruitful avenue to reassess the processes behind the patterns we identify as horizons in different regions of the Andes.

ANDEAN HISTORICAL CONSCIOUSNESS AND IDEOLOGIES OF TIME

Despite considerable diversity in premodern Andean representations of time ("Trouillot's historicity II"), it is worth identifying some general trends in Andean time reckoning and historical consciousness. This brief survey should facilitate comparison with temporal ideologies documented in other cultures but also provides a basic framework to interpret archaeologically recorded differences in Andean experiences of time explored in the following chapters. Commonalities in Andean constructions of time and history are evident in at least four domains of philosophy and practice: (1) (meta)physical/linguistic notions of time; (2) ideologies of duration and change (including the agents or movers of "history"); (3) calendrics and time measurement (delimitation of phases, and astronomical and seasonal revolutions); (4) and the politicoreligious institutions of memory, myth, divination, and archive.

(Meta)physical/Linguistic Notions of Time

In considering the first domain, Andean peoples generally recognized that space and time were inextricably bound, similar to the thinking of contemporary philosophers (see Aveni 2015a: 2; Lefebvre 2004; Nair 2015; Zuidema 2010: 209). For phenomenologists, space and time "come together in place . . . and arise in social action" (Adams 2007: 396). This (meta)physical understanding is captured in the oft-discussed Quechua concept of *pacha*, signifying at once "earth, time, world, and place." Frank Salomon and George Urioste define the term as "a moment or interval in time and a locus or extension in space" (Salomon and Urioste 1991: 14). In Inca mythology, *kay pacha* created by the Viracocha (or the sun) at Tiwanaku, refers to the present world, current temporal order, and the (middle) earth of the here and now (Cobo 1990).

As reflected in the polysemy of the linguistic term *pacha*, time in the Andes was (and in many cases still is) physically inscribed and remade in the ritual construction of place and landscape (Arnold 1992; Bouysse-Cassagne 1986). As Wilkinson and D'Altroy discuss in chapter 4, political-religious "monuments," including everything from the shrines of *wak'as* (*huacas*) to the extraordinary geocalendar of the Inca *zeq'e* (*ceque*) system, cannot be viewed as "memorials" of bygone events but were vital to the creation of the present and future. In supporting this argument, we are not simply asserting the oft-invoked archaeological abstraction that the physical persistence of the material world constrains and enables present actions—comparable to Oakeshott's "encapsulated past" (Oakeshott 1983; Olivier 2011). Instead, Andean people understood the past as inhering in and animating the here-and-now, and the storied agents of primeval times remained living social actors that largely determined the unfolding of future events (see Chase [chapter 5], Spence Morrow [chapter 7], and Seoane and Culquichicón [chapter 8] this volume).

Such thinking is immediately apparent in the authority of *malquis* and mummified Inca emperors, who were continually consulted by their followers. The wak'as of individual *ayllus* or *llactas* (a larger grouping of people that identified with a powerful wak'a) were celebrated as the movers of a primordial time of world creation (Salomon and Urioste 1991: 23). However, they usually remained active in the present affairs of their devotees, who fed, cared, and communicated with their wak'as (see Bray 2015). The power of these divinities was signaled by their ability to talk and prognosticate future events, while sacrificial offerings reciprocally ensured the material and spiritual well-being of their respective ayllus (Curatola Petrocchi 2008). The sustained influence of "ancestors" as stewards of etiological time has been documented in many premodern societies (Bloch 1971; Nyord 2013), but the Andean case is remarkable for the degree to which wak'as and culture heroes were involved in day-to-day social practices and decision making.

Ideologies of Duration and Change

Drawing from the work of Gosden and Gary Lock (Gosden and Lock 1998), Charles Cobb and Adam King note: "Many societies employ concepts of both genealogical history (where the past is created through links to ancestors) and mythical history (where a more distant past is evoked), and the ways in which specific groups articulate these histories relative to one another foster different notions of continuity and change" (Cobb and King 2005: 172).[4] In the Andean context, the mythical and genealogical were largely isomorphic, and the living landscape compressed the distance between the here-and-now, primordial origins, and generational reckonings of time (see Seoane and Culquichicón [chapter 8] this volume, Wilkinson and D'Altroy, this volume [chapter 4]). In his exploration of Quechua ontological categories of being, substance, and transformation, Salomon explains (Salomon 1998: 9, 10): "The huacas have in some contexts, individuality and properties, but in others they are seemingly imaged as long-term overarching sequences of phenomena or deeds . . . the accumulation of eventual being."

As architects of cosmogony, wak'as were the movers of history and social change (the second domain outlined above), and political figures, including Inca kings, drew inspiration from wak'as in their projects to remake the world (space-time) (Gose 1996; MacCormack 1991). Therefore, to write history was tantamount to creating place; the battles of deities in the Huarochirí manuscript carved out mountains and valleys, and the Inca "inscribed their place in history" by obsessively modifying, sculpting, renaming, and reconfiguring the natural and political landscape (Aveni 2015a: 5; Kosiba 2010; Nair 2015: 119–20; Swenson 2013). Wak'as were often associated with heroic ancestors who emerged from *paqarinas* (dawning places), usually caves and rivers that became important shrines. The lithification of ancestors created wak'as as sacred places, and rites and offerings ensured the continued flow of animating energies between divine progenitors and their mortal descendants (and thus between a spatially interconnected "past and present") (Bray 2015; Dean 2010).

This is not to argue that Eliade's "eternal return" (the constant ritual reenactment of cosmogony to ensure the continuation of world order) accurately captures Andean ideologies of historical origins and the perpetuation of space-time (see Seoane and Culquichicón this volume [chapter 8]). As calculating and animate social beings, the "creative" force of wak'as was far from predictable or repetitive but could lead to the complete reconfiguration of sociopolitical, geographic, and economic relations (Dulanto 2015: 157; Salomon 1998: 16; Salomon and Urioste 1991). Their acts did much more than revitalize the world by rebooting time or reactivating and circulating primal energies, and much would be lost in describing Andean temporal ideologies as simply "cold" (strategies that deny or neutralize change and discontinuity; but see Seoane and Culquichicón this volume [chapter 8]) (González-Ruibal 2014; Gose 1996; Hill 1988; Turner 1988). As both Wilkinson and D'Altroy (chapter 4) and Seoane and Culquichicón [chapter 8] discuss in their analysis of linguistic markers of time in Quechua and Aymara, the past and present were fused, but the pacha of the future was perceived as distant, precarious, and unpredictable.

The anxiety surrounding the uncertainty of future events is exemplified by the preeminence of oracles in Andean political history and religious philosophy (Curatola Petrocchi 2008; Gose 1996; MacCormack 1991). For instance, the *sapa* Inca would never embark on military campaigns without consulting the oracles, an act that entailed prodigious sacrifices, fasting, taboos, and back-up divinatory procedures (Curatola Petrocchi 2008; MacCormack 1991). The arts of divination in the Andes were varied and complex, and they offer a fascinating point of comparison with other premodern civilizations (see Bahrani 2008). Auguries were obtained from observing the flight of birds, the movement of spider legs, and the patterned arrangement of tossed coca leaves, and by soliciting the spoken prognostications of oracles (Curatola Petrocchi 2008; Huertas Vallejos 2008). Oneiromancy and extispicy were also highly developed, and the entrails of sacrificed guinea pigs and llamas were read to predict the future (Cobo 1990; Huertas Vallejos 2008; Hyland 2010; Mannheim 1987). Variants of divinatory dice games, such as *pichca*, widespread in the Andes at the time of the conquest, have even been identified in the iconography of the Moche (Gentile 2008; Salomon 2002).

It would be wrong simply to portray Andean ideologies of history as cyclical, and cultural understandings of "process" were highly complex and context-specific (Aveni 2015a: 4; Dulanto 2015; see Chase [chapter 5] and Seoane and Culquichicón [chapter 8] this volume). In their analysis of the Huarochirí manuscript, Salomon and Urioste (1991) describe Andean conceptions of history not so much as cyclical but as "metamorphic" (see also Seoane and Culquichicón this volume). The world was changed and remade by acts of feeding, growth, and the circulatory exchange of biological energies (Allen 1988; Cummins and Mannheim 2011: 7; Duviols 1973; Ramirez 2005; Salomon 1998: 10–11). Alternating movements between wet and soft to dry and hard states underwrote this metamorphic (and relational) ontology of time (Allen 1988, 1998). Notions of time and regeneration are more aptly described not so much as cyclical but as circulatory—a process that entailed the kinetic and transformative flow of energy, fructifying liquids, and material substances (Cummins and Mannheim 2011: 15; Lund Skar 1994; Weismantel 2004).

This circulation of life matter did not necessarily occur in a singular or predictable way, and the teleological space-time of wak'as and divine kings differed from the temporal trajectory of mortals (Gose 1993; Salomon 1998: 11). Thus the lithification of cultural heroes not only signaled a sense of permanency and power, but reflected the decelerated and enduring temporality of certain wak'as (Salomon 1998: 9; Salomon and Urioste 1991). The latent but fertilizing potency of desiccated *malquis* (a word denoting mummy and sapling in Quechua) served to draw water to their living communities, but mummy bundles remained responsive to the needs of its people only if properly fed and propitiated (Gose 1993). Scholars of Andean religion have shown the Andean theological concept of *camay* best encapsulates this sense of circulatory time; *camac* or *camaquen*, a vitalizing being (wak'as, mountains, constellations, divine progenitors) created, sustained, and reenergized their progeny on earth (Taylor 1974–76). However, camay theory affirms that this exchange of vital being was only possible through the reciprocal circulation of sacrificial offerings. The allocentric construction of space in Quechua (the position and identity of a given thing is determined relative to other objects, peoples, and places) further underscores the relational and circulatory contingencies of time in Andean thought (Mannheim 2015: 209).

The metamorphic basis of historical change was further expressed in "the inseparability of complementarity from conflict" in Andean concepts of process which Salomon and Urioste (1991: 10) describe as the "motor force in the mutability (what we would call the historicity) of west Andean society." Thus Andean understanding of social change might distantly compare with theories of the dialectic (contradictions and their resolution), the driver of history in Marxist social theory (see Swenson 2013). In the region of Huarochirí and elsewhere in the Andes, historical dynamics were expressed in terms of both violent conflicts and the marital union of deities. The nuptial alliances of different huacas, especially from different ecological zones, explained the changing geopolitical and economic fortunes of rival social groups (Duviols 1973; Swenson 2014; but see Chase [chapter 5], this volume). Of course, structuralists have long recognized that affinal relations serve as a metaphor of social reconstitution.[5] In contrast, divine sibling bonds in the Huarochirí region were expressive of continuity and more stable forms of social integration (see Salomon and Urioste 1991).

Billie Jean Isbell's (1982: 354) notion of "reversible dualism," similar to Salomon's (2002) analysis of "alternating dualism" (*dualismo alternante*) in the community of Pacota (Huarochirí) resonates with a mode of historicity based on a dialectic of conflict and complementarity (see also Gose 1996). History is propelled through the oscillation of polar opposites including the rising and setting of celestial bodies (critical to calendrical dating) and the pairing of ayllus, architectural complexes, ecological zones, and wak'as. Salomon similarly argues that present and future are articulated in terms of the alternation of opposed but complementarily apportioned spaces, people, and activities.[6] An intervallic ritual time (a "time out of time and a space out of space") allows for the physical union and subsequent passage from present to future (see also Rappaport 1992: 11–15). In other words, ideologies of time are largely congruent with philosophies of social space; *suyu, hanan, hurin* and the continuously divisible nesting of peoples and places are activated in the temporal sequencing of ritual practice, political offices, and productive activities (see also Abercrombie 1998; Arnold 1992; Bouysse-Cassagne 1986; Gelles 1995; Gose 1993, 1996; Harris 2000; Seoane and Culquichicón [chapter 8], this volume; Spence Morrow [chapter 7], this volume).

The famed Inca zeq'e (ceque) system of the Cuzco Basin operated on just such a principle of alternation and rotation (*mit'a*, or "round"). The zeq'e system consisted of 328 huacas arrayed on forty-one sight lines or processional waves, and these numerous shrines were analogous to knots tied onto *khipu* chords, the Inca recording device (see below; Bauer 1998; Cobo [1653] 1990; Zuidema 1964). The wak'as consisted of rock formations, hydraulic installations, astronomical landmarks, or other ritual constructions that were distributed in all four of the *suyus*, the gross provincial divisions of the Inca Empire that converged at the Temple of the Sun (*Qorikancha*). This remarkable complex consisted of an integrated agricultural and water shrine, a monument to Inca conquest and mythic history, a mnemonic device, a sidereal-lunar calendar, and a materialization of the social and ethnic identities of the circum-Cuzco region (Bauer 1998). The performance of rituals by set peoples at prescribed times and places created a vast, living calendar that served as a mythic charter for social action (Aveni 2015b; Zuidema 2010). The spatial arrangement of specific types of wak'as expressed principles of homology and hierarchy, and each individual shrine was organized in relation to a particular zeq'e line according to a tripartite schema of status based on degree of genealogical (and thus temporal) relatedness to the Inca king (*Collana*, primary kin; *Payan*, subsidiary kin, and *Callao*, no relationship to the ruler). (see Aveni 2015b; Zuidema 1964, 2010).

As mentioned, the entire complex radiated out from Cuzco, the sacred navel of the empire, but the individual wak'as were maintained by separate social groups (lineages or ayllus). Among the 328 wak'as arrayed around the forty-one zeq'e lines, only a few appear to have been ritually activated on any particular day of the agricultural year, and time was thus materialized and set into motion by rotating sacrifices orchestrated at individual shrines (Cobo 1990). The timing of rituals conducted at individual wak'as within the zeq'e complex appears to have moved both clockwise and counterclockwise (depending on their location within one of the four *suyus*) and outward toward the horizon as dictated by the passage of the stars, sun, and moon (Aveni 2015b).[7] Moreover, the scheduling of agriculture, infrastructural projects, communal work activities, and the religious liturgy was precisely timed by the geoastronomical measurements made possible by the zeq'e system (Aveni 2015b; Zuidema 2010). Communities in charge of a particular wak'a were thus associated not only with a distinct place, but a distinct periodicity, both in terms of mythic historical events and calendrical phases of ritual and work (days, weeks, months). As a consequence, pilgrimage to an activated zeq'e shrine was tantamount to time travel, affording not only an intensive experience of place but of an alternate moment in time. A striking exemplar of Mikhail Bakhtin's "chronotope" (Bakhtin 1981, see Roddick this volume [chapter 3]) the zeq'e is extraordinary for the degree to which it fused time with geography and social identity (Abercrombie 1998: 321). In a sense, it delivered the calendric system from the realm of abstraction, converting it into a living, pulsating, and sensual organism. Therefore, Inca zeq'e represented a radical project to both control "the means of historical production" and to regulate the routines, memories, and identities of subject populations (Trouillot 1995; see Rice 2008: 278).

Calendrics and Time Measurement

As the zeq'e system demonstrates, conceptions of process and historical agency in Andean thought can only be understood in the context of the long traditions of calendrics and astronomical time-reckoning in the Andes, the third domain outlined in the introduction to this section. Unsurprisingly, investigations of time in the Andean context have focused primarily on calendrical systems, and there is a great deal of excellent scholarship on this subject (see Aveni 2008, 2015a; Ghezzi and Ruggles 2007; Sakai 1998; Urton 1981; Urton and Aveni 1983; see also Seoane and Culquichicón this volume [chapter 8]). In recognizing the indivisibility of the spatial and temporal, time is equally inseparable from "motion" as implied not only by the word "process" but also in Salomon and Urioste's categorization of Andean conceptions of time as metamorphic and circulatory (see also Cummins and Mannheim 2011; Nair 2015: 128). This sense of motion is further captured by the rotation of peoples and activities within the Inca zeq'e complex, which in a sense activated time in place. A cultural universal, the movement of the sun, moon, stars, and planets has been charted to plot the passage of seasons and to regulate a myriad of social and ecological rhythms, including most notably farming, hunting, herding, warfare, and ritual life.

Archaeologists have identified the astral alignments and astronomical functions of Andean religious monuments dating as early as the Preceramic Period, and these temples have been interpreted as expressing an early concern with controlling the seasonal round and with scheduling rituals, agricultural activities, fishing expeditions, and other events (Benfer 2012; Urton and Aveni 1983). Gary Urton and Anthony Aveni argued long ago that the calculation of the movement of the Pleiades as well as the rising and setting of the sun on the days of both its zenith and antizenith passage played a central role in synchronizing the ritual calendar with the agricultural year (Urton and Aveni 1983; see also Nair 2015: 127; Orlove, Chiang, and Cane 2002). Robert Benfer (2012) has also proposed that Preceramic and Initial Period ceremonial architecture in coastal Peru (3500–1750 BCE) was built in alignment with lunar standstills, constellations, and the rising and setting of the sun on the solstices. Similar arguments have recently been made for the positioning of Formative Period architecture on the Taraco Peninsula, and for large andesite pillars at the Kalasasaya enclosure at Tiwanaku (Benitez 2009, 2013). Such features, including the famed sun pillars (*sucanas*) and *intiwatanas* (hitch posts of the sun), were also central to Inca astronomy (Aveni 2008; 2015b: 109–11).

The complex time-measurement of the Maya, founded on a series of interlocking calendars—the Long Count, the calendar round of fifty-two years, the ritual *tzolk'in* (260-day almanac), and the annual 365-day calendar (*Haab*)—finds no direct parallel in the Andes. However, Andean societies clearly developed sophisticated and integrated sidereal, lunar, and solar calendars that played a central role in meteorology, divination, production, exchange, and social control (Aveni 2008; Zuidema 2010). It is not an exaggeration to claim that time became a fundamental "means of production" by the Formative Period (Rice 2008: 279); its reification in architecture and calendrics was central to the construction of elite authority and the creation of imagined communities (Benfer 2012; Roddick 2013).[8] Andean ritual specialists, including the *yancas* documented in the Huarochirí manuscript, were described as calendric authorities and technicians of time, and they often served as arbiters of oracles, the immanent agents of the past that could foretell and influence the future (Salomon and Urioste 1991: 4, 72).

The calendrical reckoning of astral revolutions, which facilitated the planning of economic and religious life in the Andes, often seemed to have conformed to the circulatory, metamorphic, and fundamentally spatial conceptions of time discussed above. Similar to the Maya, Andean people commonly viewed time as "living, liminal creatures" (comparable to Mesoamerican Year Bearers) whose physical circulation set the world in motion (Rice 2008: 293). For instance, the movement of stars among the Inca was mimetically synchronized with the containment, storage (liminal gestation), release, and exchange of peoples and things, entities that came to embody astronomical time.[9] Robert Randall (1982, 1990) provides a fascinating discussion of how the period of the Inca harvest (April/May), defined by the storage of cultigens in warehouses, corresponded with the disappearance of the Pleiades (one of its names being *qollqa*) and the sacrifice of women in the Siqllapampa zeq'e at the wak'a of Mama Raroy. The harvest constituted an important temporal juncture in many societies, a time that mediated oscillations of life and death and witnessed the reversal of social roles (Leach 1961: 129).

The numerous Inca rituals and astronomical events commemorating the harvest charted a westward and downward movement of water, stars, crops, seeds, fecundity, and the female life principle. These entities circulated from kaypacha to the ocean and the underworld, a place at once associated with death and latent fertility. Therefore, the months between April and August were considered to be a time of sterility, sickness, occultation, and bleeding (menstruation). Chosen women (*acllakuna or akllakuna*) were also likened to harvested crops and cloistered in the *acllawasi* in the month of April, described as a dangerous and liminal period marked by the disappearance of the Pleiades. Female fertility remerges in August with the commencement of the planting season, when the earth is ready to accept the seed and is impregnated with the return of the rains from the underworld (that recirculates upward through the Milky Way). This period of renewal corresponds with the release and distribution of seeds from the warehouse as well as the full emergence of the Pleiades (which slowly becomes visible for the first time in June). The Qolla Raymi festival in September (marked by the nadir of the Southern Cross) celebrates the full return of fertility and the female life principle.[10] Therefore, the past as a possible future to come was understood as stored in a latent and virtual state in *acllawasi* (sacred house of chosen women), *colcas* (*qullqa*) (storage constructions) and *machays* (caves and sanctuaries for ancestral mummy bundles) (see also Gose 1993).

This belief in a stored (underground) and latent time is further exemplified by the Andean millenarian tradition of the *Inkarrí*, the return of the hidden and buried Inca king. The resurrection would be consummated with the reunion of the Inca's body and decapitated head, symbolizing the reunification of *hanan and hurin* as well as the female and male principles (Randall 1990). Ultimately, this event would bring to an end the chaotic and unjust period of Spanish colonization. The boons that come with the rediscovery of a *purun huaca*, a lost or wild wak'a of an extinct ayllu, provides another example of how time is renewed through the activation of a latent state, as does the search typically in August for *conopa* stones in the shape of llama or alpacas (Allen 1997). These power objects acted to transfer (latent) fertility and well-being to the animals they resembled (Salomon and Urioste 1991: 101).

Politicoreligious Institutions of Memory, Myth, Divination, and Archive

A consideration of the fourth dimension of historicized time warrants a brief discussion on Andean notions of the longue dureé and institutions of official history. In chapter 4, Wilkinson and D'Altroy argue that the Inca temporal ontology eschewed a clear distinction between the past and present and that commemorative practices—memorializing past "events" as understood in the Western sense—were absent in Inca politics. We agree with much of their analysis and recognize that the Inca developed a philosophically unique understanding of the immanence of the past (or the simultaneity of multiple pasts) and the continued agency of primordial actors in chartering the future. Nevertheless, it is evident that the Inca ordered "events" sequentially as occurring in specific places and times, and they employed various recording media to make sense of temporal discontinuities. In other words, they seem to have developed historically specific "chronographic" projects to periodize their own history and make sense of (or mystify) the relatedness of events (Rosen 2004).

A classic debate in studies of Inca history concerned the degree to which Spanish and acculturated chroniclers distorted the mythic structures and cultural schemas of Andean people by flattening them within a linear sequence of historical development (Netherly 1984, 1990; Zuidema 1990; see Chase [chapter 5] and Seoane and Culquichicón [chapter 8] this volume). John Rowe (1945, 1985) and his followers adopted a documentary-historicist perspective and compared overlaps and discrepancies in Spanish records of kingly succession and related sequence of events to propose a chronology of Inca history (and Rowe relied in particular on the chronicle of Cabello Balboa 1951). Proponents of structuralist theory, most notably Tom Zuidema, critiqued this approach for imposing universalist, European notions of history onto the Andean past and for misrepresenting Inca historical consciousness and sociopolitical organization (Zuidema 1964; see also Duviols 1980; Pease 1991: 36–37; Urton 1996). Zuidema famously "read the chronicles and indigenous testimonies as metaphoric renderings of deeply seated Inca (and more generally indigenous Andean) organizational structures, institutions, and mentalities rather than as faithful records of actual individual actions and past events" (Kolata 2013: 33). For instance, Zuidema argued against the existence of a succession of twelve Inca kings; instead, the records point to the sharing of power between Hanan and Hurin Cusco. The period of Inca kingship was thus shorter than Rowe's proposed chronology, for the list of rulers should be read not in terms of the exploits of individual kings but as titles and offices that expressed diarchic co-rule and the existence of parallel descent groups. In other words, Pachakuti Yupanqui should not be understood as a great man in the tradition of Alexander the Great or Julius Cesar, but as a denotation of structural position that emphasized moments of rupture and sociopolitical reconstitution. As Gary Urton (1990: 7) notes: "Zuidema's position with respect not only to

Pachakuti Inka, but to all of the kings who supposedly ruled before the arrival of the Spaniards, is that each name was, in fact, a *title* in a complex dualistic and hierarchical structure of genealogical and administrative positions." He continues (Urton 1990: 8): The chroniclers "misrepresented political structures (and events) as chronological events, and in the process they effectively historicized what was, at the base of it, an ideology of history that was timeless, repetitive, and fully interchangeable—and integrated—with political, social, and ritual structure."

Although the authors of this volume are largely sympathetic with the structuralist approach to colonial records, Zuidema has been criticized for privileging synchronic structure over indigenous forms of historiography and for exaggerating the degree to which the Inca denied history and resisted temporal discontinuities (but see Seoane and Culquichicón [chapter 8] this volume). The twelve Inca kings may very well have symbolized twelve contemporaneous panacas and their living descendants, but as Gose (1996: 388) argues, a distinct history underwrote Inca notions of divine kinship (see also Julien 2000).

Scholars have long argued that conceptions of time differed between literate and preliterate societies (see Assmann 2006). Writing allowed for a dissemination and codification of culture memory that differed significantly in scale and kind from the identities crafted in recurring ritual observations, festivals, architectural projects, and oral presentations. Of course, identity is founded fundamentally on memory and the technologies of its transmission. As Jan Assmann (2006: 40, 120) notes: "With the emergence of codified writing and classics, the temporal form of culture undergoes a change. The festive distinction between primordial time and the present is now joined by another: the distinction between past and present, antiquity and modernity."[11] However, it is reductive to pigeonhole Amerindian societies into simply one of these two domains of memory and historical consciousness ("literate vs. non-literate").

This analytic is further complicated since a number of pre-Columbian cultures developed complex writing and semasiographic systems. The extraordinary Inca khipus and their antecedents dating as early as the Middle Horizon no doubt formed the principal "archive" of the late Andean state (Urton 2003). Although only 450 or so Inca examples exist today, many thousands were archived in temples, *tambos* (rest houses), and administration centers at the time of the conquest. The Spanish relied heavily on khipus before they were destroyed as pagan relics after 1580, and Urton has argued that one-third of the surviving corpus encoded historic and narrative information (Ascher and Ascher 1981; Hyland 2010; Urton 2003, 2012). The distinct materiality and tactility of khipus have long fascinated scholars, and they clearly created a unique "historical record." Instead of simply recording past histories or present realities, the khipus may also have been envisioned as mimetic, anticipatory tools that served to trigger a desired present (whether a census summoning people to work in the mit'a or a narrative that allowed for the revival of a specific time or previously dormant place). In this sense, the khipu as historiographic device mirrors the kinetic and virtual qualities of the zeq'e system described above.

Khipus clearly played an important role in extending and reconfiguring memory that permitted novel ways of relating past, present, and future. Similar to Moctezuma's burning of the codices at Tenochtitlan, Atahualpa's desire to burn the khipus following the defeat of Huascar is a testament to the power of these records in prescribing behavior and codifying memory (Brokaw 2003: 8). Although most studies have focused on the numeric, mnemonic, and encoding properties of the khipu, recent research has begun to consider the role they played as a distinctive medium of historiography, semiosis, and poetics (see Brokaw 2003; Quilter and Urton 2002). Furthermore, the khipu as a recording and bureaucratic device can only be properly understood in relationship to the great infrastructure projects of the Inca state. The engineering marvel of the imperial highway (*Capac Ñam*) and its supporting network of reclaimed fields, rest houses, and warehouses compressed space and time and allowed for the rapid dissemination of people, armies, tribute, and information (Nair 2015). In just days, messengers (*chaskis*) could transmit government directives over thousands of kilometers. If "the past is a foreign country" (something remote, far away, and hard to grasp), then the imperial integration of far-flung territories, facilitated by khipus and the road network, likely led to new understandings of space, time, and identity. Marco Curatola Petrocchi (2008) similarly argues that Inca patronage of local oracles permitted the surveillance of provincial peoples and the appropriation of their histories. Khipus and comparable infrastructural projects have been documented for the Middle Horizon Wari, and more attention should be paid to how radical shifts in infrastructural projects, accounting, and place making might have underwritten transformations in time reckoning and historiography in the ancient Andes (Urton 2012).

Although Eurocentric biases must be taken into account, Andean people also appear to have made a distinction between ages (major pachas) of differing temporal depths. Rowe (1987) believes that the five ages or "suns" recorded in the sixteenth and seventeenth centuries were wrongly imposed on the Andes by plagiarizing Spanish chroniclers familiar with Mesoamerican sources (mainly the chronicles of Murúa, Guaman Poma, and Blas Valera). Juan Ossio (2015), however, makes a strong case that distinct ages, inaugurated by the catastrophic destruction of the preceding era (named *pachakutis*—changes in space-time), most likely had indigenous roots in South America. For instance, contemporary Andeans view pre-Columbian structures as the houses of beings "who lived before the current sun arose" (Salomon and Urioste 1991: 51). Pachakutis referred to both destruction and renovation and could refer to cataclysmic interludes that separated more stable periods of 500 or 1,000 years. As implied by his name, the Inca king, Pachakuti Yupanqui, was celebrated as the founder of a new world order. The chronicles report that he embodied a break between two distinct periods (between chaos and renewal), and he was renowned for having creating a new calendar, reorganizing government, and rebuilding the political and sacred landscape of the Cuzco region (Ossio 2015: 222).

A belief in the succession of ages associated with distinct peoples and zeitgeists has also been documented in Amazonia, while myths from the North Coast similarly recount the rise and fall of dynasties as coinciding with great environmental cataclysms triggered by the violation of religious taboos (i.e., the Naymlap kingdom of Lambayeque) (Cabello Balboa 1951 [1586]; Ossio 2015). The Revolt of the Objects in Moche iconography, paralleling myths in the Huarochirí manuscript, has also been interpreted as expressing reversals in the world order and the dawning of a new age (Quilter 1990). Although it has been debated whether the succession of ages was cyclically recurrent (as opposed to linear or teleological), these ages clearly referenced qualitatively different periods of existence. In a similar light, Justin Jennings (2008) has interpreted the emergence of the Lambayeque polity and its sudden and violent destruction as signaling revitalization movements that sought to revive a former, golden age.

Whether the rise and fall of dynasties were consciously historicized by pre-Columbian Andean peoples awaits further research—an understanding of history perhaps comparable to the Chinese "mandate of heaven" (for a similar view of the fall of Moche, see Bawden 1996). In the end, attention to how environmental perturbations, invasions, or other disruptions affected political fortunes and religious ideologies should also consider how they may have led to transformations in temporal routines and the media of memory and historiography. Gose (2008) argues that the Spanish invaders were perceived as returning "ancestors" and that such an understanding of major historical disjunctures has deep roots in the Andes (see also Harris 1995). In fact, archaeological investigations of archaistic ideologies or of the spread of horizon styles might be fruitfully analyzed in this framework of "invaders as ancestors."

The four domains of historical consciousness discussed in this section should not be viewed as mutually exclusive types. Instead they can be applied as heuristic devices to facilitate the identification and comparison of diverse historiographic projects in the Andes and beyond. As the above analysis demonstrates, multiple representations of time were clearly in play in the ancient Andes, including phases, cycles, circulations, metamorphisms, and ages of differing durations and consequences. Although conceptualized in a unique cultural calculus, some of these categories likely found parallels with contemporary philosophers' understandings of history (as event, moyenne durée, longue-dureé, virtual time, etc.). However, other modes of time work and native historiography were clearly specific to the Andes as exemplified by the khipu and Inca zeq'e system.

CONCLUDING THOUGHTS AND ORGANIZATION OF THE VOLUME

This volume concentrates on problems of time and history with a focus on the precolonial Andes from an archaeological vantage point, focusing not simply on Andean ideologies of history but on the ingrained temporal routines of everyday practice as can be detected in the material record. The deficiencies of standard chronological and historical models employed by Andean archaeologists are also scrutinized in several of the opening chapters.

As outlined in the above discussion, the authors of the volume are critical of a strict analytical divide between history and temporality, but seek instead to explore how they variably interpenetrated in Andean societies. However, the heuristic value of these concepts is recognized by the contributors, and the volume loosely follows the structure of this introductory chapter. Thus Sayre (chapter 2) and Roddick (chapter 3) begin the volume with a criticism of standard chronological models and the Andean horizon concept. By considering new theories and approaches, they present novel methods and analytical frameworks to interpret experiences of time in Chavín and the Lake Titicaca Basin. Sayre considers the well-established narratives of the "Early Horizon," exploring the continuing impact of Julio Tello in our understanding. Explicitly engaging with Trouillot's historicities at Chavín de Huántar, both the traces of historical processes and historical narratives, Sayre presents a reconsideration of the site's occupation history through the perspective of daily life, highlighting previous silences in our archive and in turn producing a distinct temporal narrative. Roddick also critiques normative narratives and established chronotopes in the Lake Titicaca Basin. He argues that our visual tools, or "chronographics," are having unintended effects in our writing. He teases apart the visual models of chronology employed by archaeologists, highlighting specifically the role of chronological charts, ceramic sequences, and stratigraphic profiles in our narrative constructions. Ultimately, Roddick underscores the potentials and pitfalls of both traditional and emerging methodologies in plotting temporal continuities and ruptures in the south-central Andes and beyond.

The ensuing chapter by D'Altroy and Wilkinson (chapter 4) explores linguistic, metaphysical, and spatial mediations of time in the Inca period. This contribution not only demonstrates the inextricability of space and time in Andean thought but proves that the relational temporality of the Andes departed dramatically from linear constructions defining modern notions of time. The chapter is of value in identifying the unique, "aniconic" technologies of history developed by the Inca that differed from the reliance on iconic historical images so characteristic of other world empires, including pre-Inca states in the Andes. In their analysis of the enclosure of stone wak'as in the Amaybamba Valley east of Cuzco, Wilkinson and D'Altroy show that the Inca did not remember, record, or manipulate an absent past but made it kin and family in the present by housing and feeding powerful wak'as. In the end, Wilkinson and D'Altroy's critique of social memory theory reveals that Inca engagements with history were far removed from archaeological *re-presentations* of remote pasts.

In chapter 5, Zach Chase shows that the "canonical prehistory" of Huarochirí, based on the chronological sequencing of waves of highland invasions, conveniently corresponds with the horizon schema employed by historians. However, this narrative is contradicted by recent archaeological excavations at the important Checa and Inca ceremonial center of Llacsatambo. At the same time, Chase argues that the mobilization of empirical archaeological data with theories on the ritual constructions of memory and the past (what he calls "performative historicity") illuminates Huarochirano conceptions of history and time. His questioning of canonical histories and the silencing of pasts (*sensu* Trouillot) complements Sayre [chapter 2] and Roddick's [chapter 3] critique of taken-for-granted historiographical approaches and chronological models.

In chapter 6, Swenson compares the settlement data of the Late Moche center of Huaca Colorada with earlier Formative and Middle Moche sites in the Jequetepeque Valley. The comparison illustrates how conceptions of time changed significantly in the region between the Formative, Early Intermediate, and Middle Horizon Periods, a point also made by Seoane and Culquichicón in chapter 8. Ultimately, his chapter serves to highlight considerable diversity in Andean ideologies of time. Swenson further applies the theoretical framework of topology to highlight the active architectural regulation of time and creation of history. The concept of topology further permits interpretation of alternate modalities of time as engineered in space, including embodied and "affective" time. In chapter 7, Spence Morrow considers how time was literally made in place through formal acts of architectural renovation also at the Moche site Huaca Colorada in the Jequetepeque Valley. He further interprets architectural models of ritual precincts interred in the priestess center of the nearby cult center of San José de Moro as power objects that effectively stored a latent or virtual space-time. The occultation of these *maquetas* in the realm of the dead was comparable to the destruction (by burial) and renewal of architectural space documented at the site of Huaca Colorada. These practices point to a particular Moche conception of time as physical, vital, and synecdochal. By synecdochal history, Spence Morrow stresses that time was materially set in motion by constantly renewing the whole by manipulating or sacrificing the part. This regulation of a nested or synecdochal time, in which the part engenders the whole and vice-versa, is evident in both the incremental renovations of the huaca and the burial of architectural representations of sacred structures at the cult center at San José de Moro

In chapter 8, Seoane and Culquichicón examine the calendric function of Andean religious architecture from the Formative to Middle Horizon Periods.

Their identification of shifts in astronomical alignments at the beginning of the Middle Horizon provides an example of how sociopolitical changes underwrote transformations in historical consciousness and the measurement of time in a pre-Inca polity. Complementing the arguments of D'Altroy and Wilkinson (chapter 4), Chase (chapter 5), and Spence Morrow (chapter 7), they contend that historical projects in later Andean civilizations were concerned with organizing peoples in space as opposed to constructing teleological narratives of sequential events. By at least the beginning of the Middle Horizon, Seoane and Culquichicón contend, time was understood not as linear but as a continuous present. The timework undertaken by astronomers and ritual specialists intended to fix time in such a way to ensure cosmic and social equilibrium and to stave off the chaos of a dreaded nonpresent (or a space without time). Finally, the volume concludes with an insightful assessment of the eight chapters by Bray. She presents some of her own interpretations on time and history in Andean archaeology and offers advice for future research on the topic.

In the introduction of a recent edited volume, Aveni notes (Aveni 2015a: 2): "Few words in the dictionary have managed to acquire as many definitions as the word *time*." Gosden (1994: 6) also writes that "there must be as many different forms of time as there are societies." In light of these statements, the chapters of the edited volume can only provide a select study of Andean temporalities and historicities. Nevertheless, the contributions show that if we wish to improve our understanding of past Andean cultures, we need to seriously question and rework normative chronological frameworks. The contributions demonstrate that a proper understanding of historical process—a long-time obsession in archaeology—must critically reflect on the history of our own temporalizing practices, as well take into account emic experiences and ideologies of time.

NOTES

1. At the beginning of the twentieth century, Uhle was the first to propose a Peruvian chronology based on alternating periods of regional and global styles. Rice explains that "the horizon concept [has generally] served two functions . . . as a chronological marker that served to integrate the local chronologies of disparate regions [as exemplified in the work of Rowe], and as an indication of cultural processes resulting in stylistic coherence over a broad region [mainly the perspectives of Willey, but also proposed in the earlier research of Bennett and Bird]. The two are not necessarily mutually reinforcing" (Rice 1993: 9).

2. In contrast, Alfredo Torero (2002: 48) argues that the Wari spoke a dialect of Aymara (see also Urton 2012: 325).

3. Unless direct reference is made to Trouillot's historicity, the authors of this volume tend to employ the terms "history" or "historiography" to refer to the production of ideologies of the past and the creation of historical consciousness more generally speaking.

4. See also Catherine Julien (2000) for a discussion of the meaning of *capac* and genealogical historiography among the Inca.

5. This Andean understanding of historical causality finds parallels in other cultures and structuralist theories more generally. For instance, Edward Leach (1961: 130) makes the general observation that "myths about sex reversals are representations of time" and draws examples from Greek and Egyptian cosmology to show that the origins of time and "beginning of becoming" are activated through the creation of contraries, "the creation of male and female not as brother and sister but as husband and wife" (Leach 1961: 131).

6. In his critical appraisal of theories of Inca diarchy, Peter Gose concludes that the Early Inca kings (their mummies) belonged to Lower Cuzco, whereas latter kings were associated with Upper Cuzco. He further notes (Gose 1996: 386): "Not only were these moieties spatialized . . . but they are also temporalized, such that Lower Cuzco was primordial and Upper Cuzco was recent." In Gose's interpretation, when an Inca king died, he entered the fertilizing domain of Lower Cuzco associated with the high priest, Willaq Umu.

7. Based on the monumental research of Tom Zuidema (see Zuidema 2010), Anthony Aveni (2015b: 112) interprets each wak'a of the zeq'e as materializing 1 of 328 days of a twelve-month sidereal-lunar calendar, with an average month consisting of 27.3 days. He suggests that 328 days may possibly have coincided with the gestation period of the llama. Moreover, Aveni proposes that each zeq'e corresponded to a week and that a grouping of zeq'es, usually three in number, materialized a month within the zeq'e calendar. As recorded in other calendric systems (Bali), the number of days in a week likely varied, as dictated by the growing season or ritual round.

8. In his discussion of annual cycles and commemorations Trouillot (1995: 116) writes: "As rituals that package history for public consumption, commemorations play the numbers game to create a past that seems both more real and more elementary . . . Numbers matter also as items in the calendar. Years, months, and dates present history as part of the natural cycle of the world."

9. For an interesting parallel in ancient Greece and the writings of Hesiod, see Purves (2004).

10. Randall (1982, 1990) further explains that things stored—the harvest, the Pleiades, the *mama sara* (sacred corn effigy), the *akllakuna*—represent the female force. At the end of the agricultural year, "Pachamama, having yielded her crops, becomes sterile, retreating into the underworld" (Randall 1990: 38): In August, the male and female principles are reunited, and this period is celebrated with rituals of fertility, sexual license, sacrifice, and water rites.

11. Assmann is especially interested in how the development of monotheism and text-based religion led to radical shifts in history and ideologies of remembering and forgetting. Abrahamic monotheism and philosophies including Platonism and Aristotelianism envisioned a world created by a distant and transcendent god, unmoved by the actions of mortals. In contrast, in many societies, including the Andes, the continued existence

of the world was dependent on sustained regimes of ritual practice. Of course, such profound theological differences can explain in part gross distinctions in traditions of memory and historical reckoning. Indeed, the Judeo-Christian underpinnings of Western temporal schemes have long bedeviled our comprehension of alternate constructions of history and time (Swenson 2014; Zuidema 1990; see also Seoane and Culquichicón this volume [chapter 8]).

REFERENCES

Abercrombie, Thomas A. 1998. *Pathways of Memory and Power: Ethnography and History among an Andean People*. Madison: University of Wisconsin Press.

Adams, Eleanor. 2007. "Time and Chance: Unraveling Temporality in North-Central Neopalatial Crete." *American Journal of Archaeology* 111 (3): 391–421. https://doi.org/10.3764/aja.111.3.391.

Adams, William Y. 1979. "On the Argument from Ceramics to History: A Challenge Based on Evidence from Medieval Nubia." *Current Anthropology* 20 (4): 727–44. https://doi.org/10.1086/202373.

Alcock, Susan E. 2002. *Archaeologies of the Greek Past: Landscapes, Monuments, and Memories*. Cambridge: Cambridge University Press.

Allen, Catherine J. 1988. *The Hold Life Has: Coca and Cultural Identity in an Andean Community*. Washington, DC: Smithsonian Institution Press.

Allen, Catherine J. 1997. "When Mountains Move Pebbles: Iconicity and Symbolism in Quechua Ritual." In *Creating Context in Andean cultures*, ed. R. Howard-Malverde, 73–84. Oxford: Oxford University Press.

Allen, Catherine J. 1998. "When Utensils Revolt: Mind, Matter, and Modes of Being in the Pre-Columbian Andes." *RES* 33 (Spring):18–27.

Arnold, Denise Y. 1992. *Hacia un orden andino de las cosas: Tres pistas de los Andes meridionales*. La Paz: Hisbol.

Ascher, Marcia, and Robert Ascher. 1981. *Code of the Quipu: A Study in Media, Mathematics and Culture*. Ann Arbor: University of Michigan Press.

Assmann, Jan. 2006. *Religion and Cultural Memory*. Stanford, CA: Stanford University Press.

Aveni, Anthony. 2008. *Foundations of New World Cultural Astronomy: A Reader and Commentary*. Boulder: University of Colorado Press.

Aveni, Anthony. 2015a. "Timely Themes: An Introduction to the Measure and Meaning of Time in Mesoamerica and the Andes." In *The Measure and Meaning of Time in Mesoamerica and the Andes*, ed. Anthony F. Aveni, 1–8. Washington, DC: Dumbarton Oaks.

Aveni, Anthony. 2015b. "Understanding Time, Space, and Social Organization in the Inca Ceque System of Cuzco." In *The Measure and Meaning of Time in Mesoamerica and the Andes*, ed. Anthony F. Aveni, 105–18. Washington, DC: Dumbarton Oaks.

Bahrani, Zainab. 2008. *Rituals of War: The Body and Violence in Mesopotamia*. New York: Zone Books.

Bailey, Geoff. 2007. "Time Perspectives, Palimpsests, and the Archaeology of Time." *Journal of Anthropological Archaeology* 26 (2): 198–223. https://doi.org/10.1016/j.jaa.2006.08.002.

Bakhtin, Mikhail. 1981. *The Dialogic Imagination: Four Essays by M. Bakhtin*, ed. M. Holquist. Austin: University of Texas Press.

Barrett, John C. 2004. "Temporality and the Study of Prehistory." In *Time and Temporality in the Ancient World*, ed. R. M. Rosen, 11–27. Philadelphia: University of Pennsylvania Museum of Archaeology and Anthropology.

Bauer, Brian S. 1998. *The Sacred Landscape of the Inca: The Cusco Ceque System*. Austin: University of Texas Press.

Bawden, Garth. 1996. *The Moche*. Malden, MA: Blackwell.

Benfer, Robert A., Jr. 2012. "Monumental Architecture Arising from an Early-Astronomical Religious Complex in Peru, 2200–1750 BC." In *Early New World Monumentality*, ed. R. L. Burger and R. M. Rosenswig, 313–63. Gainesville: University of Florida Press. https://doi.org/10.5744/florida/9780813038087.003.0012.

Benitez, Leo. 2009. "Descendants of the Sun: Calendars, Myth, and the Tiwanaku State." In Tiwanaku: *Papers from the 2005 Mayer Center Symposium at the Denver Art Museum*, ed. Margaret Young-Sanchez, 49–81. Denver: Denver Art Museum.

Benitez, Leo. 2013. "What Would Celebrants See? Sky, Landscape, and Settlement Planning in the Late Formative Southern Titicaca Basin." In *Advances in Titicaca Basin Archaeology II*, ed. Alexei Vranich, Elizabeth Klarich, and Abby Levine, 89–104 Los Angeles: UCLA Cotsen Institute of Archaeology.

Beresford-Jones, David, and Paul Heggarty. 2012. "Broadening our Horizons: Towards an Interdisciplinary Prehistory of the Andes." In *Archaeology and Language in the Andes: A Cross-Disciplinary Exploration of Prehistory*, ed. Paul Heggarty and David Beresford-Jones, 57–84. Oxford: British Academy and Oxford Press. https://doi.org/10.5871/bacad/9780197265031.003.0003.

Bintliff, John. 1991. "The Contribution of an *Annaliste*/Structural History Approach to Archaeology." In *The Annales School and Archaeology*, ed. John Bintliff, 1–33. New York: New York University Press.

Bloch, Maurice. 1971. *Placing the Dead: Tombs, Ancestral Villages and Kinship Organisation in Madagascar*. London: Seminar Press.

Boone, Elizabeth Hill. 1993. "Preface." In *Latin American Horizons*, ed. Don S. Rice, vii–viii. Washington, DC: Dumbarton Oaks.

Bouysse-Cassagne, Thérèse. 1986. "Urco and Uma: Aymara Concepts of Space." In *Anthropological History of Andean Polities*, ed. John Murra, Nathan Wachtel, and Jacques Revel, 201–27. Cambridge: Cambridge University Press. https://doi.org/10.1017/CBO9780511753091.018.

Bradley, Richard. 1991. "Ritual, Time, and History." *World Archaeology* 23 (2): 209–19. https://doi.org/10.1080/00438243.1991.9980173.
Bradley, Richard. 2002. *The Past in Prehistoric Societies*. New York: Routledge.
Braudel, Fernand. 1970. "History and the Social Sciences: The Long Term." *Social Sciences Information. Information Sur les Sciences Sociales* 9 (1): 144–74. https://doi.org/10.1177/053901847000900107.
Bray, Tamara. 2015. *The Archaeology of Wak'as: Explorations of the Sacred in the Pre-Columbian Andes*. Boulder: University of Colorado Press.
Brokaw, Galen. 2003. "The Poetics of Khipu Historiography: Felipe Guaman Poma de Ayala's 'Nueva corónica' and the 'Relación de los quipucamayos.'" *Latin American Research Review* 38 (3): 111–47. https://doi.org/10.1353/lar.2003.0029.
Burkholder, JoEllen. 1997. "Tiwanaku and the Anatomy of Time: A New Ceramic Chronology from the Iwawe Site, Department of La Paz, Bolivia." PhD dissertation. State University of New York, Department of Anthropology, Binghamton.
Cabello Balboa, Miguel. [1586] 1951. *Miscelánea antártica*. Lima: Instituto de Etnología, Universidad Nacional Mayor de San Marcos.
Castillo, Luis Jaime, and Christopher B. Donnan. 1994. "Los mochica del norte y los mochica del sur." In *Vicús*, ed. Krzysztof Makowski and Chrisopher B. Donnan, 143–81. Lima: Colección Arte y Tesoros del Perú, Banco de Crédito.
Chapdelaine, Claude. 2000. "Struggling for Survival: The Urban Class of the Moche Site." In *Environmental Disasters and the Archaeology of Human Response*, ed. Garth Bawden and Martin Reycraft, 121–42. Albuquerque: Maxwell Museum of Anthropology.
Chapdelaine, Claude. 2001. "The Growing Power of a Moche Urban Class." In *Moche Art and Archaeology in Ancient Peru*, ed. Joanne Pillsbury, 69–87. New Haven: National of Gallery of Art and Yale University Press.
Chapdelaine, Claude. 2009. "Domestic Life in and around the Urban Sector of the Huacas of Moche Site, Northern Peru." In *Domestic Life in Prehispanic Capitals: A Study of Specialization, Hierarchy, and Ethnicity*, ed. Linda R. Manzanilla and Claude Chapdelaine, 191–96. Ann Arbor: University of Michigan Press.
Chapdelaine, Claude. 2011. "Recent Advances in Moche Archaeology." *Journal of Archaeological Research* 19 (2): 191–231. https://doi.org/10.1007/s10814-010-9046-6.
Cobb, Charles, and Adam King. 2005. "Re-Inventing Mississippian Tradition at Etowah, Georgia." *Journal of Archaeological Method and Theory* 12 (3): 167–93. https://doi.org/10.1007/s10816-005-6927-y.
Cobo, Bernabe. [1653] 1990. *Inca Religion and Customs. Translated and Edited by Roland Hamilton and Foreword by John Howland Rowe*. Austin: University of Texas Press.
Cook, Anita G. 2004. "Wari Art and Society." In *Andean Archaeology*, ed. Helaine Silverman, 146–66. Oxford: Blackwell.

Connerton, Paul. 1989. *How Societies Remember*. Cambridge: Cambridge University Press. https://doi.org/10.1017/CBO9780511628061.
Cummins, Tom, and Bruce Mannheim. 2011. "The River around Us, the Stream within Us: The Traces of the Sun and Inka Kinetics." *Res: Anthropology and Aesthetics* 59/60 (Spring/Autumn): 5–21. https://doi.org/10.1086/RESvn1ms23647779.
Curatola Petrocchi, Marco. 2008. "La función de los oráculos en el imperio inca." In *Adivinación y oráculos en el mundo andino antiguo*, ed. Marco Curatola Petrocchi and Mariusz S. Ziółkowski, 15–70. Lima: Fondo Editorial, Pontificia Universidad Católica del Perú.
Dawdy, Shannon L. 2010. "Clockpunk Anthropology and the Ruins of Modernity." *Cultural Anthropology* 51 (6): 761–93.
Dean, Carolyn. 2010. *A Culture of Stone: Inka Perspectives on Rock*. Durham, NC: Duke University Press. https://doi.org/10.1215/9780822393177.
Dillehay, Tom D. 2004. "Social Landscape and Ritual Pause: Uncertainty and Integration in Formative Peru." *Journal of Social Archaeology* 4 (2): 239–68. https://doi.org/10.1177/1469605304042396.
Donnan, Christopher B. 2001. "Moche Burials Uncovered." *National Geographic* (March): 58–73.
Donnan, Christopher B. 2007. *Moche Tombs at Dos Cabezas*. Los Angeles: Costin Institute of Archaeology Press.
Donnan, Christopher B. 2009. "The Gallinazo Illusion." In *Gallinazo: An Early Cultural Tradition on the Peruvian North Coast*, ed. Jean François Millaire and M. Morlion, 17–23. Los Angeles: Costin Institute of Archaeology Press.
Dulanto, Jahl. 2015. "Time and the Other: The Early Colonial Mythohistorical Landscapes of the Huarochirí Manuscript." In *The Measure and Meaning of Time in Mesoamerica and the Andes*, ed. Anthony F. Aveni, 141–64. Washington, DC: Dumbarton Oaks.
Duviols, Pierre. 1973. "Huari y Llacuaz: Agricultores y pastores. Un dualismo pre-hispánico de oposición y complementariedad." *Revista del Museo Nacional. Museo Nacional (Perú)* 34:153–91.
Duviols, Pierre. 1980. "Algunas Reflexiones acerca del la Tesis de la Estructura Dual del Poder Incaico." *Histórica (Lima)* 4 (2): 183–96.
Febvre, Lucien. 1947. *La problème de l'incroyance au XVI siècle*. Paris: Albin Michel.
Finlay, Robert. 2010. *The Pilgrim Art: Cultures of Porcelain in World History*. Berkeley: University of California Press.
Flaherty, Michael G. 2011. *The Textures of Time: Agency and Temporal Experience*. Philadelphia: Temple University Press.
Gell, Alfred. 1992. *The Anthropology of Time: Cultural Constructions of Temporal Maps and Rhythms*. Oxford: Berg.

Gelles, Paul. 1995. "Equilibrium and Extraction: Dual Organization in the Andes." *American Ethnologist* 22 (4): 710–42. https://doi.org/10.1525/ae.1995.22.4.02a00040.

Gentile, Margarita E. 2008. "Espacio y tiempo de un oráculo andino relacionado con el agro y pesca." In *Adivinación y oráculos en el mundo andino antiguo*, ed. Marco Curatola Petrocchi and Mariusz S. Ziółkowski, 221–50. Lima: Fondo Editorial, Pontificia Universidad Católica del Perú.

Ghezzi, Ivan, and Clive Ruggles. 2007. "Chankillo: A 2300-Year-Old Solar Observatory in Coastal Peru." *Science* 315 (5816): 1239–43. https://doi.org/10.1126/science.1136415.

Goldstein, Paul S., and Bruce Owen. 2001. "Tiwanaku en moquegua: Las colonias altiplánicas." In *Huari y Tiwanaku: Modelos vs. evidencias, segunda parte*, ed. Peter Kaulicke and William H. Isbell, 169–88. Boletín de Arqueología No. 5. Lima: Pontifica Universidad Católica del Perú.

González-Ruibal, Alfredo. 2014. *An Archaeology of Resistance: Materiality and Time in an African Borderland*. New York: Rowman and Littlefield.

Gosden, Christopher. 1994. *Social Being and Time*. Oxford: Blackwell.

Gosden, Chris, and Gary Lock. 1998. "Prehistoric Histories." *World Archaeology* 30 (1): 2–12. https://doi.org/10.1080/00438243.1998.9980393.

Gose, Peter. 1993. "Segmentary State Formation and the Ritual Control of Water under the Incas." *Comparative Studies in Society and History* 35 (30): 480–514. https://doi.org/10.1017/S0010417500018557.

Gose, Peter. 1996. "The Past is a Lower Moiety: Diarchy, History, and Divine Kingship in the Inka Empire." *History and Anthropology* 9 (4): 383–414. https://doi.org/10.1080/02757206.1996.9960887.

Gose, Peter. 2008. *Invaders as Ancestors: On the Intercultural Making and Unmaking of Spanish Colonialism in the Andes*. Toronto: University of Toronto Press.

Gosselain, Olivier. 2015. "Roads, Markets, Migrants: The Historical Trajectory of a Male Hausa Pottery Tradition in Southern Niger." In *The Distribution of Technological Knowledge in the Production of Ancient Mediterranean Pottery*, ed. W. Gauß, G. Klebinder-Gauß, and C. Von Rüden, 277–96. Vienna: Sonderschriften ÖAI.

Harding, Jan. 2005. "Rethinking the Great Divide: Long-Term Structural History and the Temporality of the Event." *Norwegian Archaeological Review* 38 (2): 88–101. https://doi.org/10.1080/00293650510032707.

Harris, Olivia. 1995. "The Coming of the White People: Reflections on the Mythologisation of History in Latin America." *Bulletin of Latin American Research* 14 (1): 9–21.

Harris, Olivia. 2000. *To Make the Earth Bear Fruit: Ethnographic Essays on Fertility, Work, and Gender in Highland Bolivia*. London: Institute of Latin American Studies, University of London.

Heidegger, Martin. 1962. *Being and Time*. Oxford: Blackwell.

Hertog, François. 2015. *Regimes of Historicity: Presentism and Experiences of Time*. Trans. Saskia Brown. New York: Columbia University Press.

Herzfeld, Michael. 1991. *Place in History: Social and Monumental Time in a Cretan Town*. Princeton, NJ: Princeton University Press.

Hill, Jonathon D. 1988. *Rethinking History and Myth: Indigenous South American Perspectives on the Past*. Chicago: University of Illinois Press.

Hocquenghem, Anne Marie. 2008. "Sacrifices and Ceremonial Calendars in Societies of the Central Andes: A Reconsideration." In *The Art and Archaeology of the Moche: An Ancient Andean Society of the Peruvian North Coast*, ed. Steve Bourget and Kimberly L. Jones, 23–42. Austin: University of Texas Press.

Huertas Vallejos, Lorenzo. 2008. "Los oráculos en la historia andina (siglos XV–XVII)." In *Adivinación y oráculos en el mundo andino antiguo*, ed. Marco Curatola Petrocchi and Mariusz S. Ziółkowski, 251–72. Lima: Fondo Editorial, Pontificia Universidad Católica del Perú.

Husserl, Edmund. 1966. *The Phenomenology of Internal Time-Consciousness*. Bloomington: Indiana University Press.

Hyland, Sabine. 2010. *Gods of the Andes: An Early Jesuit Account of Inca Religion and Andean Christianity*. University Park: Pennsylvania State University.

Hyslop, John. 1993. "Factors Influencing the Transmission and Distribution of Inka Cultural Materials throughout Tawantinsuyu." In *Latin American Horizons*, ed. Don S. Rice, 337–56. Washington, DC: Dumbarton Oaks.

Ingold, Tim. 2000. *The Perception of the Environment: Essays in Livelihood, Dwelling, and Skill*. New York: Routledge. https://doi.org/10.4324/9780203466025.

Isbell, Billie Jean. 1982. "Culture Confronts Nature in the Tropics." In *Ethnoastronomy and Archaeoastronomy in the American Tropics*, ed. Anthony F. Aveni and Gary Urton, 353–63. Annals of the New York Academy of Sciences 385. New York: New York Academy of Sciences.

Isbell, William H. 2013. "Nature of an Andean City: Tiwanaku and the Production of Spectacle." In *Visions of Tiwanaku*, ed. Alexei Vranich and Charles Stanish, 167–210. Los Angeles: Cotsen Institute of Archaeology Press.

Isbell, William, and Patricia Knobloch. 2008. "ISAIS: The Origin, Development, and Dating of Tiahuanaco-Huari iconography." In *Tiwanaku—Papers from the 2005 Mayer Center Symposium at the Denver Art Museum*, ed. Margaret Young-Sánchez, 165–210. Denver: Denver Art Museum.

Isbell, William, and Helaine Silverman. 2006. "Rethinking the Central Andean Co-Tradition." In *Andean Archaeology III: North and South*, ed. William Isbell and Helaine Silverman, 497–518. New York: Springer. https://doi.org/10.1007/0-387-28940-2_20.

Janusek, John W. 2003. "Vessels, Time, and Society: Toward a Chronology of Ceramic Style in the Tiwanaku Heartland." In *Tiwanaku and Its Hinterland: Archaeological and*

Paleoecological Investigations of an Andean Civilization, vol. 2. ed. Alan L. Kolata, 30–92. Washington, DC: Smithsonian Institution Press.

Janusek, John, and Sonia Alconini. 1994. "Social Diversity and Historical Change in Tiwanaku Ceramics: Steps towards a Tiwanaku IV–V Chronology." Paper presented at the 59th Annual Meeting of the Society for American Archaeology. Anaheim, CA.

Jennings, Justin. 2008. "Catastrophe, Revitalization, and Religious Change on the Prehispanic North Coast of Peru." *Cambridge Archaeological Journal* 18 (2): 177–94. https://doi.org/10.1017/S0959774308000243.

Joyce, Rosemary A. 2000. *Gender and Power in Prehispanic Mesoamerica*. Austin: University of Texas Press.

Julien, Catherine. 2000. *Reading Inca History*. Iowa City: University of Iowa Press.

Kaulicke, Peter. 1991. "El período intermedio temprano en Alto Piura: Avances del Proyecto Arqueológico Alto Piura 1987–1990." *Bulletin de L'Institut Français d'Etudes Andines*. 21 (2): 381–422.

Knapp, A. Bernard. 1992. *Archaeology, Annales, and Ethnohistory*. Cambridge: Cambridge University Press. https://doi.org/10.1017/CBO9780511759949.

Kolata, Alan L. 2013. *Ancient Inca*. Cambridge: Cambridge University Press.

Koons, Michele L., and Bridget A. Alex. 2014. "Revised Moche Chronology Based on Bayesian Models of Reliable Radiocarbon Dates." *Radiocarbon* 56 (3): 1039–55. https://doi.org/10.2458/56.16919.

Kosiba, Steve. 2010. "Becoming Inka: The Transformation of Political Place and Practice during Inka State Formation (Cusco, Perú)." PhD dissertation. Department of Anthropology, University of Chicago, Chicago.

Kubler, George. 1970. "Period, Style, and Meaning in Ancient American Art." *New Literary History* 1 (2): 127–44. https://doi.org/10.2307/468624.

Last, Jonathon. 1995. "The Nature of History." In *Interpreting Archaeology: Finding Meaning in the Past*, ed. I. Hodder, M. Shanks, A. Alexander, V. Buchli, J. Carment, J. Last, and G. Lucas, 141–57. London: Routledge.

Leach, Edward. 1961. *Rethinking Anthropology*. London: Althone.

Lefebvre, Henri. 2004. *Rhythmanalysis: Space, Time, and Everyday Life*. London: Bloomsbury.

Lightfoot, Kent. 2001. "Traditions as Cultural Production: Implications for Contemporary Archaeological Research." In *The Archaeology of Traditions: Agency and History Before and After Columbus*, ed. Timothy Pauketat, 237–52. Gainesville: University Press of Florida.

Lockard, Greg. 2009. "The Occupational History of Galindo, Moche Valley, Peru." *Latin American Antiquity* 20 (2): 279–302. https://doi.org/10.1017/S1045663500002649.

Lucas, Gavin. 2005. *The Archaeology of Time*. London: Routledge.

Lucero, Lisa J. 2003. "The Power of Ritual: The Emergence of Maya Classic Rulers." *Cultural Anthropology* 44 (4): 523–58.

Lund Skar, S. 1994. *Worlds Together, Lives Apart: Quechua Colonization in Jungle and City.* Oslo: Scandinavian University Press.

MacCormack, Sabine. 1991. *Religion in the Andes: Vision and Imagination in Early Colonial Peru.* Princeton, NJ: Princeton University Press.

Mannheim, Bruce. 1987. "A Semiotic of Andean Dreams." In *Dreaming: Anthropological and Psychological Interpretations*, ed. Barbara Tedlock, 132–53. Cambridge: Cambridge University Press.

Mannheim, Bruce. 2015. "All Translation is Radical Translation." In *Translating Worlds: The Epistemological Space of Translation*, ed. Carlo Severi and William F. Hanks, 199–220. Chicago: Hau Books.

Marsh, Erik J. 2012. "A Bayesian Re-Assessment of the Earliest Radiocarbon Dates from Tiwanaku, Bolivia." *Radiocarbon* 54 (2): 203–18. https://doi.org/10.2458/azu_js_rc.v54i2.15826.

McEwan, Gordon. 2012. "Indicators of Possible Driving Forces for the Spread of Quechua and Aymara Reflected in the Archaeology of Cuzco." In *Archaeology and Language in the Andes: A Cross-Disciplinary Exploration of Prehistory*, ed. Paul Heggarty and David Beresford-Jones, 246–64. Oxford: British Academy and Oxford Press. https://doi.org/10.5871/bacad/9780197265031.003.0010.

McGlade, James. 1999. "The Times of History: Archaeology, Narrative, and Non-linear Causality." In *Time and Archaeology*, ed. Tim Murray, 139–63. London: Routledge.

Menzel, Dorothy. 1964. "Style and Time in the Middle Horizon." *Ñawpa Pacha* 2 (1): 1–105. https://doi.org/10.1179/naw.1964.2.1.001.

Millaire, Jean François. 2009. "Gallinazo and the Tradición Norcosteña." In *Gallinazo: An Early Cultural Tradition on the Peruvian North Coast*, ed. Jean François and M. Morlion, 1–16. Los Angeles: Costin Institute of Archaeology Press.

Mills, Barbara J., and William H. Walker. 2008. *Memory Work: Archaeologies of Material Practice.* Albuquerque: SAR Press.

Munn, Nancy. 1992. "The Cultural Anthropology of Time: A Critical Essay." *Annual Review of Anthropology* 21 (1): 93–123. https://doi.org/10.1146/annurev.an.21.100192.000521.

Murray, Tim. 1999. *Time and Archaeology*. New York: Routledge. https://doi.org/10.4324/9780203203019.

Nair, Stella. 2015. "Space and Time in Architecture of the Inca Royal Estates." In *The Measure and Meaning of Time in Mesoamerica and the Andes*, ed. Anthony F. Aveni, 119–40. Washington, DC: Dumbarton Oaks.

Netherly, Patricia. 1984. "The Management of Late Andean Irrigation Systems on the North Coast of Peru." *American Antiquity* 49 (02): 227–54. https://doi.org/10.2307/280017.

Netherly, Patricia. 1990. "Out of Many, One: The Organization of Rule in the North Coast Polities." In *The Northern Dynasties: Kingship and Statecraft in Chimor*, ed. Michael Moseley and Alana Cordy-Collins, 461–88. Washington, DC: Dumbarton Oaks.

Nyord, Rune. 2013. "Memory and Succession in the City of the Dead: Temporality in the Ancient Egyptian Mortuary Cult." In *Taming Time, Timing Death: Social Technologies and Rituals. Studies in Death, Materiality and Time 1*, ed. by Dorthe Refslund Christensen and Rane Willerslev, 195–211 Farnham: Ashgate.

Oakeshott, Michael. 1983. *On History and Other Essays*. Oxford: Basil Blackwell.

Olivier, Laurent. 2004. "The Past of the Present: Archaeological Memory and Time." *Archaeological Dialogues* 10 (2): 204–13. https://doi.org/10.1017/S1380203804001254.

Olivier, Laurent. 2011. *The Dark Abyss of Time: Archaeology and Social Memory*. Lanham, MD: Altamira.

Olsen, Bjørnar. 2010. *In Defense of Things: Archaeology and the Ontology of Objects*. Lanham, MD: Altamira.

Orlove, Benjamin, John C. H. Chiang, and Mark A. Cane. 2002. "Ethnoclimatology in the Andes: A Cross-Disciplinary Study Uncovers a Scientific Basis for the Scheme Andean Potato Farmers Traditionally Use to Predict the Coming Rains." *American Scientist* 90 (5): 428–35. https://doi.org/10.1511/2002.33.791.

Ossio, Juan M. 2015. "Ages of the World in the Andes." In *The Measure and Meaning of Time in Mesoamerica and the Andes*, ed. Anthony F. Aveni, 211–38. Washington, DC: Dumbarton Oaks.

Owen, Bruce D. 2001. "From Sequence to Social Organization: Tiwanaku Multicomponent Society in Moquegua, Perú." Paper presented at the 66th Annual Meeting of the Society for American Archaeology, New Orleans.

Owen, Bruce D. 2005. "Distant Colonies and Explosive Collapse: The Two Stages of the Tiwanaku Diaspora in the Osmore Drainage." *Latin American Antiquity* 16 (1): 45–80. https://doi.org/10.2307/30042486.

Pease, Franklin G. Y. 1991. *Los incas: Una introducción*. Lima: Pontificia Universidad Católica del Perú, Fondo Editorial.

Picazo, Marina. 1997. "Hearth and Home: The Timing of Maintenance Activities." In *Invisible People and Processes: Writing Gender and Childhood into European Archaeology*, ed. Jenny Moore and Eleanor Scott, 59–60. Leicester, England: Leicester University Press.

Ponce Sanginés, C. 1981. *Tiwanaku: Espacio, tiempo y cultura*. La Paz: Editorial Los Amigos del Libro.

Purves, Alex. 2004. "Topographies of Time in Hesiod." In *Time and Temporality in the Ancient World*, ed. Ralph M. Rosen, 147–68. Philadelphia: University of Pennsylvania Museum of Archaeology and Anthropology.

Quilter, Jeffrey. 1990. "The Moche Revolt of the Objects." *Latin American Antiquity* 1 (01): 42–65. https://doi.org/10.2307/971709.

Quilter, Jeffrey, and Luis Jaime Castillo, eds. 2010. *New Perspectives on Moche Political Organization*. Washington, DC: Dumbarton Oaks.

Quilter, Jeffrey, and Gary Urton. 2002. *Narrative Threads: Accounting and Recounting in Andean Khipu*. Austin: University of Texas Press.

Ramirez, Susan Elizabeth. 2005. *To Feed and Be Fed: The Cosmological Bases of Authority and Identity in the Andes*. Stanford, CA: Stanford University Press.

Randall, Robert. 1982. "Qoyllur Rit'i, an Inca Fiesta of the Pleiades: Reflections on Time and Space in the Andean World." *Bulletin de L'Institut Français d'Etudes Andines* 11 (1–2): 37–81.

Randall, Robert. 1990. "The Mythstory of Kuri Qoyllur: Sex, Seques, and Sacrifice in Inka Agricultural Festivals." *Journal of Latin American Lore* 16 (1): 3–45.

Rappaport, Roy A. 1992. "Ritual, Rime, Eternity." *Zygon* 27 (1): 5–30. https://doi.org/10.1111/j.1467-9744.1992.tb00996.x.

Rice, Don Stephen. 1993. "The Making of Latin American Horizons: An Introduction to the Volume." In *Latin American Horizons*, ed. Don S. Rice, 1–14. Washington, DC: Dumbarton Oaks.

Rice, Prudence. 2008. "Time, Power, and the Maya." *Latin American Antiquity* 19 (3): 275–98. https://doi.org/10.1017/S1045663500007951.

Richards, Colin, and Julian Thomas. 1984. "Ritual Activity and Structured Deposition in Later Neolithic Wessex." In *Neolithic Studies: A Review of Current Research*, ed. Richard Bradley and Julie Gardiner, 189–218. BAR International Series 133 Oxford: British Archaeological Reports.

Roddick, Andrew. 2009. "Communities of Pottery Production and Consumption on the Taraco Peninsula, Bolivia, 200 BC–300 AD." PhD dissertation. Department of Anthropology, University of California, Berkeley.

Roddick, Andrew. 2013. "Temporalities of the Formative Period Taraco Peninsula, Bolivia." *Journal of Social Archaeology* 13 (3): 287–309. https://doi.org/10.1177/1469605313485396.

Roddick, Andrew. 2015. "Geologies in Motion: Itineraries of Stone, Clay and Pots in the Lake Titicaca Basin." In *Things in Motion: Object Itineraries in Anthropological Practice*, ed. Rosemary A. Joyce and Susan D. Gillespie, 123–46. Santa Fe: SAR Press.

Roddick, Andrew. 2016. "Scalar Relations: A Juxtaposition of Craft Learning in the Lake Titicaca Basin." In *Knowledge in Motion: Constellations of Learning across Time and Place*, ed. Andrew P. Roddick and Ann B. Stahl, 126–54. Tucson: University of Arizona Press.

Roddick, Andrew P., Maria C. Bruno, and Christine A. Hastorf. 2014. "Political Centers in Context: Depositional Histories at Formative Period Kala Uyuni, Bolivia." *Journal of Anthropological Archaeology* 36 (December): 140–57. https://doi.org/10.1016/j.jaa.2014.09.010.

Roddick, Andrew P., and Christine A. Hastorf. 2010. "Tradition Brought to the Surface: Continuity, Innovation and Change in the Late Formative Period, Taraco Peninsula,

Bolivia." *Cambridge Archaeological Journal* 20 (2): 157–78. https://doi.org/10.1017/S0959774310000211.
Rosen, Ralph M. 2004. "Ancient Time across Time." In *Time and Temporality in the Ancient World*, ed. Ralph M. Rosen, 1–10. Philadelphia: University of Pennsylvania Museum of Archaeology and Anthropology.
Rowe, John H. 1945. "Absolute Chronology in the Andean Area." *American Antiquity* 10 (3): 265–84. https://doi.org/10.2307/275130.
Rowe, John H. 1956. "Cultural Unity and Diversification in Peruvian Archaeology." In *Men and Cultures: Selected Papers of the Fifth International Congress of Anthropological and Ethnological Sciences*, 627–31. Philadelphia.
Rowe, John H. 1962. "Stages and Periods in Archaeological Interpretation." *Southwestern Journal of Anthropology* 18 (1): 40–54. https://doi.org/10.1086/soutjanth.18.1.3629122.
Rowe, John H. 1985. "La constitución inca del Cuzco." *Histórica (Lima)* 9 (1): 35–75.
Rowe, John H. 1987. "La mentira literatura en la obra de Martín de Murúa." In *Homenaje a Aurelio Miró Quesada Sosa*, ed. Héctor López Martínez, 753–61. Lima: Villanueva.
Sakai, Masato. 1998. *Reyes, estrellas y cerros en Chimor: El proceso de cambio de la organización espacial y temporal de Chan Chan*. Lima: Editorial Horizonte.
Salomon, Frank. 1998. "How the Huacas Were: The Language of Substance and Transformation in the Huarochirí Quechua Manuscript." *Res: Anthropology and Aesthetics* 33 (Spring): 7–17. https://doi.org/10.1086/RESv33n1ms20166998.
Salomon, Frank. 2002. "'*Huayra Huayra Pichcamanta!*' Augurio, risa y regeneración en la política tradicional (Pacota, Huarochirí)." *Boletín del Instituto Francés de Estudios Andinos* 31 (1): 1–22. https://doi.org/10.4000/bifea.6900.
Salomon, Frank, and George L. Urioste. 1991. *The Huarochirí Manuscript: A Testament of Ancient and Colonial Andean Religion*. Austin: University of Texas Press.
Silverman, Helaine. 2004. "Introduction: Space and Time in the Central Andes." In *Andean Archaeology*, ed. Helaine Silverman, 1–15. Oxford: Blackwell. https://doi.org/10.1016/B978-0-08-051337-9.50006-7.
Stone-Miller, Rebecca. 1993. "An Overview of the 'Horizon' and 'Horizon Style' in the Study of Ancient American Objects." In *Latin American Horizons*, ed. Don S. Rice, 15–40. Washington, DC: Dumbarton Oaks.
Swenson, Edward. 2013. "Interpreting the Political Landscape of Early State Religions." In *A Companion to the Anthropology of Religion*, ed. Michael Lambek and Janice Boddy, 471–88. London: Wiley Blackwell.
Swenson, Edward. 2014. "Los fundamentos cosmológicos de las interacciones Moche Sierra durante el Horizonte Medio en Jequetepeque." In *Los rostros de Wari: Perspectivas interregionales sobre el Horizonte Medio*, ed. Justin Jennings and Luis Jaime Castillo, 79–104. Boletín de Arqueología PUCP 16 (2012). Lima: Fondo Editorial Pontificia Universidad Católica del Perú.

Swenson, Edward. 2015. "The Materialities of Place-Making in the Ancient Andes: A Critical Appraisal of the Ontological Turn in Archaeological Interpretation." *Journal of Archaeological Method and Theory* 22 (3): 677–712. https://doi.org/10.1007/s10816-014-9202-2.

Swenson, Edward. 2017. "Timing Is Everything: Religion and the Regulation of Temporalities in the Ancient Andes." In *Religion and Politics in the Ancient Americas*, ed. S. Barber and A. Joyce, 210–33. New York: Routledge.

Taylor, Gerald. 1974–76. "Camay, Camac et Camasca dans le Manuscrit Quechua de Huarochirí." *Journal de la Société des Américanistes* 63 (1): 231–44. https://doi.org/10.3406/jsa.1974.2128.

Thompson, Edward P. 1967. "Time, Work-Discipline, and Industrial Capitalism." In *Essays in Social History*, ed. M. W. Flinn, M. W. and T. C. Smout, 56–97. Oxford: Clarendon Press.

Torero, Alfredo. 2002. *Idiomas de los Andes: Lingüística e historia*. Lima: IFEA-Editora Horizonte.

Trouillot, Michel R. 1995. *Silencing the Past: Power and the Production of History*. Boston: Beacon Press.

Turner, Terrence. 1988. "Ethno-Ethnohistory: Myth and History in a Native South American Representations of Contact with Western Society." In *Rethinking History and Myth: Indigenous South American Perspectives on the Past*, ed. Jonathan D. Hill, 235–81. Chicago: University of Illinois Press.

Uceda, Santiago. 2010. "Theocracy and Secularism: Relationship between the Temple and Urban Nucleus and Political Change at the Huacas de Moche." In *New Perspectives on Moche Political Organization*, ed. Jeffrey Quilter and Luis Jaime Castillo, 132–58. Washington, DC: Dumbarton Oaks.

Urton, Gary. 1981. *At the Crossroads of the Earth and the Sky: An Andean Cosmology. Institute of Latin American Studies, Monograph 55*. Austin: University of Texas Press.

Urton, Gary. 1990. *The History of a Myth: Pacariqtambo and the Origins of the Inkas*. Austin: University of Texas Press.

Urton, Gary. 1996. "R. Tom Zuidema, Dutch Structuralism, and the Application of the 'Leiden Orientation' to Andean Studies." *Journal of the Steward Anthropological Society* 24 (2-2): 1–36.

Urton, Gary. 2003. *Signs of the Inka Khipu: Binary Coding in the Andean Knotted-String Records*. Austin: University of Texas Press.

Urton, Gary. 2012. "The Herder-Cultivator Relationship as Paradigm for Archaeological Origins, Linguistic Dispersals, and the Evolution of Record-Keeping in the Andes." In *Archaeology and Language in the Andes: A Cross-Disciplinary Exploration of Prehistory*, ed. Paul Heggarty and David Beresford-Jones, 320–44. Oxford: British Academy and Oxford Press. https://doi.org/10.5871/bacad/9780197265031.003.0013.

Urton, Gary, and Anthony Aveni. 1983. "Archeoastronomical Fieldwork on the Coast of Peru." In *Calendars in Mesoamerica and Peru: Native American Computations of Time*, ed. Antony F. Aveni and Gordon Brotherston. BAR International Series 174. Oxford: British Archaeological Reports.

Weismantel, Mary. 2004. "Moche Sex Pots: Reproduction and Temporality in Ancient South America." *American Anthropologist* 106 (3): 495–505. https://doi.org/10.1525/aa.2004.106.3.495.

Whitehead, Neil L. 2003. "Introduction." In *Histories and Historicities in Amazonia*, ed. Neil L. Whitehead, vii–xx. Lincoln: University of Nebraska Press.

Willey, Gordon R. 1945. "Horizon Styles and Pottery Traditions in Peruvian Archaeology." *American Antiquity* 11 (01): 49–56. https://doi.org/10.2307/275530.

Willey, Gordon R. 1951. "The Chavín Problem: A Review and Critique." *Southwestern Journal of Anthropology* 7 (2): 103–44. https://doi.org/10.1086/soutjanth.7.2.3628619.

Willey, Gordon R., and Philip Phillips. 1955. "Method and Theory in American Archaeology II: Historical Developmental Interpretation." *American Anthropologist* 57 (4): 723–819. https://doi.org/10.1525/aa.1955.57.4.02a00060.

Willey, Gordon R., and Philip Phillips. 1958. *Method and Theory in American Archaeology*. Chicago: University of Chicago Press.

Zuidema, Tom. 1964. *The Ceque System of Cuzco: The Social Organization of the Capital of the Inca*. Leiden: E. J. Brill.

Zuidema, Tom. 1990. *Inca Civilization in Cuzco*. Austin: University of Texas Press.

Zuidema, Tom. 2010. *El calendario inca: Tiempo y espacio en la organización ritual del Cuzco, la idea del pasado*. Lima: Fondo Editorial Del Congreso del Perú y Fondo Editorial de la Pontificia Universidad Católica del Perú.

TWO

The Historicity of the "Early Horizon"

MATTHEW SAYRE

INTRODUCTION

Archaeological chronologies have always been produced in order to document change over time and across space, but their emphasis on the material remains of the past has often led to a conflation of ceramic styles with cultural types (Hodder 1992; Trigger 2006). This in turn has led to a criticism of archaeology as simply documenting "pots marching across space" (Hodder 1982: 37). However, even in early archaeological approaches there was some flexibility inherent in this model, and cultures were rarely presented as static and unchanging relics of past societal forms.

Today, when archaeologists consider the changes that occur over millennia, many turn to the concepts developed by historians working in the Annales School. The vocabulary developed to account for the different scales of time (*longue durée*, *moyenne durée*, and the *événement*) give us a sense of how the dynamics of geography, climate change, and individual action shape and transform culture and history (see also introduction [chapter 1], this volume). However, this approach focuses on material processes that influence history, and it can limit discussions

DOI: 10.5876/9781607326427.c002

of narratives and the silencing that occurs in history (see also Roddick [chapter 3], Chase, this volume [chapter 5]). One of the means of considering how these forms interact together and how history and fiction are created was discussed by Michel-Rolph Trouillot (1995: 29): "What happens leaves traces, some of which are quite concrete—buildings, dead bodies, censuses, monuments, diaries, political boundaries—that limit the range and significance of any historical narrative. This is one of the many reasons why not any fiction can pass for history: the materiality of the sociohistorical process (historicity I) sets the stage for future historical narratives (historicity II)."

The material structuring of history differs from the narrative representations of history that are inevitably produced to define and manage our sense of this past time. In this discussion of history, the silences, or what is not considered to be a historical fact, are just as important as what is discussed. This sense of history is a palpable means of engaging with the past for archaeologists. After all, the changes in "material culture" and technology are the primary artifacts of the past that can be interpreted by archaeologists. Pots, buildings, tools, means of subjection, means of transportation, and means of production all shift over time. In current times the graphic illustrations of rising global carbon levels, documented in the Keeling Curve, provide a direct example of how material forces continue to shape future historical narratives. One concept that the archaeological project at Chavín de Huántar (Peru) has investigated is how to connect the radiocarbon record with the broader issue of change over time and space and how this change in cultural material connects to variation in the lives of the people who populated the past at Chavín (Rick et al. 2011). An exclusive focus on ceramic seriations can lead to a silencing of how aspects of daily life change over time.

This chapter follows this idea and critiques the standard narrative surrounding the "Early Horizon" and presents a reconsideration of the site's occupation history that reveals additional evidence about daily lives, changing material culture, and different relationships with the landscape and outside people. Here I focus on how the archaeological site of Chavín de Huántar, Peru, came to be known as the "mother culture" of the Andes and how the indigenous archaeologist Julio C. Tello used it as a basis to contradict outsiders' biases about the origins of "civilization" and monumental architecture in the Andes. After considering some of the critiques of master sequences in the Andes, in particular how they are poorly applied in the case of the "Early Horizon of Chavín," I turn to examine how alternative data sets and perspectives can be used to diversify our notions of the Andean past. Thus, an examination of the archaeological data from Chavín will permit a more nuanced analysis of the everyday temporal and material realities for the ancient residents and visitors of the center ("historicity I"). In turn, the analysis will allow for new considerations of historical narratives ("historicity II").

The production of history in archaeology both relates and differs from other genres of historiography (Lucas 2012). As Alison Wylie (2008: 190) stated: "To understand how silences arise, Trouillot attends to four moments in historical production, each of which has an archaeological counterpoint: the generation of textual traces, the compilation of these traces as an archive, the retrieval of traces as facts to be built into historical narratives, and the construction of narratives that have retrospective significance. The factors symmetrically shaping ignorance and knowledge figure at each of these junctures as they have been engaged by archaeologists in the context of ongoing debate about disciplinary status and standards."

HISTORICITY II: CULTURE HISTORY, REGIONAL PERSPECTIVES, AND THE SILENCING OF DAILY LIFE

The historical narratives archaeologists create can silence aspects of the past. In the case of South American archaeology there is a clear history of silencing aspects of the past, and these narratives often originate from the point of contact. The Spanish conquest forever marked the people and history of the Andes. While the brutality of the conquest still lingers to this day, the documents created by the early conquistadors also continue to shape our perceptions of the Andean past. Chroniclers like Bernabé Cobo ([1653] 1979) and Martín de Murúa ([1605] 1986) provided scholars with a picture of the Inka Empire, and the Inka elite who often spoke to the chroniclers through translators heavily influenced this portrait (see also Chase [chapter 5] and Seoane and Culquichicón [chapter 8], this volume).

The early chroniclers noted that the Inkas seemed to take credit for most of the monumental past in the areas they had conquered (Cobo [1653] 1979). There was little acknowledgment of the accomplishments of other cultures and peoples. This initially created the mistaken impression that there was not a deep tradition of building monumental architecture in the region. It would be left predominately to archaeology to rectify this situation.

Julio C. Tello first visited the site of Chavín de Huántar in 1919 as part of his larger survey of the Marañon River (Tello 1922, 1943). Tello was conducting surveys of the highland region and searching for evidence for the independent development of Andean civilization, something he believed to have begun in the highland region of Peru (Tello 1943). Chavín fell into the preconceived notions of what an early megalithic center should look like. The physical monumentality of Chavín provided such clear evidence of elaborate planning that it could easily refute foreign archaeologists, such as Max Uhle (1902), who believed that Andean civilization diffused from Mesoamerica.

The formal rediscovery of the temple did not initially lead to extensive excavations; rather, the first visits to the temple were devoted to cleaning and mapping.

After the site was cleared, a long parade of archaeologists would begin to excavate and analyze the site. Little of this early work challenged the assumption that Chavín was the primordial site of Andean civilization. In recent years (Shady Solís 2005) this view has been debunked, but modern-day Andean archaeologists are often times just as concerned with finding the oldest cities or cultures in the region. The search for the earliest site recapitulates social evolutionary biases.

While most Andean scholars are familiar with the claims and writings of Julio C. Tello, many archaeologists are not familiar with the literature that informed his intellectual writings. His pronouncements on early ritual and the basis of Chavín have been repeated time and again. Yet returning to his own scholarship, it is clear that much of it was based on intuitive formulations derived from the writings of theologians, philosophers, and chroniclers (Tello 1922).

The early chroniclers wrote their texts in an age when divine influence on the material world was assumed. Much of Guaman Poma de Ayala's ([1615] 1956) texts and drawings depict sacred images and try to correlate Inka rites with Christian themes. While modern scholars may gloss over this relationship and attempt to use Guaman Poma's work without any mention of its religious background, he lived at a time and place in which religious thought still pervaded much of the social sciences. While there is great chronological and cultural difference between the two, Tello himself was also predominately educated in religious institutions (Tantaleán 2014).

Tello's work contained profound insights, and many archaeologists continue to follow his thoughts on the nature and timing of Chavín. One of his most important contributions was his idea that the site functioned as a ritual center (e.g., Tello 1943). This claim is now part of common wisdom, most likely because archaeologists have found no irrefutable evidence to contradict it or determine its correctness. In addition, there is little evidence to support claims of military rule or extensive control over other sites outside of the Callejón de Conchucos, the valley where Chavín de Huántar is located. Tello espoused a theory that continues to exert influence on Andean constructions of chronology and social change. His concept of Chavín as an enduring Andean essence whose primary purpose was religious continues to exert a strong influence over our conceptions of Chavín and the florescence of Andean civilization (Burger 1995). This emphasis on religion and pilgrimage led to a historical narrative that silences other aspects of Chavín, such as the role of trade, markets, and common people in the construction of this monument.

Tello ensured that Chavín is generally granted its own unique place in Andean history. While many other important archaeological sites are clearly associated with armies and militaristic functions, Chavín has been considered to be a site of ritual pilgrimage. As a ceremonial place of convergence, the site could anchor a narrative that foregrounded religion and fine art, material traces associated with a certain form of advanced culture. John Rowe (1967) expanded on Tello's publications and he created a master sequence ceramic for the Andes, divided into horizons and periods, which provoked critiques. His statements acknowledged some of the inherent weaknesses of these approaches.

As Rowe (1967: 12) stated: "The negative advantage of using periods rather than stages to organize archaeological data is that the investigator runs much less danger of assuming what he set out to be trying to prove. Because of the close association of stages with the theory of cultural evolution, virtually every archaeologist who uses stages to organize his data thereby builds into them certain assumptions about cultural development without being aware that he is doing so."

The metanarratives of preconquest history constructed by Andean archaeologists often cover enormous geographical distances. In many instances, entire stretches of Argentina, Bolivia, Chile, Ecuador, and Peru are included in the descriptive narrative. While the Andes are thought to be one of the five regions of the world that gave rise to pristine states (Burger 1995; Moseley 2001; Stanish 2001), many of the metanarratives of the Andes are profoundly influenced by what we know about the Inka. The Inkas conquered communities in all the above-listed countries (Rowe 1944), and this territorial expanse is often uncritically projected into the past when analyzing pre-Inka civilizations (see for instance Schreiber 1992 and the "Wari Horizon"; see also the introduction [chapter 1] to this volume).

My discussion of Chavín's history will refer to various chronologies, but I will focus on the ones listed below (tables 2.1 and 2.2), as these are the more relevant for this specific Andean region. Thus, the previous metanarratives created for the region stand in contrast to the more nuanced understanding of people's lives in the past that is possible if we examine both material and historical traces. The tables below document chronologies of particular importance to Peru. The Peter Kaulicke (1998) chronology uses similar terminology to that of Rowe, however it emphasizes the Formative, somewhat resonating with the chronologies currently employed in Bolivia and Ecuador (Hastorf 2005). The use of the word "Formative" is becoming more common across the Andes, but as will be discussed below, the use of a term with such evolutionary meaning comes with embedded assumptions about the social-cultural evolutionary stage and lives of the people in a time period that makes some archaeologists uncomfortable.

North American Archaeologists generally used to rely on Rowe's chronology. The two chronologies presented above use different dates, partially because Kaulicke's work (table 2.2) was written later in time when many more radiocarbon dates were published.

In the Andes there is a tendency to define time periods based on ceramic stylistic change. Rowe (1967: 2) begins by defining stages: "Complex stages are stages defined on the basis of several different features which are supposed to

TABLE 2.1. Rowe's (1967) Andean chronology

Time Period	*Dates*
Preceramic	3000–2100 BC
Initial Period	2100–1400 BC
Early Horizon	1400–400 BC
Early Intermediate Period	400 BC–AD 550
Middle Horizon	AD 550–900
Late Initial Period	AD 900–1476
Late Horizon	AD 1476–1532

TABLE 2.2. Kaulicke's (1998) Formative Andean chronology

Time Period	*Dates*
Arcaico Final	1800–1500 BC
Formativo Temprano	1500–1000 BC
Formativo Medio	1000–600 BC
Formativo Tardío	600–400 BC
Formativo Final	400–200 BC
Epiformativo	200 BC–AD 100/200

occur together. If, for example, we note a frequent correlation of pottery with evidence for farming, animal husbandry, and loom weaving in the area we are studying, we may be tempted to use all four of these features together as criteria for a single cultural stage."

While this is a useful guide into how a variety of criteria can be employed to describe the past, as we will see, the inclusion of cultural forms (material culture) other than ceramics has rarely been undertaken in the Andes. As Wylie (2008: 191) stated, "What material archaeologists recover depends not only on what is visible, accessible, and technologically tractable but also on what archaeologists find interesting, puzzling, and relevant to current concerns." In Rowe's original formulation of Andean chronology, he did attempt to list some of the technological advances that were associated with each time period (Rowe and Menzel 1967), and it should be noted that these foundational scholars were not initially aiming for nuance; they were attempting to document broad historical changes. In a later section I will discuss how the inclusion of various types of cultural material could broaden our sense of how people lived their daily lives at different points in time. There is a general trend to ignore these other material traces (Trouillot's historicity I) that silence other narratives of Andean history (Trouillot's historicity II). As Rowe notes: "The inherent weaknesses of the method of using complex stages as a framework for interpretation appear only much later when the relative chronology can be made more precise and other local sequences are established. Unfortunately, by this time everyone is accustomed to thinking in terms of the traditional stages, and it is very difficult to give them up and start afresh with a more productive system" (Rowe 1967: 3).

Here we see another profound insight, as it has become tremendously difficult to convince scholars that thousands of square kilometers cannot be defined by the ceramics encountered in a single river valley. In the statement below, Rowe lays out how his master sequence was created:

> The use of a master sequence to control a system of periods can be illustrated from Peruvian data. In applying the idea to Peru I chose the Ica Valley as the most convenient point of reference, because we had made the best start toward the establishing a detailed chronology of the pottery styles of the valley. We could then say that the Initial Period begins at the time of the introduction of pottery in Ica. The Early Horizon is the time from the first appearance of Chavín influence at Ica until polychrome slip painting replaces resin painting in that valley. The Middle Horizon is the time from the beginning of Phase 9 of the Nasca style at Ica to the beginning of Phase A of the Chupalca style. The Late Horizon is the period from the beginning of Phase A of the Tacaraca style at Ica until the beginning of Phase B of the same style. Since we believe that the beginning of Tacaraca A style coincided in time with the beginning of the Inka occupation of the Ica valley, we assign the traditional date of the Inka conquest of those areas, about 1476 AD, to the beginning of the Late Horizon. (Rowe 1967: 10)

The problems and biases inherent in a method that relies exclusively on ceramic styles should now be apparent in that it silences other historical narratives. The above citation thus encapsulates Rowe's exact basis for the timeframes used in the Andes. While the language justifying the master sequence is technical and seemingly precise, later analysis has revealed that this precision mechanically equated shifts in ceramic styles with historical developments. In addition, while Rowe did acknowledge that other technologies and changes in lifestyles occurred over time, his chronology focuses almost exclusively on ceramic styles and their evolution. The Rowe chronology also raised the problems inherent in assuming that cultural evolution is uniform.

Rowe's (1967: 12) critique of circular reasoning is a salient reminder that not only stages can become self-reinforcing. The same can be true of periods, and this issue is of particular importance at the site of Chavín. It is possible to argue from the ceramic and architectural data that there was never a culture as materially sophisticated as Chavín to exist in the Callejón de Huaylas or the Callejón de Conchucos after the site of Chavín ceased to function as a major ceremonial center. The major post-Chavín sites, such as Willkawaiyan, are often viewed as pale imitations of the site itself. However, there is ample literature to challenge the notion that singular examples of monumental architecture should serve as the chief marker of cultural sophistication (Lau 2014).

One aspect of Rowe's six-stage chronology that is rarely discussed is his reliance on schema developed by Flinders Petrie (Tantaleán 2014: 87). Petrie attempted to create a cohesive chronology of the Egyptian Dynasties, and there were three major kingdoms known to archaeologists in the early twentieth century, generally referred to as the Old, Middle, and Late Kingdom. Petrie also created chronological periods of transition and disruption between these kingdoms. These intermediary periods designated times when major dynasties had not united the entire Egyptian realm. This top-down interpretation of history, in which the dominant power of royalty is used to mark all peoples' experience of history, is one means of creating a unified and homogeneous sense of time for a macroregion.

The decision to regard the Early Horizon as the first important time marker in the Andes meant that the thousands of years of human history that existed before this period were glossed over in a few spaces in the chronological chart. While Rowe and colleagues were able to identify three major horizons that had evidence for the importation of ceramic styles from different monumental centers, these horizons may have been constructed differently if the type-site had been located on the North Coast of Peru (Isbell and Silverman 2006; Koons and Alex 2014).

Tello (1943) was predominantly responsible for bringing Chavín into the public eye. As discussed above, the site meant much more to him than other sites. As an example of early monumental architecture in the highlands, it was proof that there were local antecedents to the sophistication of the Inkas. This point had clear political implications in the early twentieth century, when the Peruvian economy was dominated by coastal elites who often held derogatory ideas about people native to the highlands (Tantaleán 2014: 52).

The original vision of Chavín, seen by early colonial visitors, was that of an enormous castle. While the townspeople were not clear on who built the site, some said that it was built by the Inka (Mesia 2007: 3). These vague statements on Chavín's origins, as well as the impressions of early colonial visitors, formed the foundation upon which later work would be established. The postconquest history of archaeological work at the site is well known (Contreras 2007: 14; Mesia 2007: 1–14).

The Chavín Horizon

The Chavín Horizon is stylistically defined by its distinct and surreal iconography, the Lanzón sculpture, beings with agnathic (having no jaws) teeth, the pendant eye, and the Janabarriod ceramic style. These traits are all based on findings at the site of Chavín de Huántar. The staff god and some of the core Chavín imagery may appear earlier at other sites west and south of Chavín, suggesting that this site is more of a culmination of this style than the instigator of it (Shady Solís and Leyva 2003). As Richard Burger (1985) and Williams Leon (1985) noted, there were clear antecedents, such as Cardal and Garagay, on the coast for many of the architectural forms and iconography at Chavín.

The focus on monumentalism and aesthetics has been so prevalent that there have been few attempts to discuss how the changing nature of food production or gender relationships could be brought into broader discussions of chronologies over time. As these aspects of past lives have not been the center of much research, they have rarely been included in discussions of what cultural material, beyond iconography, marks the Early Horizon.

There are many problems with using the term "Horizon" to describe a time when its type-site is not considered to be a state or the sole major center by most archaeologists (Stanish 2001). While the issue of whether the term "Early Horizon" should continue to be employed will not be resolved here, it is worth stressing the problematic history of this term.

The occupation history of Chavín is still being debated (Burger 1995; Rick et al. 2011). As this is the case, we need to consider referring specifically to the site chronology rather than pan-Andean chronology. The Janabarriu phase is at the center of the debate surrounding Chavín's iconography and chronology. Burger (1984) first defined the Janabarriu ceramic style. That definition emphasized polished red and black wares and "the frequency of designs made by stamps and seals. Circles, circle-dots, S's, and other stylized designs . . . are impressed in rows on the exteriors of bowls, cooking pots, plates, and cups" (Burger 1995: 170). While this definition is broad, it is useful and can be used to describe a wide variety of ceramic types found in the Andes at sites outside of the Callejón de Conchucos.

The Janabarriu phase was associated with Rowe's New Temple architectural phase. The exact linkage of three ceramic styles with three architectural phases may have been an interesting chronological find, but there is ample reason to believe that the Janabarriod ceramic style had a longer lifespan than that initially proposed of 390–200 BC (Rick et al. 2011). This finding overturns the proposed contemporaneity of the ceramic style with the architectural phase. Therefore, when I refer to Janabarriu or Janabarriod, I am referring strictly to a ceramic style and not the dates discussed above. This attempt to link two forms of material evidence—ceramics and architecture—was problematic, but it is relevant to this discussion of the diversity of experiences that were prevalent during the time when the temple was built and occupied.

More recent work at Chavín has established that the Black and White phase coincided with a major construction initiative of the ceremonial core of the site that lasted from approximately 1000 to 550 BC (Rick 2008: 11). This phase is associated with Janabarriod ceramics. During this phase, almost all of the galleries in

Buildings A and B, as well as in the East Area, were constructed (Kembel 2008: 44). "In the final monumental stage, the Black and White Stage, site-wide additions, including plazas, terraces, and open staircases, are built with high levels of symmetry, decorated fine stonework, and standardized galleries (Kembel 2008, 45)." This is the time when Chavín reached the height of its supraregional appeal. This change in the spatial layout of the site reflects an increasing importance in controlling how visitors experienced the temple as there was a transition from communal to enclosed spaces.

One useful attribute to the phasing is that it does not postulate a necessary one-to-one correlation between ceramic styles and building construction events or styles. In addition, this timeframe covers the majority of the occupation of the sectors of La Banda that will be discussed below.

Culture historical approaches thus silence various aspects of the past. The material roles of plants, animals, and domestic artifacts rarely inform these broad historical narratives. Rather, we are left with a depiction of ideologies portrayed on pottery that serve to illustrate the entire histories of regions and people across broad swathes of time and space. This limits our understanding of everyday life and how people reacted to and modified the material world around them.

HISTORICITY I: TRACES OF PROCESS AT CHAVÍN DE HUÁNTAR

In general, critical discussions of chronologies acknowledge that creating broad frameworks of change over space and time inevitably end up minimizing variability across sites. A large monumental site such as Chavín de Huántar has many discrete sectors (table 2.3) with varied architecture and life histories (figure 2.1). Below I discuss some of Chavín's sectors and what the radiocarbon dates from these areas reveals about their use over time. The Monumental Center of Chavín has been the focus of most of the research conducted at the site over nearly the past 100 years. The work has focused on the "Old Temple," "New Temple," Circular Plaza, Square Plaza, and buildings surrounding these areas (Contreras 2007: 14; Kembel 2008; Lumbreras 1989; 2007; Mesia 2007). The Stanford Project at Chavín has challenged some of the older assumptions about the construction sequence at the site and the division between the "Old" and "New" Temple. The recent radiocarbon dating of the Circular Plaza has placed the construction and use of this feature of the site between 900 and 500 BCE (Rick et al. 2011: 115). Galleries located near the circular plaza, such as the Caracoles Gallery, fall well within this timespan.

The West Field has been the subject of less archaeological work than the Monumental Center. However, this sector may offer greater possibilities in the future as it has been subject to less modification and cleaning. The area undoubtedly contains terraces and smaller structures, but it may also be associated with

TABLE 2.3. Sectors of Chavín and approximate radiocarbon range of concentrated occupation (from Rick et al. 2011)

Sector	*Dates*
West Field	1000–600 BC
Wacheqsa	1100–550 BC
La Banda SM	900–500 BC
Black and White phase of the Monumental Center	900–500 BC

monumental constructions and galleries (Contreras 2007). Daniel Contreras led recent excavations in this region, which uncovered a *mito*-style hearth and associated constructions (Contreras 2010). This hearth and associated deposits also date to between 900 and 500 BCE. This is also within the time range of the Black and White phase, when Janabarriod-style ceramics were in use at Chavín. The presence of a ritual structure and the lack of evidence for production in the sector reveal that this space was used and experienced in a significantly different fashion from the Wacheqsa or La Banda sector.

The Wacheqsa zone is an area located to the north of the Monumental Center. This sector is located near the Wacheqsa River, and it has been the subject of intermittent research. The excavations led by Christian Mesia uncovered a complex sector that had tremendous stratigraphic variation (Mesia 2007). Terraces, platforms, middens, and canals have been uncovered in this sector. Its location near canals and the river likely means that parts of the Wacheqsa zone were constantly in use, even before the construction of the Monumental Center. A series of radiocarbon dates were obtained from materials deriving from this sector dating between 800 and 500 BCE, but there are also dates that fall outside of this range (Rick et al. 2011: 118). It has also been argued that the middens in Wacheqsa contain evidence for feasting or communal food consumption events (Mesía Montenegro 2014).

The La Banda zone is located to the east of the Monumental Zone, across the Mosna River, and it encompasses a large area associated with occupations from the preceramic times up until the present (Sayre 2010). Recent excavations in the sector have uncovered a concentrated domestic area with evidence for houses, production areas, patios, and well-organized pathways (Sayre 2010; Sayre, Miller, and Rosenfeld 2016). The upslope area (Sector 3, Unit 5A) east of the central occupation (Sector SM) is characterized by long occupations and socially stratified settlements. This upslope sector dates to between 1250 and 400 BCE, whereas the concentrated domestic settlement in Sector SM dates to between 900 and 400 BCE. It is likely that this sector represents a single cohesive occupation, and the dates concentrate to between 800 and 600 BCE. It deserves mention that a high quantity of Janabarriod ceramics were recovered from La Banda.

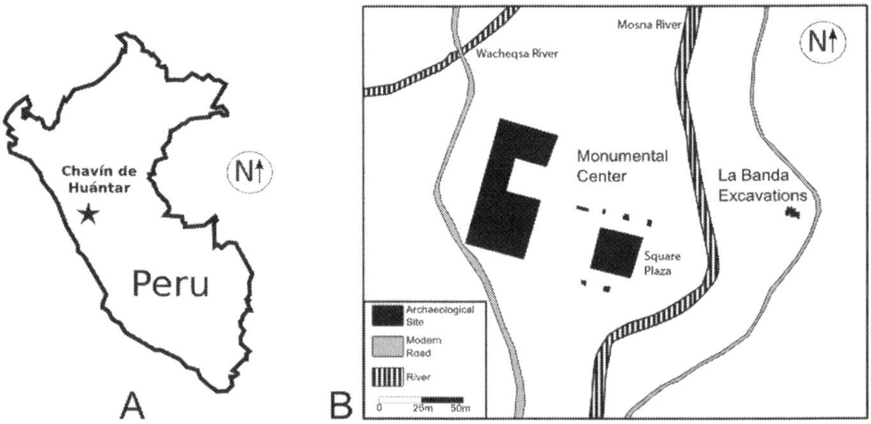

FIGURE 2.1. *Location of Chavín de Huántar in Peru and location of site sectors. (Modified from Sayre 2010: 65. Thank you to M. Miller and D. Stansbury for assistance with the initial image.)*

A reanalysis of carbon dates within changing contexts offers new possibilities for an investigation of historical change in Chavín. Charles Stanish (2003: 85–89) has written that historical chronologies based off of radiocarbon dates do not presuppose gradual transitions between periods and avoid some of the developmental assumptions inherent to evolutionary chronologies (see Roddick this volume [chapter 3]). Two quotes, in particular, are worth analyzing in order to understand how the Stanford Project at Chavín de Huántar is attempting to understand change over time at the site.

As John Rick et al. (2011: 125) stated:

> Moreover, we expect to have shown that it is necessary and productive to pay careful attention to the radiocarbon date. A radiocarbon date is a proposal, not a conclusion; it has individual implications, and in groups they can yield reliable information but only if it is on their own terms by taking advantage of their relationship with time, not due to an alleged exact identity with a time period. Researchers must become accustomed to, and celebrate, ambiguity and blurred borders over time. In Fig. 25 the authors of this paper attempt to demonstrate a summary of the contentions of, to the extent possible, the inaccuracies as well as the position that much of the changes over time cannot be defined by rigid divisions. This is not only because the methodology used does not allow such precision, but rather, because the original phenomena, changes in time, were probably gradual and a serious study should approach the reality of the past with the realization that the past should not be limited by inadequate methodologies and perspectives.[1]

As stated above, the material record limits some interpretative possibilities. It also impacts conclusions about relationships between different materials that are our guides into activities in the past:

> The observations presented here lead to a chronological summary of what is known of Chavin that will serve as a basis for future work and will be corrected and changed in the future. The relationship between dating, ceramics, and architecture will always be difficult, and we should be cautious in inferences that are made from the dating of residential areas, monumental areas and, especially, the construction of monumental buildings. Within these observations you can find a good part of the recent history of problems of dating Chavin. The authors argue that there are large and varied perspectives that can take advantage of these historical lessons about archaeological time.[2]

The recent dating work reveals how the project is less focused on creating an absolute chronology of what used to be known as the Early Horizon; rather, they focus on local temporalities and domestic rhythms. Our understanding of small-scale material transitions informs our understanding of the broader changes that were occurring in the larger society.

The radiocarbon record permits greater nuance in understanding how people lived in and constructed sites. The gaps in key portions of the record remind us that time periods are not uniform; rather, they are defined by both moments of heightened activity and long lulls in monumental construction. We can thus see that phases are not monotonous blocks of time that lump all lives together in singular activities. For instance, our evidence for the production of ceremonial goods, possibly employed by priests, is not uniform. The concentrated evidence for this type of activity gives insights into how rituals and their associated paraphernalia changed over time, sometimes in conjunction with changing architectural forms, and sometimes occurring independently of those changes.

ANOTHER KIND OF HISTORICITY II. CONSTRUCTING DAILY LIFE

The question of what can we learn about peoples' everyday lives at Chavín is an important one. In the end, macroscale social and political change can only be understood in terms of microscale reconstitutions in gender roles, food consumption craft production, and so forth. Following Trouillot, Wylie (2008: 196) states, "These strategies of repositioning evidence and critical historiography, supplemented by Trouillot's third strategy—that of exploiting dissonance among interpretive conventions to identify points at which projective conventions may be operating—all play a role in identifying and countering simplistic models of social organization that have dominated fourth moment interpretive and explanatory theorizing." Thus, the material traces of this history allow us to propose different narratives, historicity II, that are presented about quotidian life at Chavín.

The evidence for food production in the domestic La Banda sector comes from the analysis of macrobotanical and microbotanical remains recovered at Chavín (Sayre 2010). While nonwood macrobotanical remains were recovered in the sector, these remains were sparse. It is evident that the majority of the plants consumed in the La Banda sector were highland plants native to the area. It appears that much of the food at La Banda was produced in the fields to the east of the settlement, but there is limited evidence that the inhabitants of La Banda spent the majority of their time working in these fields. These fields would have been accessed by various members of the community, and most likely by people of all ages. The limited evidence for farming implements in the sector may be because the La Banda inhabitants removed them from their homes, or posttemple occupants may have removed them, as these tools would have been useful for any highland farmer.

To provide another example, the residents of La Banda and greater Chavín were able to manage their fuel use over time. One striking find from this project was the lack of dung in the macrobotanical samples. While some dung may have been used in fires and later destroyed by taphonomic factors, dung is commonly identified in the archaeobotanical samples of other areas of the Andes, such as Xauxa and the Bolivian altiplano (Bruno 2008; Hastorf 1993; Whitehead 2007). At Chavín, wood appears to have been plentiful and well managed, presumably as an ecological strategy similar to those discussed in field rotation and fallow systems. The dung from camelids and other animals was likely used as a fertilizer in fields with wood being the preferred fuel item. The long-term management of forest resources not only points to successful ecological practices but also means that further work is needed regarding the political and social processes that led to the rise and decline of the main temple.

There is limited domestic evidence of changing gender roles over time at Chavín. The lack of burials and human remains makes this area of inquiry difficult to follow, as does the fact that some degree of cultural essentialism would be necessary in order to analyze the role that certain domestic artifacts, such as clod breakers and weaving awls, played in the lives of people at Chavín. These types of artifacts have been recovered at Chavín, but they have rarely been found in primary contexts or burials. The art from the Monumental Center does provide some evidence of men and women using different styles of dress, but this area of research has yet to be expanded (Lumbreras 2007).

These new data sets on domestic life are often ignored or underemphasized even when they are a part of the research project; however, they can reveal a more nuanced understanding of social change at Chavín. These data sets (dung, faunal remains, wood use, etc.) are rarely the material traces of the past that are interpreted as chronological indicators, historicity I, but they can be used in such a fashion (see Asch and Asch Sidell 1988). The macrobotanical record from the site does not indicate that diets and household configurations changed in lockstep with major shifts in temple use (Sayre 2010). This silencing of the narratives surrounding daily life is important to consider as we examine how peoples' lives changed over time.

The team at Chavín has collected the fine-grain evidence that provides insights into the history of production, consumption, and trade at the site. Some of the evidence allows insights into the quieter aspects of daily life, such as the production of artifacts for religious specialists to wear in the temple (Sayre, Miller, and Rosenfeld 2016). The dates from La Banda and the West Field reveal that these spaces, outside of the Monumental Center proper, have their most marked occupation in the 800–600 BCE time period. This is the time that the Black and White phase of temple construction, the Janabarriod ceramic style, and Middle Formative Time Period were all at their height (Rick et al. 2011: 127). It demonstrates that there can be discontinuities between the early monumental construction, which occurred before these dates, and the construction of outside domestic and semiprivate ceremonial spaces. The patterning of occupations expands and contracts in relation to events and construction in the main monument, yet they did not directly coincide with the initial significant architectural events. There is significant evidence that goods produced in La Banda were for priestly use in the temple area, yet the La Banda sector appears to have only been significantly occupied for a portion of the time in which the monument functioned. This expanded analysis allows us to see different use patterns and occupation histories at Chavín. In many instances, this type of analysis requires intensive lab work and analysis, the type of work that is not the immediate focus of rapid interpretation. The results of future research in this area will enrich our understanding of how daily life changed over time for the inhabitants of Chavín.

DISCUSSION: SCALES OF CHANGE AT CHAVÍN DE HUÁNTAR

In this chapter I have attempted to present sequences of traces of change—an issue that links earlier interpretations of Chavín (Burger 1995; Rick et al. 2011; Rowe and Menzel 1967; Tello 1943) to new narratives about life at the site. This research now focuses on more fine-grained interpretations of differences across Chavín and changing patterns of site use across time. This movement away from a sole focus on temple construction has exciting chronological implications.

The Early Horizon is a case study in how one type-site and one ceramic style can be taken as emblematic of the experience of an entire macroregion. These temporal metanarratives can limit our sense of how people lived their lives in the ancient Andes. That being said, there is value in examining large-scale processes and how they impact societal change over time. Two prominent examples that are well documented in the current anthropological literature are the case of increasing caloric (energy) consumption by different societies over time and the Keeling Curve. Ian Morris (2010) has demonstrated that many preindustrial empires reached a technological hard barrier, beyond which it was not possible to expand production further. Energetic and material limits constrained the size and ambitions of empires, yet surges in energy production could lead to imperial expansion. The industrial era and the consumption of fossil fuels have dramatically changed the amount of energy available to people on earth. The Keeling Curve is one record of the carbon pollution that has resulted from this production of energy. While calories and carbon may not immediately appear to be material parts of the sociohistorical process (historicity I), they are nonetheless material markers of human action that dramatically alter future historical narratives and possibilities (historicity II). The local changes that we will all experience in the future—whether conflict, drought, migration, or changing notions of what it means to be a citizen—will be shaped by these material processes. Many of the debates about how to mark the beginnings of the Anthropocene (or whether to state that it coincides with the Holocene) focus on the material traces of our current era, whether that is the inclusion of plastic in geological strata or the radioactive fallout in soils, revealing just how much material forces continue to shape history (Certini and Scalenghe 2015).

When we look to the past, it should be clear that while horizons and stages have inherent problems, the issue of categorizing change across time and space will always be a challenge for archaeology and anthropology. The material record from the Andean past clearly demonstrates that time periods should be delineated through a consideration of a variety of different and overlapping material forms—including architecture, food, settlement patterns, ceramics, energy use, and so forth.

Rowe's horizon chronological system in the Andes began with the Early Horizon of Chavín. The later horizons, middle and late, are defined by the rise of states or empires—the Wari, Tiwanaku, and the Inka. However, there has always been debate about the limited role of the polities of the North Coast of Peru in these chronologies. While some people argue that Pampa de las Llamas-Moxeke or Chavín were states (see Stanish 2001: 52), most scholars do not agree with this point of view. The Moche were considered to be an example of an early state in the Andes, but they are generally not referred to as an archaeological horizon marker. The varied application of the term "horizon" is an example of the nuance that is lost when applying this term across an entire region.

The chronology of Chavín de Huántar has long been used to mark an entire time period across a broad part of the ancient Andes. The complexity of the forms of cultural material has led the Stanford Project to return to a focus on radiocarbon dating deriving from a distinct historical context of practice, rather than stylistic-based, chronologies. The new chronology focuses on the site and builds from across a variety of contexts. As of yet, this bottom-up chronology is not being developed in order to refine our sense of the entire Formative Period in the Central Andes. Rather, this chronology is constructed with the knowledge that new dates will force reevaluations of life at the site.

A local focus on the occupation at the temple of Chavín and the coeval occupations of sectors surrounding the Monumental Center—such as the construction of ritual space in the West Field, middens in the Wacheqsa Sector, or production areas in the domestic areas of the La Banda Sector—has led to a heightened sense of diversity in the material patterns and temporal rhythms of life across the site. It is clear that Chavín has little in common with the other "horizons" that are used to mark the reach and power of ancient Andean empires. As such, it serves as a prime example of how we can begin to think differently about how people experienced time in the Andes. The passage of generations, and the changes in the lives of people at Chavín over that time, force us to consider just how little knowledge we gain from placing cultures in boxes. Rather, by examining how life was experienced across the site, we reveal a more complete picture of the temporality and historicity of Andean peoples.

This chapter presented evidence on how early scholars who worked at Chavín attempted to look at large-scale processes and linked these events to particular ceramic forms. While a broad time and scale approach based on the most ubiquitous archaeological material was perhaps necessary at the beginning, a current examination of this early work reveals that it does not provide a nuanced picture of life in the past. This is not to state that ceramics do not provide valuable insights into past histories, or that overt material culture finds are not significant. After all, the carbon record of the Anthropocene, as seen in the Keeling Curve, shapes our own past and future. However, our microapproach to the past and the examination of traces of past daily life enrich this experience. We are beginning to engage in scholarship that investigates the rhythms and patterns of domestic life at Chavín, and there will be more of this work in the future (Sayre, Miller, and Rosenfeld 2016; Rosenfeld and Sayre 2016). There are values to both approaches, but here I have presented evidence about how a focus on certain aspects of material culture (mainly ceramic styles) forces us to think in terms of horizons. In the case of the Early Horizon, it is clear that this silences many aspects of history.

The stakes to this argument are high. The silenced aspects of past life do not have to remain silent. Although all narratives inevitably highlight certain experiences and concepts, a continued effort to expand our knowledge of past life beyond ideas embedded in ceramic chronologies will allow scholars to try to understand Andean temporalities and historicities through a more emic lens. These new considerations permit us to consider both the material history of the past as well as the traces that shape future historical narratives.

NOTES

1. Translation by author. "Por otra parte, se espera haber demostrado que es necesario y productivo prestar atención detenida y explícita a los fechados. Un fechado radiocarbónico es una proposición, no una conclusión; tiene implicancias individuales, y en grupo puede rendir información confiable, pero solo si se lo trata en sus propios términos por medio del aprovechamiento de su relación con el tiempo, no con una supuesta identidad exacta con el tiempo. Los investigadores deben acostumbrarse y hasta celebrar la ambigüedad y fronteras fofas en el tiempo. En la Fig. 25 se intenta una síntesis de lo que sostienen los autores del presente trabajo y se demuestra, en la medida posible, la inexactitud y, también, su posición respecto a que mucho de los cambios en el tiempo no pueden ser definidos por divisiones rígidas no solo porque la metodología utilizada no permite tal precisión, sino, más bien, porque los fenómenos originales, los cambios en el tiempo, probablemente fueron graduales y la comprensión del estudioso debe acercarse a la realidad del pasado y no que el pasado se vea limitado por metodologías y perspectivas inadecuadas."

2. Translation by author. "Las observaciones presentadas aquí llevan a una síntesis cronológica de lo conocido de Chavín que servirá como base para ser corregido y, sin duda, cambiado en el futuro. La relación entre fechados, ceramic y arquitectura siempre será difícil, y realmente se debe tener cautela en las inferencias que se hacen de los fechados de zonas residenciales, zonas monumentales y, todavía más, de la construcción de edificios monumentales. Dentro de estas observaciones se puede encontrar una buena parte de la reciente historia problemática de los fechados de Chavín. Los autores sostienen que hay grandes y variadas perspectivas que se pueden aprovechar de estas lecciones históricas sobre el tiempo arqueológico."

REFERENCES

Asch, David L., and Nancy Asch Sidell. 1988. "Archaeological Plant Remains: Applications to Stratigraphic Analysis." In *Current Paleoethnobotany: Analytical Methods and Cultural Interpretations of Archaeological Plant Remains*, ed. Christine A. Hastorf and Virginia S. Popper, 86–96. Chicago: University of Chicago Press.

Bruno, Maria C. 2008. "Waranq waranqa: Ethnobotanical Perspectives on Agricultural Intensification in the Lake Titicaca Basin (Taraco Peninsula, Bolivia)." PhD dissertation. Washington University, St. Louis.

Burger, Richard L. 1984. *The Prehistoric Occupation of Chavín de Huántar, Peru*. Berkeley: University of California Press.

Burger, Richard L. 1985. "Concluding Remarks: Early Peruvian Civilization and Its Relationship to the Chavín Horizon." In *Early Ceremonial Architecture in the Andes*, ed. Christopher B. Donnan, 262–89. Washington, DC: Dumbarton Oaks.

Burger, Richard L. 1995. *Chavín and the Origins of Andean Civilization*. London: Thames and Hudson.

Certini, Giacomo, and Riccardo Scalenghe. 2015. "Is the Anthropocene Really Worthy of a Formal Geologic Definition?" *Anthropocene Review* 2 (1): 77–80. https://doi.org/10.1177/2053019614563840.

Cobo, Bernabé. [1653] 1979. *History of the Inca Empire: An Account of the Indians' Customs and Their Origin, Together with a Treatise on Inca Legends, History, and Social Institutions*. Trans. Roland Hamilton. Austin: University of Texas Press.

Contreras, Daniel A. 2007. "Sociopolitical and Geomorphologic Dynamics at Chavín de Huántar, Peru." PhD dissertation. Stanford University, Stanford, CA.

Contreras, Daniel A. 2010. "A Mito-Style Structure at Chavín de Huántar: Dating and Implications." *Latin American Antiquity* 21 (1): 3–21. https://doi.org/10.7183/1045-6635.21.1.3.

Guaman Poma de Ayala, Felipe. [1615] 1956. *La nueva crónica y buen gobierno*. Lima: Editorial Cultura, Dirección de Cultura, Arqueología e Historia del Ministerio de Educación Pública del Perú.

de Murúa, Martín. [1605] 1986. *Historia General del Perú*, ed. Manuel Ballesteros. Historia 16. Madrid: Historia.

Hastorf, Christine A. 1993. *Agriculture and the Onset of Political Inequality before the Inka*. Cambridge: Cambridge University Press.

Hastorf, Christine A. 2005. "The Upper (Middle and Late) Formative in the Titicaca Region." In *Advances in Titicaca Basin Archaeology*, vol. 1, ed. Charles Stanish, Amanda B. Cohen and Mark S. Aldenderfer, 65–94. Monograph 54. Los Angeles: Cotsen Institute of Archaeology at UCLA.

Hodder, Ian. 1982. *Symbols in Action: Ethnoarchaeological Studies of Material Culture*. Cambridge: Cambridge University Press.

Hodder, Ian. 1992. *Theory and Practice in Archaeology*. London: Routledge.

Isbell, William, and Helaine Silverman. 2006. "Rethinking the Central Andean Co-Tradition." In *Andean Archaeology III: North and South*, ed. William Isbell and Helaine Silverman, 497–518. New York: Springer. https://doi.org/10.1007/0-387-28940-2_20.

Kaulicke, Peter. 1998. "Perspectivas regionales del Periodo Formativo en el Perú: Una introducción." *Boletín de Arqueología PUCP* 2:1–15.

Kembel, Silvia R. 2008. "The Architecture at the Monumental Center of Chavín de Huántar: Sequence, Transformations, and Chronology." In *Chavín: Art, Architecture, and Culture*, ed. William J. Conklin and Jeffrey Quilter, 35–81. Monograph 61. Los Angeles: Cotsen Institute of Archaeology, University of California.

Koons, Michele L., and Bridget A. Alex. 2014. "Revised Moche Chronology Based on Bayesian Models of Reliable Radiocarbon Dates." *Radiocarbon* 56 (3): 1039–55. https://doi.org/10.2458/56.16919.

Lau, George F. 2014. "Intercultural Relations in Northern Peru: The North Central Highlands During the Middle Horizon." *Boletín de Arqueología PUCP* 16:23–51.

Leon, Williams C. 1985. "A Scheme of the Early Monumental Architecture of the Central Coast of Peru." In *Early Ceremonial Architecture in the Andes*, ed. Christopher B. Donnan, 227–40. Washington, DC: Dumbarton Oaks.

Lucas, Gavin. 2012. *Understanding the Archaeological Record*. Cambridge: Cambridge University Press.

Lumbreras, Luis G. 1989. *Chavín de Huántar en el nacimiento de la civilización andina*. Lima: Instituto Andino de Estudios Arqueológicos.

Lumbreras, Luis G. 2007. *Chavín: Excavaciones arqueológicas*. Lima: Universidad Alas Peruanas.

Mesia, Christian. 2007. "Intrasite Spatial Organization at Chavín de Huántar during the Andean Formative: Three Dimensional Modeling, Stratigraphy and Ceramics." PhD dissertation. Stanford University, Stanford, CA.

Mesía Montenegro, Christian. 2014. "Festines y poder en Chavín de Huántar durante el período Formativo Tardío en los Andes Centrales." *Chungara (Arica)* 46 (3): 313–343. https://doi.org/10.4067/S0717-73562014000300002.

Morris, Ian. 2010. *Why the West Rules—For Now: The Patterns of History, and What They Reveal about the Future*. New York: Farrar, Straus, and Giroux.

Moseley, Michael E. 2001. *The Incas and Their Ancestors: The Archaeology of Peru*. 2nd ed. New York: Thames and Hudson.

Rick, John W. 2008. "Context, Construction, and Ritual in the Development of Authority at Chavín de Huántar." In *Chavín: Art, Architecture, and Culture*, ed. William J. Conklin and Jeffrey Quilter, 3–34. Monograph 61. Los Angeles: Cotsen Institute of Archaeology, University of California.

Rick, John W., Christian Mesia, Daniel A. Contreras, Silvia R. Kembel, Rosa M. Rick, Matthew P. Sayre, and John Wolf. 2011. "La cronología de Chavín de Huántar y sus implicancias para el Periodo Formativo." *Boletín de Arqueología PUCP* 13:87–132.

Rosenfeld, Silvana A., and Matthew P. Sayre. 2016. "Llamas on the Land: Production and Consumption of Meat at Chavín de Huántar, Peru." *Latin American Antiquity* 27 (04): 497–511. https://doi.org/10.7183/1045-6635.27.4.497.

Rowe, John H. 1944. *An Introduction to the Archaeology of Cuzco*. Cambridge, MA: Museum.

Rowe, John H. 1967. "An Interpretation of Radiocarbon Measurements on Archaeological Samples from Peru." In *Peruvian Archaeology: Selected Readings*, ed. John H. Rowe and Dorothy Menzel, 16–30. Palo Alto, CA: Peek Publications.

Rowe, John H., and Dorothy Menzel, eds. 1967. *Peruvian Archaeology: Selected Readings*. Palo Alto, CA: Peek Publications.

Sayre, Matthew P. 2010. "Life across the River: Agricultural, Ritual, and Production Practices at Chavín de Huántar, Perú." PhD dissertation. University of California at Berkeley, Berkeley, CA.

Sayre, Matthew P., Melanie J. Miller, and Silvana A. Rosenfeld. 2016. "Isotopic Evidence for the Trade and Production of Exotic Marine Mammal Bone Artifacts at Chavín de Huántar, Peru." *Archaeological and Anthropological Sciences* 8 (2): 403–17. https://doi.org/10.1007/s12520-015-0230-y.

Schreiber, Katharina J. 1992. *Wari Imperialism in Middle Horizon Peru*. Anthropological Papers 87. Ann Arbor: Museum of Anthropology, University of Michigan.

Shady Solís, Ruth. 2005. *Caral Supe, Perú: La civilización de Caral-Supe. 5000 años de identidad cultural en el Perú*. Lima: Instituto Nacional de Cultura, Proyecto Especial Arqueológico Caral-Supe.

Shady Solís, Ruth, and Carlos Leyva. 2003. *La ciudad sagrada de Caral-Supe: Los orígenes de la civilización andina y la formación del Estado prístino en el antiguo Perú*. Lima: Instituto Nacional de Cultura, Proyecto Especial Arqueológico Caral-Supe.

Stanish, Charles. 2001. "The Origin of State Societies in South America." *Annual Review of Anthropology* 30 (1): 41–64. https://doi.org/10.1146/annurev.anthro.30.1.41.

Stanish, Charles. 2003. *Ancient Titicaca: The Evolution of Complex Society in Southern Peru and Northern Bolivia*. Berkeley: University of California Press. https://doi.org/10.1525/california/9780520232457.001.0001.

Tantaleán, Henry. 2014. *Peruvian Archaeology: A Critical History*. Walnut Creek, CA: Left Coast Press.

Tello, Julio César. 1922. *Introducción a la historia antigua del Perú*. Lima: Sanmarti y cia. impresores.

Tello, Julio César. 1943. "Discovery of the Chavin Culture in Peru." *American Antiquity* 9 (1): 135–60. https://doi.org/10.2307/275457.

Trigger, Bruce G. 2006. *A History of Archaeological Thought*. 2nd ed. Cambridge: Cambridge University Press. https://doi.org/10.1017/CBO9780511813016.

Trouillot, Michel-Rolph. 1995. *Silencing the Past: Power and the Production of History*. Boston: Beacon Press.

Uhle, Max. 1902. "Types of Culture in Peru." *American Anthropologist* 4 (4): 753–59. https://doi.org/10.1525/aa.1902.4.4.02a00080.

Whitehead, William T. 2007. "Exploring the Wild and Domestic: Paleoethnobotany at Chiripa, a Formative Site in Bolivia." PhD dissertation. University of California at Berkeley, Berkeley, CA.

Wylie, Alison. 2008. "Mapping Ignorance in Archaeology: The Advantages of Historical Hindsight." In *Agnotology: The Making and Unmaking of Ignorance*, ed. Robert Proctor and Londa L. Schiebinger, 183–205. Stanford, CA: Stanford University Press.

THREE

Disordering the Chronotope and Visualizing Inhabitation in the Lake Titicaca Basin

ANDREW P. RODDICK

INTRODUCTION

In this chapter I discuss chronology and temporality in the Lake Titicaca Basin, exploring how our visual tools are associated with a particular narrative structure. Archaeological narratives are the sequence within which we order our data, and the way our empirically driven stories achieve an overall coherence (Pluciennik 1999; Ricoeur 1984). In analyzing our narratives of the past, we can strip away complex language and reveal unquestioned assumptions and a variety of genres, and the ways that time and space are configured by particular "chronotopes" (Ballard 2003; Joyce 2002; Pluciennik 1999: 654; Terrell 1996). The term "chronotope" refers to the fusing of time and space into a concrete whole, and directs our attention to how such structures are enacted through our narrative practices (Bakhtin 1981). Mikhail Bakhtin defined the term specifically for literary analysis, but anthropologists have developed this term to analyze broader social phenomena, including cultural landscapes and performances where time and space are drawn together (Basso 1984: 43–44; Crossland 2014: 38–39; Thornton 2012: 17). Archaeologists have drawn on this concept to reflect on how we write

DOI: 10.5876/9781607326427.c003

history and construct our narratives (Joyce 2002; Lucas 2005: 60–61; see also Swenson and Roddick [chapter 1], this volume).

In her exploration of the languages and genres of archaeology, Rosemary Joyce (2002: 34) discusses the dominance of the progression chronotope, which transforms time "into a chain of cause and effect with an underlying directionality toward complexity or improvement." This particular way of conceiving time and space relies on a single linear temporality: the universal, or homogeneous time of chronology (Lucas 2005: 9–15). The use of chronology does not necessitate the progressive chronotope. Indeed, Gavin Lucas shows that there are other ways of mapping temporal rhythms and experiences while not abandoning the advantages to such a form of universal time, which might include the ability to track macroscale processes. Yet our discipline has grown with both this dominant chronotope and efforts at periodization, and certainly both define the majority of approaches to time in the Andes. As the chapters of this volume demonstrate, an unwavering emphasis on a totalizing, homogeneous time masks the range of historically contingent temporalities of particular places and times. The progressive story of state development is only one kind of narrative of the past, and even within this narrative there are likely many distinct kinds of temporalities, several of which might play important roles both in larger social process and lived experiences of pre-Columbian inhabitants. Drawing all material culture into a directional temporality therefore constrains our explanatory models. A homogeneous time masks the lived realities of the experience, representation and manipulations of time, in favor of the temporal pegs of periods and horizons, which are admittedly useful, but in some cases may be "arbitrary fabrications" and "given a fictitious unity" (Collingwood 1927: 324; discussed in Lucas 2005: 52–53).[1]

Andeanists have begun to develop new research trajectories, drawing on a diversity of theoretical orientations and growing data sets to move beyond such totalizing temporal schemes. They are exploring relational forms of time inherent in particular social relationships, objects, and landscapes (Abercrombie 1998; Dulanto 2015; Ossio 2015; Sillar 2009; Swenson in press; Weismantel 2004; and this volume). I recently attempted to destabilize the dominant chronotope of linear progression by developing an archaeology of inhabitation (sensu Barrett 1999) for the Lake Titicaca Basin, exploring how particular temporal tensions might have been relevant to longer-term processes (Roddick 2013). I suggested that a focus on the multitemporal landscape (or the "taskscape") of the Taraco Peninsula might produce new research inroads (see also Roddick 2015). Despite my explicit attention to various forms of temporality and explicit use of concepts meant to resist the dominant chronotope, I often slipped back into a narrative of progress. My continued use of standard period names, weighed down with evolutionary expectations, and my ongoing interest in the development of the state certainly impacted this project. But I am convinced that my *visualization of time* played an equally important role. Archaeological narratives rely not just on text but also on "chronographics," those charts, figures, and illustrations used to represent time.

In this chapter, I suggest that the chronographics we deploy to serve larger nuanced arguments are unwittingly constraining our abilities to develop distinct narratives. I argue that increased attention to such visual tools might push our analyses and interpretations in new directions. After a brief discussion of some recent work on visualization, I explore several visual conventions that are central to our normal science. These visualizations, from quick sketches to formalized charts, have their own contingent histories within our discipline. Indeed, even the simplest of doodles can become essential components of archaeological investigation, a "matter of fact" rather than a problem to consider and justify (Moser 2014; see also Moser 1998; Moser and Gamble 1997; Smiles and Moser 2005). Although these forms were once heavily debated, today they are rarely reflected upon. In this chapter I explore the chronological chart, ceramic illustrations, section drawings, and Harris Matrices in the Lake Titicaca Basin.

VISUALIZING TIME AND MOVING LINES

One of the enduring results of the postprocessual turn in Anglo-American archaeology is an increased reflexivity in regards to our practices, including the craft of fieldwork (Hodder 1997; Leighton 2015; Lucas 2001; Shanks and McGuire 1996). Although it wasn't until the 1990s that a number of (primarily British) archaeologists began to attend to the visual languages of archaeology (Bradley 1997; Gamble 1992; Molyneaux 1997; Moser and Gamble 1997; Van Reybrouck 1998; but see Piggott 1978), this research has developed into a mature scholarship (Cochrane and Russell 2007; Fyfe, Caseldine, and Gillings 2010; Morgan 2014; Moser 2001, 2014; Perry 2009, 2014; Pillsbury 2012; Swogger 2000). This growing literature demonstrates that the "imaging of archaeological knowledge" is as much a theoretical enterprise as the written language, and is much more integral to our narrative construction than we often realize (Smiles and Moser 2005: 1). "Serial orders of representation" play a key role in getting us from the archaeological site or artifact to the final written form and can (intentionally or unintentionally) preordain certain research results while occluding other interpretive possibilities (Shanks and Webmoor 2010: 89, 107). The history and circulation of particular kinds of archaeological images play a key role in our chronotopes, where time, "thickens, takes on flesh, becomes artistically visible; likewise, space become charged and responsive the movement of time, plot, and history" (Bakhtin 1981: 84). Archaeologists have recently joined historians and art historians in unpacking the black boxes (sensu Latour and Woolgar 1979) of these chronographics. They are also exploring new conceptualizations of time through innovative forms of visualization.

An example from Mesoamerica demonstrates the questions and approaches possible through a reevaluation of our modes of representation. Susan Gillespie's critical analysis into early excavations and later circulations of artifacts from the Olmec site of La Venta highlights the practices and temporalities that lay behind linear chronological narratives and a number of iconic yet "passive" images (Gillespie 2008: 110). She explores a range of deposits and caches at Complex A that were revisited over many years, suggestive of a kind of "memory work." Careful analysis of stratigraphic profiles from the original excavations shows how construction practices permitted the past and present to be mutually constituted, eventually producing a new temporal horizon and system of referents for those inhabiting this changing landscape (Gillespie 2008). Gillespie (2011) demonstrates how the maps and perspective drawings from the original 1959 excavations of Complex A from La Venta were simplified, homogenizing the various strands of time embedded in the original illustrations, and thus shaping current interpretations of the site. In a more recent effort, Gillespie (2015) highlights how the distinct temporalities of particular objects are often missed due to our modes of representation. The famous jade figures from "Offering 4" have clear evidence of distinct social lives and varying tempos, or "itineraries," but since they have been consistently photographed from the same viewpoint, these temporalities have been masked. Some scholars have begun to explore new forms of visualization, which provides important ways to disrupt some of our temporal narratives (Morgan 2009; Webmoor 2005; Van Dyke 2006; and see below). For instance, Gillespie and Michael Volk develop a more nuanced perspective of both the original excavations and the dynamics of dwelling in complex A at La Venta through 3D modeling (Gillespie and Volk 2014: 9–10).

Our older, traditional tools of visualization, however, have particular advantages, including accessibility and affordability. The most traditional visualization of time involves the timeline and the chronological chart. These visual tools help to order cultures, but are also complicit in our very conceptualization of space-time. They are the "organizing center" for our narrative structure, "where the knots of narrative are tied and untied" (Bakhtin 1981: 250). Although the familiar timeline and chronological chart form "such a familiar part of our mental furniture that it is sometimes hard to remember that we ever acquired them in the first place" (Rosenberg and Grafton 2013: 10), their history encapsulates the resilience and dominance of the progressive chronotope.

Daniel Rosenberg and Anthony Grafton trace the history of the timeline, in part, as a story of Western conceptions of history and the grammar of graphical historical representation (Rosenberg and Grafton 2013). Although the timeline is less common in modern archaeological practice (but see below), many of their findings are pertinent to our own visual practices. For example, they position their study of these 250-year-old "cartographies of time" as an analysis of the line, the basic component of most historical diagrams. Lines played the role of temporal metaphor, but they also were a physical, graphical choice that changed the way that history was conceived (Rosenberg and Grafton 2013: 13). "By the middle of the nineteenth century, a wide and inventive range of charts, including some that were quite technical, could be found in general publications . . . But through all of this, the line itself straight, crooked, curved, or branching—remained the principal visual metaphor by which historical chronologies were envisioned" (Rosenberg and Grafton 2013: 138). Rosenberg and Grafton demonstrate that when deployed in chronographs, lines have "the power to change the world" (Ingold 2007: 3).

CHRONOLOGICAL CHARTS AND THE PRODUCTION OF HOMOGENEOUS TIME

Like the timeline, the utility and pervasiveness of the chronological chart have "made us blind to the underlying assumptions they convey about the past" (Stone-Miller 1993: 20–21). Chronological charts perform several complex tasks: they simplify several sets of more complex visual data, homogenize different rates of change, and compress generations of significant variability into a form of 'fossilized' and universal time (McGlade 1999: 143; Roddick and Hastorf 2010). The graphical choices made by archaeologists in these charts—the lines and tidy symmetrical boxes on the pages of our publications—convey particular kinds of statements of the past. In her innovative analysis of the graphical conventions in chronological charts, Rebecca Stone-Miller (1993) pays particular attention to the line. She maps out the diverse ways of visualizing time, beginning with Alfred Kroeber's (1944: 112) careful use of labels and lines to represent a rather open-ended temporalization schema, a model that was adaptable to new information.[2]

Stone-Miller also discusses the role of John Rowe's (1962) chronology building. Rowe's importance in contemporary Andean chronologies is, of course, well known to Andean archaeologists as it continues to structure our discipline today (see also Sayre [chapter 2] this volume). Rowe combined pan-regional styles or horizons (Chavín, Tiwanaku, and Inca) with periods of local differentiations, all of which were associated with the newly available radiocarbon dates creating a unified, homogeneous time for the central Andes. Certainly Rowe's approach was a systematic and rigorous form of data collection (Isbell and Silverman 2006a: 10), yet Stone-Miller highlights his role in our chronographics. Rowe (1962: 41) argued that our temporal boundaries should be more strongly stated, "shown as a straight line on the chart of relationships."[3] In fact, Rowe's enduring contribution to Andean archaeology is perhaps best *illustrated* in the ubiquitous chronological charts. These charts are characterized by unbroken flat lines (which have only moved in the subtlest of ways in recent decades), conveying the false premise that "change occurs across a region simultaneously and irrevocably" (Stone-Miller 1993: 25).

The Lake Titicaca Basin, and in particular the site of Tiwanaku, plays a particularly important role in the history of the Andean chronological chart. For example, Stone-Miller points to Max Uhle (1902), whose work ordering Andean style depended on distinguishing Tiwanaku-style ceramics from pre-Tiwanaku, post-Tiwanaku, and Inca materials. In his extensively annotated translation of Fernando Montesinos, Philip Means (1920) produced one of the earliest chronological tables explicitly dealing with the Titicaca Basin.[4] In an effort to develop a sense of universal time, Means (1920: xxxv) constructs several chronological charts assembling Maya historical development (using the Maya Long Count as a cultural yardstick) with "the folklore material hidden in Montesinos" and the emerging archaeological data of the central Andes. His presentation of the archaeological data surmises that Tiwanaku culture likely dated from AD 200 to AD 500, and Middle and Late Tiwanaku from AD 500 to AD 900. This table is remarkable in that this sequence, produced thirty years before the development of radiocarbon dating across the Andes, is well aligned with currently accepted chronometric dates for the Formative through Tiwanaku V related phases. Still, it is a far cry from the chronological chart common to south-central Andean archaeology today, and it wasn't until later in the twentieth century that archaeologists began to produce more recognizable graphic representation of this homogeneous time.

We might skip ahead thirty years, when Wendell Bennett (1950) produced one of the early chronological charts for the northern and southern Lake Titicaca Basin. Bennett's (1936) excavation of the Formative Period Monticulo (or mound) at Chiripa and his seriation of ceramic material at this site and at Tiwanaku resulted in a more data-oriented representation of time. Figure 3.1 depicts some of the pioneering work to develop a localized chronology, highlighting a distinct series of phases fenced off by the hard lines of culture historical assumptions. This research was developed out of earlier nationalism-oriented work of Tello, and the subsequent culture area research by Kroeber and by Bennett himself (Bennett 1948; Bennett and Bird 1949; see also Isbell and Silverman 2006a: 7–10 and Sayre [chapter 2], this volume). Bennett's "Central-Andean co-tradition," stressed "the over-all unit of culture history of an area within which the component cultures have been interrelated over a period of time" (Bennett 1948: 1), and he argued that these changes were uniform across the entire central Andes (Isbell and Silverman 2006a: 8). In other words, we see the first clear articulation of homogeneous time, and little evidence of the uncertainty that Stone-Miller noted in Kroeber's earlier visualization.

Since the first development of these early chronological charts, our understandings of time-space systematics have changed, the number of surveyed and excavated sites has risen exponentially, our corpus of material culture has grown, and the quantity of publications relying on chronological charts has increased accordingly. We might categorize the charts currently in circulation as either evolutionary or historical in nature (Stanish 2003: 85). Charles Stanish suggests

SOUTH TITICACA BASIN				NORTH TITICACA BASIN				
PERIODS	COPACABANA	TIAHUANACO	ACHACACHE	MOHO	TARACO	PUNO	JULI	PERIODS
INCA	9	8	2+	+	2+ 8+ 3		1 2	Chucuito INCA Taraco
KHONKHO	2	4	20+	5	2+ 9	3+ 2 1	+	Sillustani COLLAO Allita Amaya
DECADENT TIAHUANACO	4	7	2		1	4	1	DECADENT TIAHUANACO
CLASSIC TIAHUANACO	6	6						
EARLY TIAHUANACO	1	2			4	1	2	PUCARA
CHIRIPA	3	3						

FIGURE 3.1. *Modified version of Bennett's (1950: fig. 35) chronology of the Lake Titicaca Basin.*

the advantage of evolutionary chronologies, which rely upon developmental stages, is that they place societies into a broader explanatory narrative. A disadvantage, however, is that they do not deal adequately with variability. In contrast, historical chronologies, which depend on absolute dates, don't presuppose gradual transitions between periods yet can be overly particularistic (Stanish 2003: 85–89 and fig. 5.1). Stanish argues that the Rowe horizon system does not work well in the Titicaca Basin, as developments on the coast happen at a different time period and at different rates of change. "For instance, the first monumental architecture on the Peruvian coast dates to perhaps 3000 BC, but the first corporate architecture in the basin was not built until at least 1500 BC or, probably, even later. The dates for the Early Horizon and Early Intermediate period likewise do not correspond to cultural developments in the area. Tiwanaku expansion occurred around AD 650 and ended around AD 1100, a century or two different from the Middle Horizon and Late Intermediate periods." (Stanish 2003: 88–89). Instead Stanish advocates a "dual chronological system," which incorporates a broad evolutionary chronology, some aspects of the Rowe system, and a variety of local historical chronologies around the basin (Stanish et al. 1997). The major graphical change in the visualization of time in Stanish's system is the inclusion of several extra columns (Stanish 2003: 90, fig. 5.2). In his initial graphical presentation, no hard bold lines are employed. Yet in in subsequent uses of this chart, Stanish's students and colleagues return to using bold lines, emphasizing the significance of particular evolutionary shifts. Despite the careful inclusion of local historical chronologies, the expectations of the dominant chronotope remain (Isbell 2004: 235–36).

Work on periodization continues (Janusek 2003; Klarich 2005; Lemuz Aguirre 2001; Marsh 2012; Roddick 2009; Stanish et al. 2014; Steadman 2007, among others), but there has been little change in the presentation of time. I surveyed fifty-six master's and PhD theses produced over the past twenty-five years and found little chronographic variation. Several earlier iterations are somewhat more careful in their graphical representation. See, for instance, Anita Cook's (1985) use of dotted lines or gaps (the deliberate absence of lines) to sketch out chronological divisions and annotations noting whether phases were based in style or radiocarbon assays (Cook 1985: 317, table 3.3). Other frameworks, such as Marc Bermann's (1990: 522), draw in a number of regional phases as a means to highlight distinct regional processes that often do not line up with Rowe's broader Ica sequence. Citational practices in more recent theses suggest the reliance on a limited number of chronological charts, with those by Alan Kolata (1983), John Janusek (2003), and Christine Hastorf (2003) being the most commonly employed.[5] Most of these iterations use a version of Stanish's dual chronological system, with little discussion of the logic or rational behind this system (but see Klarich 2005: 42–43; Levine 2012: 61). Isbell and Silverman (2006b) find this approach problematic: "Many of the archaeologists working in the Titicaca Basin are employing site-specific, or sub-area specific chronologies that inter-digitate with the Rowe chronology in only the broadest fashion . . . While the development of a unique chronology is generally more sensitive to local issues than a chronology linked to a master sequence created elsewhere, the negatives of such an approach must be recognized" (Isbell and Silverman 2006b: 503). They worry that approaches that stray from Rowe's master chronology will not address macroscale processes, resulting in a temporally disjointed narrative. Nevertheless, Isbell and Silverman agree with Stanish that our temporalization should primarily illuminate the development of evolutionary trajectories, despite Isbell's (2004) earlier critical stance on empirical research and evolutionary expectations.

Almost all of the recent theses about the Titicaca region share this evolutionary focus. The dominant theme across the sample of Titicaca theses stresses political complexity, with particular attention to the development of the state. Many recent theses deploy the word "emergence"—whether of complexity, the state, elites, or leaders—in the title, and almost all in their thesis abstracts. Researchers employ their chronological charts to identify when elites make their debut, usually thought to correspond with the Middle and Late Formative, at which point they become central to most narratives (Roddick 2009: 33–36). It is rare that chronological charts are used explicitly to either generate new hypotheses or new kinds of stories about the Lake Titicaca Basin. Christine Hastorf and I, in our attempt to think about traditions in a social and process-based manner, recently attempted such a project. In mapping out the standard sequence for the basin, we added a column that estimated the approximate number of generations for each stage (Roddick and Hastorf 2010: 160, fig. 2). This was a small effort that endeavored to highlight lived lives and cross-generational processes.

There are two striking examples of distinct chronologies, both found outside of academic theses, that nevertheless draw from and contribute to the dominant chronotope. Arthur Griffin and Stanish (2007) use spatial agent-based modeling (ABM) and simulation to explore the factors relating to the development of complexity in the region. Their publication is particularly rich in info-graphics, and they include a range of provocative figures to visualize population growth rates and overall polity development. Their modeling draws in a range of factors, including ecology and trade, but focuses in particular on traits associated with the progressive narrative, such as agricultural intensification and the emergence of ranking. Their timeline of the region (Griffin and Stanish 2007: 5, fig. 2) visually aligns stages (Archaic, Formative, etc.) with agriculture (nonintensive farming, terraces, raise fields), and with levels of political complexity and leadership (foragers, hamlets, small chiefdoms, large chiefdoms, etc.). Much like other chronographics, this timeline is "closely calibrated to both the interests and the vision of their users" (Rosenberg and Grafton 2013: 12). The hypothetical chiefs mapped on the timeline are given further agency in their broader model parameters, where values are calculated for potential chief's "strength" and "kill ratio" (Griffin and Stanish 2007: 14).

Another timeline is currently on display at the Museo de Cerámicas at Tiwanaku, an image that demonstrates the impact that the material manifestations of our narrative constructions has on ongoing sociopolitical processes (El-Haj 1998). Tiwanaku has become a major tourist site since the early excavations by Bennett and reconstructions by Ponce Sanginés, and thousands of Bolivian and international visitors stream through this museum each year. Tiwanaku is the most important heritage site in Bolivia and "has come to symbolize and embody the evolutionary rise of 'complex societies' . . . and has ultimately played a key role in the creation of Bolivia's national patrimony" (Kojan and Angelo 2005: 388). As Kojan and Angelo demonstrate, our archaeological narratives in the Altiplano are not separate from larger political projects and nationalist campaigns; the progressive chronotope is very active on the current political landscape (Joyce 2002: 31). Drawing on historian Carlos Mamani Condori (1996), Kojan and Angelo suggest that the pre-Hispanic history of the country is not viewed or discussed as a history of indigenous populations, but as part of a larger homogenized national narrative. This process can be traced back

FIGURE 3.2. *Timeline on display in the Tiwanaku ceramics museum, August 2015. (Photograph courtesy of John Janusek.)*

to the nationalistic efforts of Carlos Ponce Sanginés, but it was also on display at the inauguration of Evo Morales at the site in 2006, an event involving a number of complex overlapping narratives (Kojan 2008).

Shortly after Morales's inauguration, a new timeline appeared in the ceramic museum (figure 3.2). Many familiar elements are seen in this image, from the common Titicaca Basin period names, to the hard divisions between periods and their alignment with chronological dates. There are, however, some striking differences. While Tiwanaku is often a primary focus in our modern chronographs, this timeline highlights the modest remains of Chiripa, perhaps due to the recognizable artifacts from this well-researched site.[6] In an effort to simplify the chronology, and presumably to link to objects in the rest of the museum, the timeline offers representations of artifacts and places, a series of index fossils for the viewer. Finally, the timeline extends beyond the Spanish conquest, the usual endpoint of the archaeological past, up into the present. The timeline highlights the election of Evo Morales, the redefining of Bolivia as the Estado Plurinacional de Bolivia, and the recent launch of the Tupac Katari satellite. Here we see the nationalist entanglements of the progressive chronotope and representations of homogeneous time (Lucas 2004: 113).

This visually arresting timeline is the first image that a tourist sees upon entering the museum. As s/he circulates among the displays, s/he may very well overhear Bolivian or visiting foreign archaeologists critiquing and or mocking this timeline. The selection of questionable artifacts as indexes and the somewhat inaccurate dates are certainly matters of concern. Moreover, the identification of a sitting president with a chronology of the *longue durée* strikes many as too politically heavy-handed. The election of Morales was a great victory for indigenous people across South America and Left-leaning people globally, but as his current presidency extends into its tenth year, his position in both the larger progressive narrative and this particular chronographic might be seen as an effort to marginalize contesting political voices. Moreover, one wonders what the lifespan of this representation of time will be; if Morale's political party loses the next election, this image could very well be taken down. Yet this celebration of Morales as an individual, but more specifically what he represents to many Bolivians and outside observers (political progression and the emergence of a new kind of indigenous political elite), is not far removed from the attention currently being paid to emerging elites in the Formative periods. In other words, this is just another, albeit explicitly political, representation of the dominant chronotope.

There is much more to be explored vis-à-vis the relationship between our dominant chronotope, visual practices, and the politics of contemporary Bolivia. The power of these images goes far beyond heuristic devices, and experimentation with ways of reckoning of time might have unexpected outcomes. Certainly Evo Morales's government is aware of the power of reckoning and displaying time. For example, a new clock was installed in the Plaza Murillo (the central Plaza in La Paz where the presidential palace is located) that runs backward. Marcelo Elio, the president of Bolivia's Congress, stated to the Associated Press that this modification is "a clear expression of the de-colonization of the people" (Domonoske 2014; see Mamani Condori 1989 and Verdesio 2013 for Andean perspectives on decolonizing archaeological time).

Let me finish my brief overview of the Titicaca Basin chronological chart with a pragmatic reason that archaeologists might also want to further experiment with their representations of time.

Although few scholars working in the region question the overall progressive narrative, empirical evidence strongly suggests that the processes of the development of the Tiwanaku polity was not all that gradual or progressive. For instance, in a series of recent papers exploring the regional distribution of the Andean Iconographic Series and through a recalibration of radiocarbon dates, Isbell and Patricia Knobloch have suggested that we reconfigure our periodization for the Late Formative and the subsequent growth of Tiwanaku (Isbell and Knobloch 2006, 2008). They believe Tiwanaku urban developments happened much later than previously believed. Janusek's ongoing work at Tiwanaku also demonstrates a shifting sense of the timing and tempo for the coalescence of Tiwanaku. He has recently argued that while elements such as monumental architectural forms have a longer developmental process in the region, other elements occur quite suddenly including volcanic stone construction, urban expansion, and ceramic assemblages (Janusek 2015, and personal communication, November 2015). Recent work on Late Formative ceramics (Roddick 2009) and the application of Bayesian models to radiocarbon assays from Tiwanaku (Marsh 2012; see below) challenge traditional evolutionary models that the great altiplano city emerged gradually, but point instead to a rapid, "big-bang" formation of the Tiwanaku polity.

Stanish (2003: 86) has suggested that historic chronologies permit for tracking such rapid rates of change, yet the linear geometry of our current chronographics are not capturing the punctuated, quickened rates of social change. Here we might begin to consider other forms of visualization, such as that suggested by George Kubler forty-five years ago (also summarized by Stone-Miller 1993). Kubler (1970: 129–30) rejected the usual "unreal fences" of typical chronological charts and content-filled temporal containers. Instead, he argued for the merits of a chart with very few of the recognizable forms, one that explored distinct geometries and rates of change. While key organizational concepts are included—such as "Formative," "Classic" and "Post-Classic"—he eliminates the progressive and constant move toward complexity, replaced by another kind of temporality: "fast centres" and "slow margins." These patterns are visualized not in terms of hard-and-fast breaks but rather as diagonals, highlighting his emphasis on dynamic historical change (Stone Miller 1993: 30). This image is slightly disorienting (Stone-Miller 1993: 30), a discomfort that likely arises from how normalized particular chronographics have become in our field.

Kubler's image hints at important steps we might take to disorder the chronotope and to develop new imaginings and images of temporal process and the experience of time. When Kubler developed this chart in the late 1960s, it was an "empty" chronology, "filled only with the thin air of cultural theory." But the explosion of research in the Titicaca Basin since the late 1980s lends more empirical support for distinct rates of change. Kubler's efforts challenge us to follow his efforts of identifying and mapping out distinct tempos. Rather than explaining away the variability we are finding because either it fails to fit our narrative structures or because we lack the means of representing such rates, we should still try to engage Kubler's schema and alternate temporal frameworks devised outside the Titicaca Basin. Certainly such an approach might invite withering critiques or, more likely, silent rejection (Joyce 2002: 137). But this kind of image might also highlight distinct forms of temporality currently masked by our chronological charts.

PICTURES OF POTTERY TAPED ON THE WALL

Let me now turn to explore the time materialized in objects, the very things that scaffold the chronological charts that circulate through academia and public representations in museums. Early chronological work consisted of efforts to temporalize poorly recorded material encountered in museums, a problem that continues today in museums such as the ceramics museum at Tiwanaku (Stone-Miller 1993). A key approach in both early and current modes of periodization involves defining gradual changes in ceramic forms and design styles as well as defining "stoppages" (sensu Gosden 2006: 431–38) to permit the drawing of the hard lines of the chronological charts discussed above. Early approaches relied on careful modes of visualization, often assembling the entire corpus, and varying periods of potting traditions on a single page (Jones 2001: 352). Here we could turn to representations of ceramics by a variety of scholars, whether Bennett's (1934) early analysis, Stig Rydén's (1947) richly illustrated monograph, or Sonia Alconini Mujica (1995) and Janusek's (2003) more recent reevaluations of pottery and time at Tiwanaku.

I would like to focus instead on Dwight Wallace's (1957) doctoral thesis, which clearly shows how the visualization of the sequential transformation of pottery forms the Titicaca Basin. Wallace plays a central role in chronology building in the Titicaca Basin, and his dissertation represents a detailed and rigorous early attempt to extract time from objects. As Janusek (2003: 33) stresses, "to this day, his analysis remains among the most comprehensive seriation of highland Andean ceramics." Drawing on the approach of his advisor Rowe and the Berkeley tradition of chronology building, his similarity seriation was meant to be tested independently against stratigraphic deposits. He seriated 215 whole vessels from published sources and 650 whole vessels from museum collections that derived primarily from grave lots. Wallace's seriation started with an ordering based on materials found in the lowest levels of Bennett's excavations, and it was then correlated with Adolph Bandelier's (1910) excavations at the site of Qeya Qollu Chica, creating a category still known today as "Qeya" (for a summary see Cook 1985: 90–102). Today this Qeya style forms part of the Late Formative II phase (Janusek 2003). However, this style of pottery appears to be much more limited across the region than early work at Tiwanaku suggested, and is thus not an ideal diagnostic of these phases (Roddick 2009: 233–38).[7]

Wallace's careful seriation of ceramic style traits across his assemblages relied heavily on a series of visualizations to map out changing design units and forms. One matrix charts each of his forty-one design types, another reconfigures those findings to order them in time, and yet another explores the relationship of design types and decorative techniques (Wallace 1957: tables 6–8). A final matrix used careful design to seriate all style traits (Wallace 1957: 119, table 9). Here he places Qeya and Tiwanaku styles on the upper horizontal axis, and shape groups are positioned running down the vertical axis; careful hatching highlights questionable dating. Although Wallace (1957: 121) suggests this figure "more or less speaks for itself," he goes on to develop a more fleshed out argument. A final

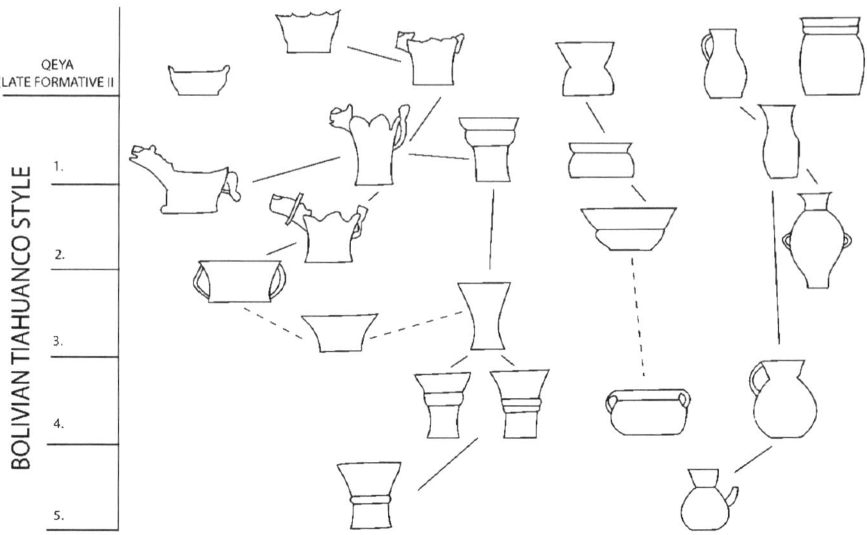

FIGURE 3.3. *A modified version of Wallace's (1957) flowchart demonstrating the shift in vessel forms from the early Qeya (Late Formative II) to Tiwanaku phase ceramics. Note that early phases are displayed at the top, and most recent at the bottom.*

chronographic, represented here (figure 3.3), shows the progression of vessel form from Qeya through the five Tiwanaku phases. This figure simplified the complex chronological series even more, but this simplicity allowed a quick reference point and became a shorthand for many basin archaeologists (see Burkholder 1997 fig. 6.16 for a similar chart, but one emerging out of careful depositional analysis and highlighting form overlaps). Much like other visualizations of data, it appears that the more data that are collected, the simpler our visualizations, whether in the form of tables (Goody 1977, discussed in Stone-Miller 1993: 28), charts, or matrices. This tendency to reductionism results in our chronographs becoming more of a visual shorthand rather than an interpretive device.

In her careful synthesis of chronology building in the region, Cook (1985) underscores the strengths of Wallace's vigorous analyses, but also points out several weaknesses. These include (1) The museum-quality fine wares may be a very biased representation of the archaeological record, and (2) the objects themselves might have more complex use-lives and temporalities than typically displayed in visualizations such as figure 3.3.

First, Cook (1985: 99) suggests that Wallace's analysis primarily stood on the use of "fancy grave lots of the kind favored by pot-hunters, whose collections were best represented in the museum collections." A primary reason for Wallace's focus on the fancier vessels is that his particular form of similarity seriation, which relies on recording form and design style variation, simply would not be possible on much of the archaeological record. Yet Wallace's attempt to order time based only on decorated whole vessels from museum collections is part of a larger problem. The simpler, more mundane sherds were either not collected, or "almost shamefully" hidden away from museum exhibits or published volumes (Lucas 2005: 127). Most of our narrative construction happens around this small sample of objects, while the other 99 percent of undecorated, broken, eroded ceramics are not analyzed. There are many problems in using design style and what scholars presume are "elite wares" in developing our chronologies (see Swenson and Roddick [chapter 1], this volume). For instance, the sequence leading up through the Middle Horizon might be thought of a process of "stylization," with the signaling potential of pottery (of status, ethnicity, and other forms of affiliation) becoming more essential to social practices such as feasting and other forms of political theater (Roddick 2009: 392–95). If so, then treating decorated pottery as our primary timepiece is extremely problematic, and they perhaps should be analyzed more narrowly to create "iconographic series" (Isbell and Knobloch 2006, 2008).

Another reason for the focus on decorated vessels, however, might be the difficulties that mid-twentieth-century scholars encountered in visualizing the variation in traits beyond shape and design. Publications from this period include beautiful plan maps and artifact illustrations, often of a quality far surpassing what we are producing today. (A comparison of the hundreds of beautiful hand-drawn and finely stippled artifact illustrations in Rydén's [1947] monograph to my own efforts [Roddick 2009] may have led to a brief bout of depression.) Yet these scholars were not always deploying shared conventions, and most primarily focused on decorated diagnostics. Today, of course, there are both expensive and innovative ways (see below), and cheap and established conventions for drawing and photographing even the most mundane of artifacts. The literature is particularly rich in the latter kind. My thesis relied on computer and camera technology rather than creative skill, but these tools allowed me to communicate ceramic variation through time by digitizing my less artistic illustrations, combining photos with these illustrations, highlighting details of paste and surface finishes through photography and a flat-bed scanner, and translating Munsell codes to RGB color values on the computer monitor (Roddick 2009: 239).

This ability to quickly produce images has had palpable impacts, and just as our number of chronological charts have increased in the region, so too have the photos and illustrations of diagnostic rims, bases, handles, motifs, and design styles. As the example from Susan Gillespie in Mesoamerica demonstrates, these are not passive images and these emerging conventions also might mask particular temporalities. Gillespie shows that such images, like the charts discussed in the previous section, do not sit quietly in the pages of dissertations, reports, and published monographs and journals, but rather have their own dynamics histories and use-lives. Ceramic representations are printed out, photocopied, and in some cases, reorganized for field labs across the region. It is quite normal to see the figures from the later seriations taped to lab walls as both experienced and neophyte ceramicists train themselves to perceive the attributes behind complex classification systems. Just as both decorated *and* undecorated ceramics once circulated in the past (Roddick 2015), these two-dimensional images now form networks through their circulation (Jones 2001; Wallis 2015).

Cook's second critique of Wallace's system relates to temporality more directly. Although she finds much to celebrate in Wallace's work, she is concerned that it may not represent great chronological spans of time (due to relying on poorly justified design units). Furthermore, Wallace's schema may not capture other kinds of temporality. Cook (1985: 99) asks what if some design units, such as the feline designs so important to his seriation, were only popular for short periods of time? In other words, Cook suggests that particular attributes of ceramics may have their own rates of change. This point has been fleshed out in a number of rich ethnographic case studies in recent years. For instance, pastes, forming techniques, and design styles often have their own historical trajectories related to particular sociopolitical contexts (Gosselain 2011; Roddick 2015, 2016; and see various case studies discussed in Roddick and Klarich 2013). Although we are not yet in a position to track such distinct temporalities and rates of change, we may be able to do so in the very near future. Due in no small part to the legacy of Rowe and the training provided by his later students such as Lee Steadman (1995; but see also Mohr 1966), Titicaca Basin scholars today rely on more than vessel shape and design, and an increasing number of scholars are exploring attributes that reflect the entire sequence of steps of production, from raw material acquisition through firing, use, and discard. We are beginning to shift our attention to the dynamics of attributes, their associated techniques, and the communities of practice implicit in their changes across time and space (Roddick 2015).

Here the problems of visualizing time return. Wallace established the importance of paste in recognizing Qeya ceramics, "with its light buff or brown micaceous paste which is relatively soft," which was in contrast to the "Tiahuanaco style" with its "hard red- or orange-brown paste" (Wallace 1957: 19). In the last twenty years there has been an increasing recognition that variation in raw materials can be more appropriate for mapping spatiotemporal variation in the region than decorative style alone. Peoples' choices appeared to change over time, but also across distinct geological units, making it a useful system for our space: time systematics (Roddick and Klarich 2013). Yet this variability continues to be *described* rather than *visualized*. Attempts to draw from these descriptions can be extremely frustrating, as the distinctions are sometimes hard to translate to what we perceive through the hand lens. The system we deploy on the

Taraco Peninsula shares this flaw, as can be seen in the description of our paste groups. For instance, this limitation is evident in our first 5 mineral tempered pastes in the classificatory system developed by Lee Steadman for the Taraco Archaeological Project:

Paste 1: Medium size, angular and subangular, black, white, translucent. Tends to have a lot of mica, subcompact, and medium texture. Fiber version would be p. 18. Inclusions in p. 17 finer than p. 1

Paste 2: Translucent subrounded and rounded, subcompact, medium size, medium texture, mica, and some blacks.

Paste 3: Fine, very fine and medium white opaque, subcompact, and medium texture.

Paste 4: Medium and fine translucent or gray angular inclusions, medium texture.

Paste 5: Fine texture paste; fine inclusions, compact/subcompact white translucent and opaque, very few blacks. For fine and semi-nice wares.

I have tried to visualize temper changes through a relatively simple flowchart (Roddick 2009: 227, fig. 7.9), an image that I have used in training my students. Yet I doubt this visualization will serve as a long-lived heuristic device or challenge deeper investigation; it is still primarily descriptive with the same problems highlighted by Matthew Bandy (2001: 56): "It is very difficult to learn to distinguish the various paste groups as I have described them here. They really cannot be conveyed satisfactorily in writing." Elsewhere I have suggested that this issue necessitates face-to-face apprenticeship in learning ceramic analysis (Roddick 2009: 204). More recently, in training and working with my own students, I have used a portable digital microscope to produce detailed paste images for both heuristic purposes and later semiquantitative analyses (Rivas-Tello 2015). I am beginning to share these images as a Filemaker database that can be accessed either on laptops or tablets in the field (figure 3.4), allowing for the temporal variability of pastes to be explored through description, but also through an explicit visual vocabulary.

Of course, visualizing changes in paste alone does little in itself to challenge the dominant chronotope. It only captures one step of the production sequence rather than assembling the range of temporalities of objects, from the multiple chronologies (based on various C14 dates), to different typologies and the varying timing of attributes discussed above (Lucas 2005: 98). For instance, a comparison of radiocarbon dates and ceramic attributes suggests that Late Formative production sequences may have changed at different rates in different regions. Some of the changes noted in operational sequences are likely

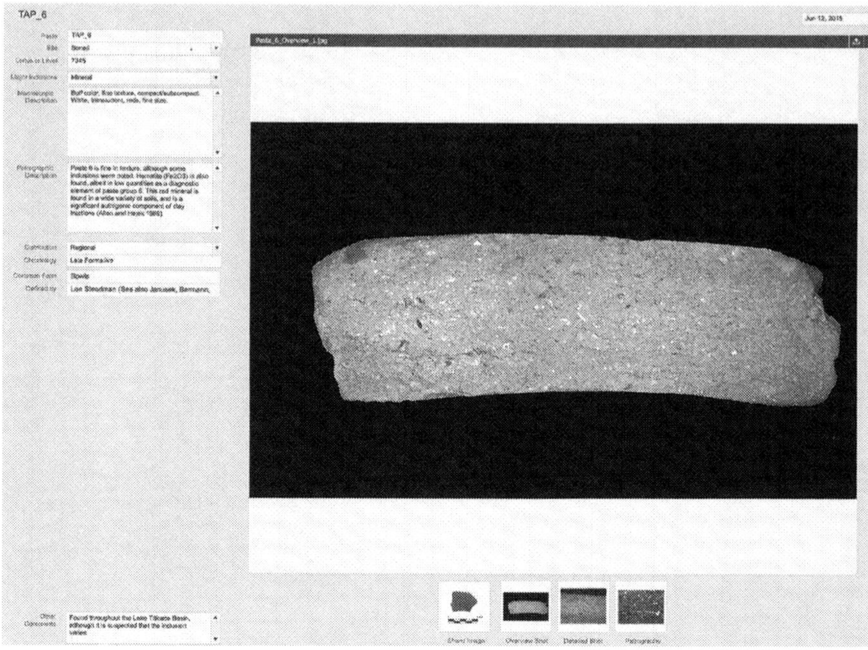

FIGURE 3.4. *A screenshot from ongoing work in the Titicaca Paste database, which assembles chronological information, written descriptions, and several images offering different levels of details from the sherd, cross-section, and petrographic visual scales. This Filemaker database is currently available on both the paid desktop/laptop and the free tablet version.*

associated with new working rhythms or what have elsewhere been called "habit time." Subtle changes in forming and finishing techniques hint at changes in habitual, embodied rhythms (Roddick 2013: 296–99). This varying rate of change in ceramic technology may be more than simply a methodological frustration, and hints at important intercommunity variability in temporal rhythms, a variability that should be represented in both maps *and* chronographics (see below). The changes in particular choices through time can be displayed by either timelines for particular artifacts (Lucas 2005: 107, fig. 4.6), or new kinds of charts and artistic illustrations for single objects (Thomas 1996 fig. 6.10) and larger assemblages (Lucas 2005: 105, fig. 4.5; Roddick and Hastorf 2010: 165, fig. 4). However, illustrating the dynamics of operational sequences of assemblages across the landscape is more difficult, a scale where the temporality of objects becomes more complex, particularly as they become more clearly entangled in a range of other kinds of things. There certainly are options. One solution might lie in the kind of matrices that Wallace was developing for his design style. Another option, more appealing in both visual qualities and potential for quantitative explorations, is the use of Geographic Information Systems. I have begun building a database of ceramic attributes across the region. Surprisingly, I have found no other examples of GIS-systems tracking ceramic attributes either in Andean archaeology or within wider archaeological scholarship.

The above discussion brings me to ask, Is the temporality of pottery actually *worth prioritizing*? Are the temporalities of pottery so important that we ought to go beyond illustrations on the wall, to search for new complex visualization schemes or to develop new matrices or GIS databases? What if all pottery, not just "elite wares," is inappropriate for periodization and the production of homogeneous time? Of course, scholars have long questioned whether pottery actually tracks larger processes of social change (Adams 1979; Kubler 1970). As a ceramicist who focuses on much more than chronology, I believe there are a wide number of social questions of the past that the ubiquitous potsherd can answer. In light of the exciting ceramic work occurring both in the Andes and beyond, these questions do little to shift my analytical focus (though obviously could have much deeper consequences!). These questions do highlight a key issue: to my knowledge we have never actually engaged in proper scholarly discussion around time and ceramics in the Lake Titicaca Basin or the greater south-central Andes.[8] This is astonishing for radiocarbon dating appeared seventy-five years ago, and ceramic sequences remains the foundation of our temporal frameworks.

One option here is to move to consider a range of material practices, not limiting ourselves to the ubiquitous pottery sherd. In fact, processual archaeologists worked to integrate the changing styles of a range of objects and features in their work on periodization. For instance, in his work at Chalchuapa (El Salvador), Robert Sharer (1978: 207, fig. 1) produced a chronological chart showing the seriation of different industries. His chronographic does not assume, a priori, the grounding chronometric function of ceramic artifacts, but rather traces the variation across a wide variety of material practices, including pottery, other clay artifacts, ground stone, bone artifacts, and architecture. While firmly planted in a chronotope of progress, this representation actually challenges this narrative. The chronographic shows that the tempo of change in various materialities do not align, that the kind of chronology articulated in El Salvadorian ceramic sequences cannot stand as a transparent measure of historical change. A similar effort in the Titicaca Basin, or indeed many regions of the Andes, could also unsettle our homogeneous time, encourage us to reevaluate the fences of our chronological charts and, at the very least, introduce a richer sense of multitemporality.

RE-EVALUATING DEPOSITIONAL TIME

In constructing his seriation of Formative and Tiwanaku ceramics, Wallace chose not to engage with stratigraphy. Had he analyzed the deposits (or at least a visual representation of them), his regional seriation might have substantially shifted and perhaps introduced other kinds of time. For instance, a stratigraphic analysis of the pots in Wallace's sample might reveal them to be primarily associated with heirloom pieces, objects that might have returned briefly into circulation, only to be buried with particular individuals living in a time long after their initial manufacture (Cook 1985: 99). In this last section, I consider how our visualizations of deposits and stratigraphic relationships have contributed to our dominant narrative convention. As will be seen, some of these stratigraphic changes play a surprisingly central role in the progressive chronotope, while others have had little impact despite their potential to tell different kinds of stories about the past.

Early excavations at Tiwanaku continue to play a key role in current understandings of the history and development of the Tiwanaku polity (Isbell and Knobloch 2006; Janusek 2003; Marsh 2012). For instance, Carlos Ponce Sanginés (1981) developed his chronological framework based on the loose interpretations of the excavation notes, drawings, and thirty radiocarbon dates of Alfred Kidder (1967). He defined a three-stage sequence with five political and economic subperiods. For instance, he hypothesized that the Late Formative lasted from 1580 BC to AD 375, a range well beyond the tighter chronology available today (see below). Ponce Sanginés believed that Bennett's red-banded vessels, often called Kalasasaya vessels today (due to their ubiquity in the Kalasasaya structure at Tiwanaku), were linked to a first phase of the Late Formative, which he called Tiwanaku I (Ponce Sanginés 1971, 1993). Ponce Sanginés defined the Tiwanaku II assemblage by a coarse undecorated ware with mica inclusions (Ponce Sanginés 1981). This problematic phase relies upon the differentiation of two strata from different types of contexts—his Tiwanaku I strata included grave lots, while his Tiwanaku II was from an occupation midden. Furthermore, the undecorated ceramics from his Tiwanaku I deposit were not described (Ponce Sanginés 1993: 53, table 1; as discussed in Janusek 2003: 37–38). Most scholars have since discarded this phase, as it could not be defined outside of Tiwanaku proper (and indeed this singular deposit!) and may not effectively represent temporal phases. His Tiwanaku III was later connected with the decorated Qeya ware through a general synthetic overview rather than systematic analysis (Janusek 2003: 39; see note 6 in this chapter). Ponce Sanginés fit these data into the dominant chronotope, with Tiwanaku I–III deposits laying the foundation for a larger narrative that conformed to Gordon Childe's model of social evolution (Marsh 2012: 204; see also Stanish 2003: 88).

Erik Marsh (2012) has recently returned to the supposed epochal break of the end of the Late Formative, exploring Ponce Sanginés's sequence and the early work of Wendell Bennett, Alfred Kidder, and Gregorio Cordero Miranda upon which it is based. Marsh explores the relationship of decorated ceramics, their depositional contexts at Tiwanaku, early radiocarbon dates, and their role in constructing our current cultural chronology. He pays particular attention to dates associated with the Late Formative, where he finds that there are few published data to evaluate the depositional context. His analysis of the stratigraphy of Bennett's pit V, a profile reproduced in his text (Marsh 2012: 208, fig. 2a), suggests that many of Bennett's dates are reliable and that in this case the "clearest temporal indicators are the marked stratigraphic boundaries" (Marsh 2012: 209). He also redraws the more complex profile from Cordero Miranda's excavations in the Kalasasaya (Marsh 2012: 209, fig. 3), a sequence with several dates associated with construction fills, and thus these are less reliable dates. Marsh uses the contextual information gleaned from these profiles and Bayesian statistical modeling to tighten up the dates and ultimately to present a more fast-paced version of the developmental narrative.

Marsh's work demonstrates the value in revisiting stratigraphy, particularly from strata that have played an essential role in constructing larger narratives (see also discussion of Gillespie [2011] earlier). There clearly are benefits to tighter date ranges in our periods through reanalysis of profiles and Bayesian statistical approaches, and our current chronology does not sufficiently draw on stratigraphic relationships to explore temporal duration (Isbell 2004: 216–17, see also Lucas 2015). There is a desperate need for the digitization of the various section drawings (and field notes) of projects over the past 100 years of excavations at Tiwanaku. Yet our current approach to time in these profiles is limited to defining breaks through soil changes, associated artifact assemblages, and radiocarbon dates. These associations are essential to the epochal shifts of our progressive chronotope. But what other kinds of temporalities, beyond the sequence of chronology, may be teased out from section drawings?

I now turn to explore some possibilities through more recent stratigraphic representations, which provide much better contextual control. I'm specifically interested here in three very distinct kinds of ash deposits found across the Lake Titicaca Basin. The archaeologists involved in the study of these strata and their interfaces have developed different interpretations of their significance to wider social relations. Yet in each case we see the relationship between material traces of deposition, our ways of visualizing these strata, and our dominant chronotope.

The first example of a significant ash deposit is seen at the site of Taraco (Stanish and Levine 2011). This large site (approximately one square kilometer in size), is located along the Rámis River in the Northern Titicaca Basin, below the modern town of the same name. Excavations in one area of the site recovered a deep, complex stratigraphy, with deposits from architectural fills; middens; buildings that were "remodeled, disassembled, or destroyed"; and, more important, a long stratum of ash measuring approximately thirty-five meters in length. Stanish and Abigail Levine include an illustration of this profile, which clearly shows the ash stratum, the various components of a destroyed adobe structure, and the precise location of samples exported for radiocarbon dating (Stanish and Levine 2011: 13905, fig. 4). A total of twelve assays were produced to date the burning deposit. Levine and Stanish (2013) pursued additional security in producing their radiocarbon dates through the careful use of multiple samples, which provided a level of statistical verification that is rare in many archaeological projects. In addition to their stratigraphic sequence, the authors explore changing patterns in artifact distributions. They find that in contexts above the ash stratum, there is an increase in the density of agricultural hoes and a decrease in both pottery and obsidian.

Stanish and Levine (2011: 13903) suggest the well-dated stratum of ash is the remains of a purposeful large-scale burning event associated with warfare and raiding during the Late Formative. The shifting artifact ratios in subsequent deposits are interpreted as evidence for the development of the state:

> Evidence of warfare is first documented around 500 B.C.E. A major, high-status section of Taraco was burned in the first century A.D., after which economic activity at the site dropped dramatically. Coincident with the conflagration, Pukara adopted many of the characteristics of state societies, including expansion at least 100 km to the south and southeast, with many of the populations falling under its powerful influence. The people residing at Taraco at the very least lost the political freedom to form their own trade networks with high-status people in other regions for at least 2 centuries. In short, our work at Taraco demonstrates that in this area of early civilization, war was a critical factor in first-generation state development. (Stanish and Levine 2011: 13906)

Here we might consider a similar process of interpretation of a burning event to a larger narrative described in the work of Nadia Abu El-Haj (1998). El-Haj also discusses the role of a particular ash deposit, here a stratum associated with the "Burnt House" of Jerusalem. Archaeologists argued for a direct association between this stratum with the destruction of Jerusalem by the Romans: "With no a priori narrative, ash could simply be evidence of accidental fire (or fires). On the basis of the material evidence alone, there is no way of determining either which possible cause the evidence of fire indexes or whether all the evidence of fire at a single site points toward a single historical cause—be that a historical event or an accident. In other words, such archaeological remains can only be interpreted as evidence of significant historical events on the basis of an a priori story" (El-Haj 1998: 171).

While Stanish and Levine demonstrate that it is very unlikely that the Taraco burn event was accidental, we are still lacking other lines of evidence for systematic regional violence across the Late Formative Titicaca Basin (see Janusek 2015: 249 on lack of evidence in southern basin, and Arkush and Stanish 2005 on evidence in greater Andes). Just as the case of the "Burnt House" was fit within a larger preexisting narrative of events (El-Haj 1998: 169–70), the Taraco stratigraphic layer, and its visual representation, fits seamlessly into a larger narrative of state development, and specifically one that involves powerful chiefs with particular "kill ratios."

The second example comes from the Taraco Peninsula in the southern Lake Titicaca Basin (not to be confused with the site above), where the Taraco Archaeological Project (TAP) has been excavating a number of sites for almost twenty-five years, including the site of Chiripa. This site is central to Formative Period narratives, with the artifacts and sequence of public architecture playing a key role in our chronology of the periods preceding Tiwanaku. The excavations by TAP in the ceremonial mound, or Monticulo, focused on the so-called Lower Houses of Chiripa, those constructions below the circular formation of Upper Houses of the more recent Formative levels. Excavations into these lower strata revealed a sequence of three structures, superimposed one on top of each other and their use and abandonment dating between 800 and 400 BC. This sequence of construction, destruction, and burning is well illustrated in a profile of these excavations (Bandy 1999: 44, figure 12). This research found that each structure had between 2 and 4 yellow floors associated with them, with a total of 8 floors through the sequence. As is often the case with pre-Columbian ceremonial structures in the Andes, the floors were carefully cleaned before being abandoned, and then covered by the next structure.

It is between these floors that the ash layers were found, a sequence of ritual deposits involving first a sand fill, then a burning event. Bandy (1998) notes that burning events are found on top of six of the eight floors in the sequence between 550 BC and 380 BC. These deposits, clearly illustrated in the stratigraphic profiles, are much less impressive in scale than the large-scale burn event discussed above by Stanish and Levine. Unlike their interpretation of the thick ash deposit as epochal marker, Bandy suggests these ash lenses are the residue of practices of ritualization "associated with the closing, or 'killing,' of an old floor, and the construction or opening of another" (Bandy 1999: 45). But in an earlier unpublished conference paper, Bandy provides a different interpretation, involving a slightly more specific form of temporality. Specifically, he suggested that these eight burning cycles were associated with the succession of a generation every twenty years or so. Unfortunately, Bandy did not further elaborate on the ritual temporality or how this process of generational succession might reveal a different perspective on Chiripa. It may be that Bandy did not wish to advance such a conjecture outside of a conference presentation, or that he felt there were not sufficient data to support such a claim. Yet his suggestion of generational succession provides an interesting avenue for thinking about other kinds of time associated with stratigraphic deposits on the Taraco Peninsula.

FIGURE 3.5. *Tiwanaku phase pits encountered in the excavations of the site of Sonaji, on the Taraco Peninsula.*

A third, and final kind of ash deposit is found in excavations throughout the southern Lake Titicaca Basin at the onset of Tiwanaku, at the same time that a new kind of material culture was circulating through the region. This involved the digging and filling of large pits, innovative spatial practices that characterized the Tiwanaku phenomenon (Roddick and Janusek 2011). Although people living in the Formative also dug pits, the scale and complexity of pitting increased. John Janusek and I have begun to explore these pits, which include basin-shaped refuse pits, to "repurposed" refuse trenches. These pits ranged in size from approximately two meters in diameter to huge extensive amorphous basins, but all include ash with a range of soil textures. Although once defined as strictly an urban phenomenon, they have recently been found at the large site of Khonkho Wankane and the smaller sites on the Taraco Peninsula, such as Sonaji (figure 3.5). At another Taraco site, that of Kala Uyuni, we found that of all the contexts excavated in the 2009 season, 33 percent consisted of Tiwanaku pit contexts. Although the pits on the Taraco Peninsula are not as large as those found at the urban center of Tiwanaku, they can be over three meters in diameter with over 1,000 liters of fill (though not all of this is ash).[9]

Due to their ubiquity in the archaeological record, these pits are found in many of the drawn stratigraphic profiles of our archaeological reports, yet they have never been explicitly discussed in academic publications. We have proposed a variety of explanations for these pits, including the use of earthen ovens, pottery firing, and waste disposal. Nevertheless, we have not ascertained a singular purpose. However, they do point to larger trends in the movement of earth (whether for dryland planting, raised-field agriculture, or the construction of monumental architecture) and new social practices of the Middle Horizon (Roddick and Janusek 2011). What might this practice mean in terms of the way that inhabitants dwelled within these sites? How might the significant practice of digging hint at the larger temporality of the landscape? Such questions require we visually foreground these deposits, but also that we consider their temporality and their relationship to our larger narratives.

Each of these examples relies upon a particular kind of visualization associated with the progressive chronotope. Stanish and Levine (2011), for instance, interpret evidence of their thick ash deposit as a purposeful burning episode correlated with state development. The cyclical resurfacing and pitting, however, support a very different kind of narrative. The ash recovered in the cyclical time of the Chiripa burn events and the Tiwanaku phase pits are not altogether useful in our larger dominant chronotope. Despite encountering many of these pits, which have destroyed a large number of otherwise well-preserved Formative deposits, and despite the ubiquity of their descriptions in our field reports, they have never been published or problematized in the scholarly literature. Bandy's example of cyclical time, briefly mentioned in a conference paper, was also never published. Instead, much of his work has focused on long-term population dynamics and their relationship to the emergence of elites, political centers, and eventually the state (Bandy 2001: 182). While Bandy's (2001, 2006) visualizations of changing populations across the landscape are particularly evocative,[10] the various temporal elements produced at the site scale are ignored, or at the very least subsumed into a larger narrative of progressive state development.

In order to explore options in visualizing temporality in these cases, we might turn to the number of researchers who have been exploring the multitemporal nature of sites and to the various pasts that make up a particular material (González-Ruibal 2014; Olivier 2011). Several of these scholars have produced rather radical chronographics to display such variability. For instance, Cornelius Holtorf (2002) suggests we track the multitemporal nature of all objects and deposits, a project that will lead to the overturning of conventional narratives. Holtorf (2002: 191) experiments in upending the "linearity of time's arrow" by producing a sketch of how the site might be visualized in a nonlinear way. He repurposes the line to represent reuses and reappropriations, rather than the steady forward march of time (Holtorf 2002: 201, fig. 7). Much like Kubler's chronographic, this image is disorienting and certainly destabilizes our chronotope. Yet I am not sure this is the solution. It would not further interpret the rich contextual detail available in all three examples discussed above. Depositional contexts and stratigraphic analyses provide our most effective reconstructions of chronology, but also the tempos and rhythms of both ritual events and everyday practice (Dillehay 2004; McAnany and Hodder 2009; Roddick, Bruno, and Hastorf 2014).

In some cases, it may be worthwhile to further explore the temporalities in preexisting visual modes of representation. In our excavation work on the Taraco Peninsula, we employ stratigraphic profile drawings in the field, but also actively develop Harris Matrices (Harris 1979), like the one of Sonaji presented in figure 3.6. On the left side of the diagram the entire sequence of stratigraphic events is illustrated, while the right hand side depicts the details of one particular structure or architectural subdivision. This complex image presents a mapping of a variety of events, from floor surfacing (and resurfacing), middens deposits, fill episodes, and a variety of pit cuts and their subsequent infilling. In creating this Harris Matrix, we transformed the stratigraphic interfaces that are illustrated as lines in section drawings into the organized cells of depositional events. Although this final version was produced in the lab setting, it requires many discussions at the trench's edge, an interpretive process highlighted by Ian Hodder (1997). The Harris Matrix itself has particularities to the Andean region, variation recently highlighted by Mary Leighton (2015: 74–78). But in all cases in which the Harris Matrix is deployed, it is not simply a way of representing time, but also a way to further enable a form of "disciplined seeing" (sensu Dennis 1989; see also Morell-Hart 2015). This visual tool trains us to "see" time at the site level while excavating stratigraphic interfaces (Brown and Harris 1993).

The matrix in figure 3.6 captures a variety of times—from the order of deposits, ceramic chronological patterns, and radiocarbon assays. Here we can see significant ordering done by such diagrams of stratigraphic sequence. Indeed, "a Harris Matrix diagram is to archaeological stratification what a normal calendar is to the days and weeks of a year" (Brown and Harris 1993: 18). There are, however, limitations in what kinds of temporalities the Harris Matrix may capture. For instance, Martin Carver (1990) has suggested that the Harris Matrix cannot explore the different duration of deposits. Marley Brown and Edward Harris, however, reply to this criticism by calling it a nonissue; the diagrams "can be lengthened, shortened, or otherwise reordered to give some indication of duration of depositions an interfaces" (Brown and Harris 1993: 19). Of course, this does not ultimately deal with the thornier issue of contemporaneity (Lucas 2012: 121; Lucas 2015). But Lucas also has a stronger critique, namely, that the Harris Matrix still relies upon a linear and abstract causal chain and thus cannot capture other significant forms of time (Lucas 2005: 39; 2012: 84). Lucas argues that the matrix is all too dependent on a simple homogeneous model of time, one that is associated with periodization and forward-moving models. Yet as the various chapters in Harris, Brown and Brown (1993) demonstrate, the matrix is actually flexible for exploring time at

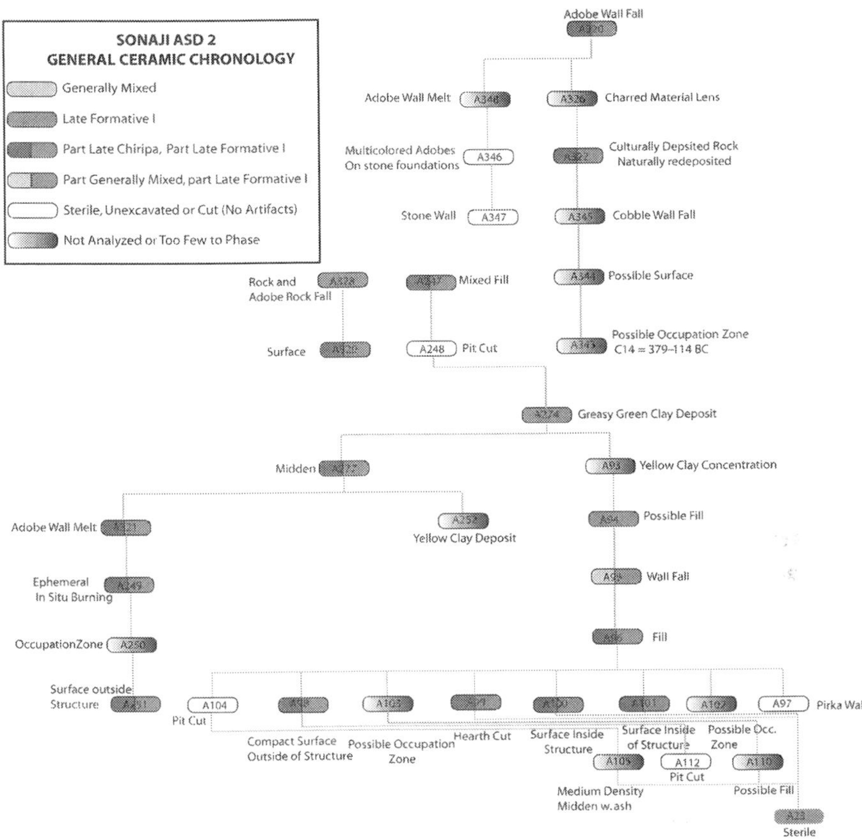

FIGURE 3.6. *Harris Matrix of the stratigraphic sequence of deposits and cuts associated with one building (ASD 2) from the site of Sonaji on the Taraco Peninsula.*

a variety of spatial and temporal scales, and might also be able to track temporalities beyond the homogenous chronological sequence.

Figure 3.6 displays the well-ordered sequence of deposits, constructions, and pitting events in our excavations at Sonaji, a complex multicomponent site at the tip of the Taraco Peninsula. The excavators of this site generated a wide number of sketches of the site while working in the field, then subsequently tightened up the specifics of the sequence once the excavations were finished. The initial use of the Harris Matrix software appropriately aligned the stratigraphic events. I then used Adobe Illustrator to realign the spatial layout of the deposits into the various terraces and buildings of this site, and color-coded the cells based on the seriation of my ceramic analyses. In some ways, the chronographic already challenges a simple progressive narrative about Sonaji. This site shows no stratigraphic evidence of occupation in the earlier phases of the

Middle Formative and the mound is not a single construction event, both of which had been hypothesized in earlier survey work (Bandy 2001). Instead it is built up through a long complex sequence of deposits and pitting, including the Tiwanaku pits, and likely occurring over several generations. The matrix also highlights the routines and cycles of daily life over many generations, a rather different chronotope (Roddick 2009: 44–51).

A closer look at figure 3.6 also reveals evidence of encounters with the past: pits associated with Tiwanaku phase occupants producing mixed ceramic deposits. Pitting, no matter the particular end goal, is also about materially engaging with the past. Inhabitants of the Taraco Peninsula digging up this landscape during the Middle Horizon would have encountered deposits from hundreds of years earlier, thus engaging with material practices of their ancestors. So how to highlight such cyclical processes on the Harris Matrix? One simple option would be another kind of graphical codification layered on top of the sequential time presented here. Another would be to generate a clone of the core matrix, but highlighting moments when the past was encountered during later occupation and sketching out lines to highlight such encounters. Either option would draw attention to the chronotope of daily life.

A similar approach to the flooring sequence at Chiripa might also reveal the social rhythms associated with early public architecture. Although such a focus has received little attention in Andean archaeology, "it is the stratigraphy of a place that perhaps best represents the empirical record of site use, its temporal construction phases, its cycles of occupation and abandonment, and its time-space compressions and extensions" (Dillehay 2004: 248). Building on Bandy's suggestion, I have argued elsewhere that this form of ritualized burning might represent a form of public time, linking practitioners to a broader social memory of the ancestors buried beneath the mound (Roddick 2013). Of course this remains a hypothesis and to be further explored would require directed attention by other scholars. Arguments such as time-space compressions of ritual use, or ancestral memory, based on single stratigraphic profile are problematic and require larger sample sizes. They also require the "disciplined seeing" provided by chronographics. Here one can imagine aligning up Harris Matrices from several sites, to see the larger patterns of shared practice across a larger region, of the changing temporal rhythms indexed by our depositional histories.

Finally, what of the role of visualization in mapping the larger regional processes, of the kinds of epochal breaks highlighted by earlier scholars such as Ponce Sanginés, and still emphasized by Stanish and Levine (2011) and Bandy (2001, 2006)? Stanish and Levine's argument of conflict is compelling for two reasons. First, their focus on long sequences across the Titicaca Basin speaks to the larger-scale narratives uniquely available to archaeology. Indeed, we must be careful in directing our attention to chronotopes often associated with ethnographic time that we don't miss the temporalities of "big histories" (Robb and Pauketat 2013: 18–19). Second, Stanish and Levine have an ideal scenario; namely, that their particular section drawing aligns with their larger argument around social evolution. But just as my argument above around habit time is simply suggestive due to sample size, so is the use of a single section drawing supporting a larger narrative of violence in state emergence. In fact, in many cases when larger data sets are visually ordered we can highlight particular temporalities. Such a case is evident in reflecting on the relationship between different kinds of data used to produce homogenous time on the Taraco Peninsula. In a recent image (Roddick et al. 2014: 144, fig. 2), we lined up Bandy's progressive narrative model with broad ceramic phasing, depositional events at Kala Uyuni, and radiocarbon dates. Although there are distinct precisions and scales, the timing of these columns clearly did not cohere easily into the homogeneous temporality often presented in our current macroscale narratives.

CONCLUSIONS: TOWARD VISUALIZING INHABITATION ON A MULTITEMPORAL LANDSCAPE

I began this chapter discussing my recent attempts at developing a different kind of narrative, one that might draw in a different kind of temporalization for the Lake Titicaca Basin. The various examples I've presented throughout this chapter demonstrate the totalizing power of our current chronotope, the slow but progressive forward march, a movement toward growth—in prestige, power, site size, and demography. I've attempted to show how we might bring about new temporal perspectives to objects, sites, and landscapes by exploring alternatives to periodization and homogeneous time. The archaeological record brims with distinct kinds of temporalities. As the chapters in this volume demonstrate, some of these narratives may be based in radically distinct ontologies. Other kinds of time, however, are easier to explore, yet are being masked by our narrative structures and particular visual schemes. Visualizations of time—whether in the form of the time chart, object and depositional histories or landscapes—are not "docile" media (Law and Whittaker 1988: 160) but rather key elements, often constraining our ability to find other kinds of emplotment in our data sets. They not only circulate and produce academic networks, but also impact larger social narratives. I've attempted here to evaluate current forms of visualizing time, to bring to the surface the rich history of Andean traditions in visualization.

R. G. Collingwood's comments from seventy-five years ago on the importance of recognizing our own past practices as we consider our future of archaeological visualization seems particularly pertinent: "No historical problem should be studied without studying . . . the history of historical thought about it" (Collingwood 1939: 132). Twenty years ago Clive Gamble (1992) found an interesting pattern in the archaeological literature: the more theoretical the paper, the fewer the visualizations. He noted that postprocessual archaeologists of the 1980s relied upon innovative ways of writing about the past, but generally avoided illustration. Stone-Miller (1993) has suggested that the more successful charts often have a manageable scale to them. Nevertheless, these same charts mask and inaccurately display "subtle, complicated, or obscure passages of history" (Stone-Miller 1993: 31). Both Gamble and Stone-Miller stress that theoretically informed archaeologies must make visualization a priority, both in terms of its production and critical stances on its consumption. Some of the most innovative archaeologists have been involved with the visual arts (Bradley 1997), scholars who developed particular ways of seeing, and it may be that further engagements with them and experts in infographics may help us advance the discipline. But as I stressed, this isn't necessarily about being artistic. Since Gamble's 1992 paper, we have seen an explosion of tools for "visualization"—databases, digital photography, GIS, and virtual reality to name a few. Even the Harris Matrix might help us explore distinct temporalities if we consider both what it displays *and* what it masks.

Although the majority of work in the Titicaca Basin, including my own, has been positioned around a story of the gradual rise and fall of Tiwanaku, our data sets suggest other possible temporal narratives. Here we might begin to consider anew the visualization of an inhabited landscape, a fairly new research trajectory, one that pushes us to consider the future possibilities of visualization. For instance, the attempts to map demographic processes (Bandy 2006), or explore the rhythms of shifting lake levels (Roddick 2013), all provide enticing possibilities for thinking about temporal disjunctures and continuities. As such, we might work toward more detailed chronographics that bring in not simply the typical "chapters" of Lake Titicaca Basin chronology, but also elements of the longue durée and processes that crosscut lived generations. If the past is made up of a variety of times, it may be foolish to think we can "harmonize [them] into a single picture" (Gosden and Kirsanow 2006: 28). Nevertheless, only by visualizing them (on paper or on the screen) can we begin to seek out temporal discordances and to work toward an understanding of temporally complex but always fully inhabited landscapes.

ACKNOWLEDGMENTS

Thanks to all members of the Taraco Archaeological Project, but especially Christine Hastorf, Matt Bandy, and Lee Steadman. John Janusek provided the photo for figure 3.2 and our discussions around Tiwanaku pits played a valuable role in this chapter. Erik Marsh offered important suggestions on the chronology in the Titicaca Basin. This chapter was originally presented at the Theoretical Archaeological Group meetings in Chicago (2013) in a session coorganized by

Ed Swenson, and I thank him for valuable suggestions and edits. Maria Bruno, Colleen Morgan, and Elizabeth Klarich also provided valuable critical feedback. Finally, thanks to Kathryn Killackey and her archaeological illustrations (killackeyillustration.com), which are always more than just support for a textual argument, but rather generate important new questions.

NOTES

1. It is, however, worth exploring whether the narratives constructed in the present link to larger kinds of lived narrative structures or experiences. As Ian Hodder (1993) has suggested, we might explore how "the material record itself was produced within a narrative experience . . . there are objective aspects of past material behaviour which can inform us about the narrative being told." In such cases it is possible to bring narrative and material culture together: "their" and "our narratives can provisionally be fused." Such an approach is explored in several of the other chapters in this volume.

2. It should be noted that Stone-Miller is an art historian, and thus brings a particular kind of attention to visualization in archaeology.

3. William Isbell and Helaine Silverman contrast Rowe's chronological chart with Bennett's earlier sketch of the Central Andean Co-Tradition: "Rowe's chronological chart included only three coastal areas, North, Central and South, but a full complement of five highland zones, although with slightly different terminology: North (Ancash), North Central (Ayacucho), South Central (Cuzco), South (Puno), and Bolivia (southern Titicaca Basin), so it is unlikely that Rowe meant to exclude Bennett's Far North Coast or Far South Coast from the Central Andean region. More likely, there simply was not enough information to fill in chronological columns for those sub-areas in the 1960s, and Rowe was loath to speculate" (Isbell and Silverman 2006a: 10). Here two points are worth highlighting: (1) The design space of a particular graph impacted how time was figured. (2) For Rowe the depiction of speculation was graphically represented by a complete absence. This clearly contrasts with Kroeber, who was more inclined to visually display uncertainty, thus opening up space for the defining of temporal phases in the future.

4. Thanks to Erik Marsh for bringing this to my attention.

5. I suspect the substantial decrease in the creation of new chronologies stems from the reluctance of archaeologists to challenge an early generation of Andean culture historians who debated the very foundations of our chronological system. In other words, "if it ain't broke, don't fix it."

6. This is despite the southern Titicaca Basin having a high density of similar Formative Period sites, not to mention the rest of Bolivia. See, for instance, Lémuz Aguirre's (2001) work in the Santiago de Huata area, for an example of a nearby region with a high density of large Middle and Late Formative sites. David Kojan and Dante Angelo also provide an insightful discussion of the rich Bolivian history that is often neglected in favor of narratives of Tiwanaku state emergence (Kojan and Angelo 2005).

7. Since Wallace's work, only a small quantity of these sherds have been found. For instance, in their regional survey in the Tiwanaku Valley, Juan Albarracin-Jordan and James Mathews found only three of 1,099 sites had Qeya diagnostic ceramics (Albarracin-Jordan and Mathews 1990: 82). Recent excavations elsewhere have also found these to be quite rare (Roddick 2009: 236), and less than a dozen possible Qeya sherds were found after two decades of excavations by the Taraco Archaeological Project.

8. The need for this discussion did emerge in a formal discussion, organized by myself and Aimée Plourde, at the 2004 SAA meetings in Montreal. Our roundtable discussion "The Evolution of Complexity in the Titicaca Basin, South-Central Andes: Evaluation and Reappraisal of the Formative Period" was, in retrospect, overtly structured by the chronotope critiqued in this chapter. Nevertheless, we discussed ceramic chronologies, different kinds of temporal scales, and the various ways of mapping them. The seed for a cross Titicaca Basin conference on ceramics and chronology was planted, but has not yet produced fruit.

9. The smaller size may also be the result our excavation methodology: since Taraco sites are deeper (see figure 3.5), we have not been able to conduct the horizontal excavation methodology deployed at both Tiwanaku and Khonko Wankane.

10. Bandy draws surface ceramics into his larger narrative of the longue durée, suggesting that they index population size and dynamics, and he generates a series of landscape images showing population densities change across our archaeological periods (Bandy 2006: 217, 231 figs. 8.4, 8.9).

REFERENCES

Abercrombie, Thomas Alan. 1998. *Pathways of Memory and Power: Ethnography and History among an Andean People*. Madison: University of Wisconsin Press.

Adams, William Y. 1979. "On the Argument from Ceramics to History: A Challenge Based on Evidence from Medieval Nubia." *Current Anthropology* 20 (4): 727–44. https://doi.org/10.1086/202373.

Albarracin-Jordan, Juan, and James Edward Mathews. 1990. *Asentamientos prehispanicos del valle de Tiwanaku*. La Paz: Producciones CIMA.

Alconini Mujica, Sonia. 1995. *Rito, Símbolo e historia en la piramide de Akapana, Tiwanaku*. La Paz, Bolivia: Editorial Acción.

Arkush, Elizabeth, and Charles Stanish. 2005. "Interpreting Conflict in the Ancient Andes." *Current Anthropology* 46 (1): 3–28. https://doi.org/10.1086/425660.

Bakhtin, Mikhail M. 1981. *The Dialogic Imagination: Four Essays*. Austin: University of Texas Press.

Ballard, Chris. 2003. "Writing(pre) History: Narrative and Archaeological Explanation in the New Guinea Highlands." *Archaeology in Oceania* 38 (3): 135–48. https://doi.org/10.1002/j.1834-4453.2003.tb00540.x.

Bandelier, Adolph. 1910. *The Islands of Titicaca and Koati*. New York: Hispanic Society of America.

Bandy, Matthew S. 1998. *Excavations on the Chiripa Mound*. Paper presented at the 63rd meeting of the Society for American Archaeology, Seattle, WA.

Bandy, Matthew S. 1999. "Montículo Excavations." In *Early Settlement at Chiripa, Bolivia: Research of the Taraco Archaeological Project*, ed. Christine A. Hastorf, 43–49. Berkeley, CA: Archaeological Research Facility.

Bandy, Matthew S. 2001. *Population and History in the Ancient Titicaca Basin*. Unpublished dissertation. Department of Anthropology, University of California, Berkeley.

Bandy, Matthew S. 2006. "Early Village Society in the Formative Period in the Southern Lake Titicaca Basin." In *Andean Archaeology III: North and South*, ed. William Isbell and Helaine Silverman, 210–36. New York: Springer Science+Business Media LLC. https://doi.org/10.1007/0-387-28940-2_10.

Barrett, John. 1999. "The Mythical Landscapes of the British Iron Age." In *Archaeologies of Landscape Contemporary Perspectives*, ed. Wendy Ashmore and Barnard Knapp, 253–65. Oxford: Blackwell Publishers.

Basso, Keith. 1984. "Stalking with Stories: Names, Places, and Moral Narratives among the Western Apache." In *Text, Play, and Story: The Construction and Reconstruction of Self and Society*, ed. Stuart Plattner, 19–53. Washington, DC: American Ethnological Society.

Bennett, Wendell C. 1934. "Excavations at Tiahuanaco." *Anthropological Papers of the American Museum of Natural History* 34 (3): 359–494.

Bennett, Wendell C. 1936. *Excavations in Bolivia*. New York: American Museum of Natural History.

Bennett, Wendell C. 1948. "A Revised Sequence for the South Titicaca Basin." *Memoirs of the Society for American Archaeology* 4:90–92.

Bennett, Wendell C. 1950. "Cultural Unity and Disunity in the Titicaca Basin." *American Antiquity* 16 (2): 89–98. https://doi.org/10.2307/276886.

Bennett, Wendell C., and Junius B. Bird. 1949. *Andean Culture History: American Museum of Natural History Handbook Series*. New York: American Museum of Natural History.

Bermann, Marc. 1990. *Prehispanic Household and Empire, at Lukurmata, Bolivia*. PhD dissertation. University of Michigan, Ann Arbor.

Bradley, Richard. 1997. "'To See Is to Have Seen': Craft Traditions in British Field Archaeology." In *The Cultural Life of Images: Visual Representations in Archaeology*, ed. Brian Molyneaux, 62–72 London: Routledge.

Brown, Marley R., III, and Edward C. Harris. 1993. "Interfaces in Archaeological Stratigraphy." In *Practices of Archaeological Stratigraphy*, ed. Edward C. Harris, 7–20. London: Academic Press.

Burkholder, JoEllen. 1997. "Tiwanaku and the Anatomy of Time: A New Ceramic Chronology from the Iwawi Site, Department of La Paz, Bolivia." Unpublished PhD dissertation. State University of New York, Binghamton.

Carver, Martin H. 1990. "Digging for Data: Archaeological Approaches to Data Definition, Acquisition and Analysis." In *Lo scavo archeologico: Dalla diagnosi all'edizione. III ciclo di lezioni sulla ricerca applicata in archeologia*, ed. R. Francovich and D. Manacorda, 45–120. Florence: Insegna del Giglio.

Cochrane, Andrew, and Ian Russell. 2007. "Visualizing Archaeologies: A Manifesto." *Cambridge Archaeological Journal* 17 (1): 3–19. https://doi.org/10.1017/S0959774307000029.

Collingwood, R. G. 1927. "Oswald Spengler and the Theory of Historical Cycles." *Antiquity* 1 (3): 311–25. https://doi.org/10.1017/S0003598X00000685.

Collingwood, R. G. 1939. *An Autobiography*. Oxford: Oxford University Press.

Cook, Anita Gwynn. 1985. "Art and Time in the Evolution of Andean State Expansionism." PhD dissertation. State University of New York, Binghamton.

Crossland, Zoë. 2014. *Ancestral Encounters in Highland Madagascar: Material Signs and Traces of the Dead*. Cambridge: Cambridge University Press. https://doi.org/10.1017/CBO9781139565882.

Dennis, Michael Aaron. 1989. "Graphic Understanding: Instruments and Interpretation in Robert Hooke's Micrographia." *Science in Context* 3 (2): 309–64. https://doi.org/10.1017/S0269889700000855.

Dillehay, Tom D. 2004. "Social Landscape and Ritual Pause: Uncertainty and Integration in Formative Peru." *Journal of Social Archaeology* 4 (2): 239–68. https://doi.org/10.1177/1469605304042396.

Domonoske, Camila. 2014. "Bolivia's De-Colonized 'Clock of the South' Turns Widdershins." *NPR News*, June 26, 2014. http://www.npr.org/sections/thetwo-way/2014/06/26/325804334/bolivias-de-colonized-clock-of-the-south-turns-widdershins.

Dulanto, Jahl. 2015. "Time and the Other: The Early Colonial Mythohistorical Landscapes of the Huarochirí Manuscript." In *The Measure and Meaning of Time in Mesoamerica and the Andes*, ed. Anthony F. Aveni, 141–64. Washington, DC: Dumbarton Oaks.

El-Haj, Nadia Abu. 1998. "Translating Truths: Nationalism, the Practice of Archaeology, and the Remaking of Past and Present in Contemporary Jerusalem." *American Ethnologist* 25 (2): 166–88. https://doi.org/10.1525/ae.1998.25.2.166.

Fyfe, Ralph, Chris Caseldine, and Mark Gillings. 2010. "Pushing the Boundaries of Data? Issues in the Construction of Rich Visual Past Landscapes." *Quaternary International* 220 (1): 153–59. https://doi.org/10.1016/j.quaint.2009.09.005.

Gamble, Clive. 1992. "Reflections from a Darkened Room." *Antiquity* 66 (251): 426–31. https://doi.org/10.1017/S0003598X00081564.

Gillespie, Susan D. 2008. "History in Practice: Ritual Deposition at La Venta Complex A." In *Memory Work: Archaeologies of Material Practices*, ed. Barbara J. Mills and William H. Walker, 109–36. Santa Fe: SAR Press.

Gillespie, Susan D. 2011. "Archaeological Drawings as Re-presentations: The Maps of Complex A, La Venta, Mexico." *Latin American Antiquity* 22 (1): 3–36. https://doi.org/10.7183/1045-6635.22.1.3.

Gillespie, Susan D. 2015. "Journey's End (?): The Travels of La Venta Offering 4." In *Things in Motion: Object Itineraries in Anthropological Practice*, ed. Rosemary A. Joyce and Susan D. Gillespie, 39–61. Santa Fe: School for Advanced Research Press.

Gillespie, Susan D., and Michael Volk. 2014. "A 3d Model of Complex A, La Venta, Mexico." *Digital Applications in Archaeology and Cultural Heritage* 1 (3–4): 72–81. https://doi.org/10.1016/j.daach.2014.06.001.

González-Ruibal, Alfredo. 2014. *An Archaeology of Resistance: Materiality and Time in an African Borderland*. Lanham, MD: Rowman and Littlefield.

Goody, Jack. 1977. "Literacy and Classification: On Turning the Tables." In *Text and Context: The Social Anthropology of Tradition*, ed. Ravindra K. Jain, 205–22. Philadelphia: Institute for the Study of Human Issues.

Gosden, Chris. 2006. "Material Culture and Long-Term Change." In *Handbook of Material Culture*, ed. Chris Tilley, Webb Keane, Susanne Kuechler, Mike Rowlands, and Patricia Spyer, 425–42. London: Sage Publications. https://doi.org/10.4135/9781848607972.n28.

Gosden, Chris, and Karola Kirsanow. 2006. "Timescales." In *Confronting Scale in Archaeology: Issues of Theory and Practice*, ed. Gary Lock and Brian L. Molyneaux, 27–37. New York: Springer.

Gosselain, Olivier P. 2011. "Fine If I Do, Fine If I Don't. Dynamics of Technical Knowledge in Sub-Saharan Africa." In *Investigating Archaeological Cultures: Material Culture, Variability, and Transmission*, ed. B. W. Roberts and M. Vander Linden, 211–27. New York: Springer. https://doi.org/10.1007/978-1-4419-6970-5_11.

Griffin, Arthur F., and Charles Stanish. 2007. "An Agent-Based Model of Prehistoric Settlement Patterns and Political Consolidation in the Lake Titicaca Basin of Peru and Bolivia." *Structure and Dynamics: EJournal of Anthropological and Related Sciences* 2 (2). http://escholarship.org/uc/item/2zd1t887.

Harris, Edward. 1979. *Principles of Archaeological Stratigraphy*. London: Academic Press.

Harris, Edward C., Marley R. Brown III, and Gregory J. Brown. 1993. *Practices of Archaeological Stratigraphy*. Academic Press: London.

Hastorf, Christine A. 2003. "Community with the Ancestors: Ceremonies and Social Memory in the Middle Formative at Chiripa, Bolivia." *Journal of Anthropological Archaeology* 22 (4): 305–32. https://doi.org/10.1016/S0278-4165(03)00029-1.

Hodder, Ian. 1993. "The Narrative and Rhetoric of Material Culture Sequences." *World Archaeology* 25 (2): 268–82. https://doi.org/10.1080/00438243.1993.9980243.

Hodder, Ian. 1997. "'Always Momentary, Fluid and Flexible': Towards a Reflexive Excavation Methodology." *Antiquity* 71 (273): 691–700. https://doi.org/10.1017/S0003598X00085410.

Holtorf, Cornelius. 2002. "Excavations at Monte Da Igreja Near Évora (Portugal), from the Life-History of a Monument to Re-Uses of Ancient Objects." *Journal of Iberian Archaeology* 4:177–201.

Ingold, Tim. 2007. *Lines: A Brief History*. Oxon: Taylor and Francis.

Isbell, William Harris. 2004. "Cultural Evolution in the Lake Titicaca Basin: Empirical Facts and Theoretical Expectations." *Revista de Antropologia* 33:209–41.

Isbell, William H., and Patricia J. Knobloch. 2006. "Missing Links, Imaginary Links: Staff God Imagery in the South Andean Past." In *Andean Archaeology III: North and South*, ed. William H Isbell and Helaine Silverman, 307–51 New York: Springer. https://doi.org/10.1007/0-387-28940-2_14.

Isbell, William H., and Patricia J. Knobloch. 2008. "SAIS—the Origin, Development, and Dating of Tiahuanaco-Huari Iconography." In *Tiwanaku—Papers from the 2005 Mayer Center Symposium at the Denver Art Museum*, ed. Margaret Young-Sanchez, 165–210. Denver: Denver Art Museum.

Isbell, William H., and Helaine Silverman. 2006a. "Regional Patterns." In *Andean Archaeology III: North and South*, ed. William H. Isbell and Helaine Silverman, 3–19. New York: Springer. https://doi.org/10.1007/0-387-28940-2_1.

Isbell, William H., and Helaine Silverman. 2006b. "Rethinking the Central Andean Co-Tradition." In *Andean Archaeology III: North and South*, ed. William Isbell and Helaine Silverman, 497–518. New York: Springer. https://doi.org/10.1007/0-387-28940-2_20.

Janusek, John W. 2003. "Vessels, Time, and Society: Towards a Chronology of Ceramic Style in the Tiwanaku Heartland." In *Tiwanaku and Its Hinterland: Archaeology and Paleoecology of an Andean Civilization*, vol. 2. ed. Alan L. Kolata, 30–92. Washington, DC: Smithsonian Institution Press.

Janusek, John W. 2015. "Tiwanaku Urban Origins: Distributed Centers and Animate Landscapes." In *The Cambridge World History.* Vol. 3.:,*Early Cities in Comparative Perspective, 4000* BCE*—1200* CE, ed. Norman Yoffee, 229–52. Cambridge: Cambridge University Press. https://doi.org/10.1017/CHO9781139035606.015.

Jones, Andrew. 2001. "Drawn from Memory: The Archaeology of Aesthetics and the Aesthetics of Archaeology in Earlier Bronze Age Britain and the Present." *World Archaeology* 33 (2): 334–56. https://doi.org/10.1080/00438240120079324.

Joyce, Rosemary A. 2002. *The Languages of Archaeology: Dialogue, Narrative, and Writing*. New York: Blackwell. https://doi.org/10.1002/9780470693520.

Kidder, Alfred V. 1967. "Digging in the Titicaca Basin." In *Peruvian Archaeology: Selected Readings*, edited by J. H. Rowe and D. Menzel, 132–45. Palo Alto: Peek.

Klarich, Elizabeth A. 2005. "From the Mundane to the Monumental: Defining Early Leadership Strategies at Late Formative Pukara, Peru." PhD dissertation. University of California, Santa Barbara.

Kojan, David. 2008. "Paths of Power and Politics: Historical Narratives at the Bolivian Site of Tiwanaku." In *Evaluating Multiple Narratives: Beyond Nationalist, Colonialist, Imperialist Archaeologies*, ed. Junko Habu, Craig Fawcett, and John M. Matsunaga, 69–85. New York: Springer. https://doi.org/10.1007/978-0-387-71825-5_6.

Kojan, David, and Dante Angelo. 2005. "Dominant Narratives, Social Violence and the Practice of Bolivian Archaeology." *Journal of Social Archaeology* 5 (3): 383–408. https://doi.org/10.1177/1469605305057585.

Kolata, Alan L. 1983. "The South Andes." In *Ancient South Americans*, ed. J. D. Jennings, 241–86. San Francisco: W. H. Freeman.

Kroeber, Alfred L. 1944. *Peruvian Archaeology in 1942: Publications in Anthropology 4*. New York: Viking Fund.

Kubler, George. 1970. "Period, Style and Meaning in Ancient American Art." *New Literary History* 1 (2): 127–44. https://doi.org/10.2307/468624.

Latour, Bruno, and Steve Woolgar. 1979. *Laboratory Life: The Social Construction of Scientific Facts*. Beverly Hills, CA: Sage.

Law, John, and John Whittaker. 1988. "On the Art of Representation: Notes on the Politics of Visualisation." In *Picturing Power: Visual Depiction and Social Relations*, ed. Gordon Fyfe and John Law, 160–83. London: Routledge.

Leighton, Mary. 2015. "Excavation Methodologies and Labour as Epistemic Concerns in the Practice of Archaeology: Comparing Examples from British and Andean Archaeology." *Archaeological Dialogues* 22 (1): 65–88. https://doi.org/10.1017/S1380203815000100.

Lemuz Aguirre, Carlos. 2001. "Patrones de Asentamiento Arqeológico en la Península de Santiago de Huata, Bolivia." Licienciatura thesis, Carrera de Arqueologia, Universidad Mayor de San Andres, La Paz.

Levine, Abigail R. 2012. "Competition, Cooperation, and the Emergence of Regional Centers in the Northern Lake Titicaca Basin, Peru." PhD dissertation. University of California, Los Angeles.

Levine, Abigail, and Charles Stanish. 2013. "The Importance of Multiple 14C Dates from Significant Archaeological Contexts." *Journal of Archaeological Method and Theory* 21 (4): 824–36.

Lucas, Gavin. 2001. *Critical Approaches to Fieldwork: Contemporary and Historical Archaeological Practice*. Abingdon: Psychology Press. https://doi.org/10.4324/9780203170007.

Lucas, Gavin. 2004. "Modern Disturbances: On the Ambiguities of Archaeology." *Modernism/Modernity* 11 (1): 109–20. https://doi.org/10.1353/mod.2004.0015.

Lucas, Gavin. 2005. *The Archaeology of Time*. London: Routledge.

Lucas, Gavin. 2012. *Understanding the Archaeological Record*. Cambridge: Cambridge University Press.

Lucas, Gavin. 2015. "Archaeology and Contemporaneity." *Archaeological Dialogues* 22 (01): 1–15. https://doi.org/10.1017/S1380203815000021.

Mamani Condori, Carlos. 1989. "History and Prehistory in Bolivia: What about the Indians?" In *Conflict in the Archaeology of Living Traditions*, ed. R. Layton, 46–59. London: Routledge.

Mamani Condori, Carlos. 1996. "History and Prehistory in Bolivia: What about the Indians?" In *Contemporary Archaeology in Theory: A Reader*, ed. Robert W Preucel and Ian Hodder, 632–45. Oxford: Blackwell Publishers.

Marsh, Erik J. 2012. "A Bayesian Re-Assessment of the Earliest Radiocarbon Dates from Tiwanaku, Bolivia." *Radiocarbon* 54 (2): 203–18. https://doi.org/10.2458/azu_js_rc.v54i2.15826.

McAnany, Patricia A., and Ian Hodder. 2009. "Thinking about Stratigraphic Sequence in Social Terms." *Archaeological Dialogues* 16 (1): 1–22. https://doi.org/10.1017/S1380203809002748.

McGlade, James. 1999. "The Times of History: Archaeology, Narrative and Non-Linear Causality." In *Time and Archaeology*, ed. Tim Murray, 139–63. London: Routledge.

Means, Philip Ainsworth. 1920. *Memorias antiguas historiales del Perú*. London: Hakluyt Society.

Mohr, Karen. 1966. "An Analysis of the Pottery of Chiripa, Bolivia: A Problem in the Archaeological Classification and Inference." MA thesis. Department of Anthropology, University of Pennsylvania, Philadelphia.

Molyneaux, Brian Leigh. 1997. "Introduction: The Cultural Life of Images." In *The Cultural Life of Images, Visual Representation in Archaeology*, ed. Brian Leigh Molyneaux, 1–10. London: Routledge.

Morell-Hart, Shanti. 2015. "The Scene of Disciplined Seeing." *Then Dig*. Http://arf.berkeley.edu/then-dig/2015/06/the-scene-of-disciplined-seeing/.

Morgan, Colleen L. 2009. "(Re) Building Çatalhöyük: Changing Virtual Reality in Archaeology." *Archaeologies* 5 (3): 468–87. https://doi.org/10.1007/s11759-009-9113-0.

Morgan, Colleen L. 2014. "Archaeology and the Moving Image." *Public Archaeology* 13 (4): 323–44. https://doi.org/10.1179/1465518715Z.00000000077.

Moser, Stephanie. 1998. *Ancestral Images: The Iconography of Human Origins*. Ithaca, NY: Cornell University Press.

Moser, Stephanie. 2001. "Archaeological Representation: The Visual Conventions for Constructing Knowledge about the Past." In *Archaeological Theory Today*, ed. Ian Hodder, 262–83. Cambridge: Polity Press.

Moser, Stephanie. 2014. "Making Expert Knowledge through the Image: Connections between Antiquarian and Early Modern Scientific Illustration." *Isis* 105 (1): 58–99. https://doi.org/10.1086/675551.

Moser, Stephanie, and Clive Gamble. 1997. "Revolutionary Images: The Iconic Vocabulary for Representing Human Activity." In *The Cultural Life of Images: Visual Representation in Archaeology*, ed. Brian L. Molyneaux, 184–212. London: Routledge.

Olivier, Laurent. [2008] 2011. *The Dark Abyss of Time: Archaeology and Memory*. Plymouth: AltaMira Press.

Ossio, Juan M. 2015. "Ages of the World in the Andes." In *The Measure and Meaning of Time in Mesoamerica and the Andes*, ed. Anthony Aveni, 211–38. Washington, DC: Dumbarton Oaks.

Perry, Sara. 2009. "Fractured Media: Challenging the Dimensions of Archaeology's Typical Visual Modes of Engagement." *Archaeologies* 5 (3): 389–415. https://doi.org/10.1007/s11759-009-9114-z.

Perry, Sara. 2014. "Crafting Knowledge with (Digital) Visual Media in Archaeology." In *Material Evidence: Learning from Archaeological Practice*, ed. Robert Chapman and Alison Wylie, 189–210. London: Routledge.

Piggott, Stuart. 1978. *Antiquity Depicted: Aspects of Archaeological Illustration*. London: Thames and Hudson.

Pillsbury, Joanne. 2012. *The Past Presented: Archaeological Illustration in the Americas*. Washington, DC: Dumbarton Oaks.

Pluciennik, Mark. 1999. "Archaeological Narratives and Other Ways of Telling." *Current Anthropology* 40 (5): 653–78. https://doi.org/10.1086/300085.

Ponce Sanginés, Carlos. 1971. "La cerámica de la época I de Tiwanaku." *Pumapunku* 2: 7–28.

Ponce Sanginés, Carlos. 1981. *Tiwanaku: Espacio, tiempo y cultura*. La Paz, Bolivia: Juventud.

Ponce Sanginés, Carlos. 1993. "La cerámica de la época I (aldeana de Tiwanaku)." *Pumapunku (Nueva Época)* 4: 48–89.

Ricoeur, Paul. 1984. *Time and Narrative*. Chicago: University of Chicago Press.

Rivas-Tello, Daiana. 2015. "Digital Image Analysis of Late Formative Micaceous Pastes in the Lake Titicaca Basin, Bolivia." Poster presented at the 34th Annual Meeting of the Northeast Conference on Andean Archaeology and Ethnohistory, Toronto, ON.

Robb, John, and Timothy R. Pauketat. 2013. "From Moments to Millennia: Theorizing Scale and Change in Human History." In *Big Histories, Human Lives: Tackling Problems of Scale in Archaeology*, ed. John Robb and Timothy R. Pauketat, 3–33. Santa Fe, NM: School for Advanced Research Press.

Roddick, Andrew P. 2009. "Communities of Pottery Production and Consumption on the Taraco Peninsula, Bolivia, 200 BC–300 AD." PhD dissertation. University of California, Berkeley.

Roddick, Andrew P. 2013. "Temporalities of the Formative Period Taraco Peninsula, Bolivia." *Journal of Social Archaeology* 13 (3): 287–309. https://doi.org/10.1177/1469605313485396.

Roddick, Andrew P. 2015. "Geologies in Motion: Itineraries of Stone, Clay and Pots in the Lake Titicaca Basin." In *Things in Motion: Object Itineraries in Anthropological Practice*, ed. Rosemary A. Joyce and Susan D. Gillespie, 123–45. School for Advanced Research Advanced Seminar Series. Santa Fe: SAR Press.

Roddick, Andrew P. 2016. "Scalar Relations: A Juxtaposition of Craft Learning in the Lake Titicaca Basin." In *Knowledge in Motion: Constellations of Learning across Time and Place*, ed. Andrew P. Roddick and Ann B. Stahl, 126–54. Tucson: University of Arizona Press.

Roddick, Andrew P., Maria C. Bruno, and Christine A. Hastorf. 2014. "Political Centers in Context: Depositional Histories at Formative Period Kala Uyuni, Bolivia." *Journal of Anthropological Archaeology* 36 (December): 140–57. https://doi.org/10.1016/j.jaa.2014.09.010.

Roddick, Andrew P., and Christine A. Hastorf. 2010. "Tradition Brought to the Surface: Continuity, Innovation and Change in the Late Formative Period, Taraco Peninsula, Bolivia." *Cambridge Archaeological Journal* 20 (2): 157–78. https://doi.org/10.1017/S0959774310000211.

Roddick, Andrew, and John Janusek. 2011. *From Profanity to Profundity: (Grudgingly) Learning to Appreciate Tiwanaku 'Ash Pits' as Tiwanaku Cultural Practice*. Paper presented at the 51st Annual Meeting of the Institute of Andean Studies, Berkeley, CA.

Roddick, Andrew P., and Elizabeth Klarich. 2013. "Arcillas and Alfereros: Clay and Temper Mining Practices in the Lake Titicaca Basin." In *Mining and Quarrying in the Ancient Andes: Sociopolitical, Economic and Symbolic Dimensions*, ed. Nicholas Tripcevich and Kevin Vaughn, 99–122. New York: Springer. https://doi.org/10.1007/978-1-4614-5200-3_5.

Rosenberg, Daniel, and Anthony Grafton. 2013. *Cartographies of Time: A History of the Timeline*. Princeton, NJ: Princeton Architectural Press.

Rowe, John H. 1962. "Stages and Periods in Archaeological Interpretation." *Southwestern Journal of Anthropology* 18 (1): 40–54. https://doi.org/10.1086/soutjanth.18.1.3629122.

Rydén, Stig. 1947. *Archaeological Researches in the Highlands of Bolivia*. Gèoteborg: Elanders Boktryckeri Aktiebolag.

Shanks, Michael, and Randall H. McGuire. 1996. "The Craft of Archaeology." *American Antiquity* 61 (01): 75–88. https://doi.org/10.1017/S0002731600050046.

Shanks, Michael, and Timothy Webmoor. 2010. "A Political Economy of Visual Media in Archaeology." In *Re-presenting the Past. Archaeology through Image and Text*, ed. Sheila Bonde and Stephen Houston, 87–110. Providence: Brown University Press.

Sharer, Robert J. 1978. *Pottery and Conclusions*. Vol. 3. The Prehistory of Chalchuapa, El Salvador. Philadelphia: University of Pennsylvania Press.

Sillar, Bill. 2009. "La saisonnalité des techniques." *Techniques and Culture* 52–53:90–119.

Smiles, Sam, and Stephanie Moser. 2005. "Introduction [to]: The Image in Question." In *Envisioning the Past: Archaeology and the Image*, ed. Sam Smiles and Stephanie Moser, 1–12. Malden MA: Blackwell. https://doi.org/10.1002/9780470774830.

Stanish, Charles. 2003. *Ancient Titicaca: The Evolution of Complex Society in Southern Peru and Northern Bolivia*. Berkeley: University of California Press. https://doi.org/10.1525/california/9780520232457.001.0001.

Stanish, Charles, Cecilia Chávez Justo, Karl LaFavre, and Aimée Plourde. 2014. *The Northern Titicaca Basin Survey: Huancané-Putina*. Vol. 1 of Memoirs of the Museum of Anthropology. Ann Arbor: University of Michigan.

Stanish, Charles, Edmundo de la Vega, Lee Steadman, Cecilia Chávez Justo, and Kirk L. Frye, L. O. Mamani, Mathew T. Seddon, and P. C. Chuquimia. 1997. *Archaeological Survey in the Juli-Desaguadero Region of Lake Titicaca Basin, Southern Peru*. Chicago: Field Museum of Natural History. https://doi.org/10.5962/bhl.title.3578.

Stanish, Charles, and Abigail Levine. 2011. "War and Early State Formation in the Northern Titicaca Basin, Peru." *Proceedings of the National Academy of Sciences of the United States of America* 108 (34): 13901–906. https://doi.org/10.1073/pnas.1110176108.

Steadman, Lee H. 1995. "Excavations at Camata: An Early Ceramic Chronology for the Western Titicaca Basin, Peru." PhD dissertation. University of California, Berkeley.

Steadman, Lee. 2007."Ceramic Analysis." In *Kala Uyuni: An Early Political Center in the Southern Lake Titicaca Basin: 2003 Excavations of the TaracoArchaeological Project*, ed. Matthew Bandy and Christine Hastorf, 67–112. Contributions of the Archaeological Research Facility No. 64. Berkeley: University of California.

Stone-Miller, Rebecca. 1993. "An Overview of the 'Horizon' and 'Horizon Style' in the Study of Ancient American Objects." In *Latin American Horizons*, ed. Don S. Rice, 15–40. Washington, DC: Dumbarton Oaks.

Swenson, Edward R. In press. "Changing Conception of Time: Moche Chronologies vs. Moche Temporalities." In *Times of Change and Changes of Time: An Inquiry about Absolute and Relative Chronologies of the Moche from Northern Peru*, ed. Luis Jaime Castillo and Claude Chapelaine. Washington, DC: Dumbarton Oaks.

Swogger, John. 2000. "Image and Interpretation: The Tyranny of Representation." In *Towards Reflexive Method in Archaeology: The Example at Çatalhöyük*, ed. Ian Hodder, 143–52. Cambridge: McDonald Institute for Archaeological Research.

Terrell, John Edward. 1996. "Lapita as History and Culture Hero." In *Oceanic Culture History: Essays in Honour of Roger Green*, ed. J. Davidson, G. Irwin, B. Foss Leach, A. Pawley, and D. Brown, 51–66. Special Publication. Dunedin: New Zealand Journal of Archaeology.

Thomas, Julian. 1996. *Time, Culture, and Identity*. Cambridge: Cambridge University Press.

Thornton, Thomas F. 2012. *Being and Place among the Tlingit*. Seattle: University of Washington Press.

Uhle, Max. 1902. "Types of Culture in Peru." *American Anthropologist* 4 (4): 753–59. https://doi.org/10.1525/aa.1902.4.4.02a00080.

Van Reybrouck, David. 1998. "Imaging and Imagining the Neanderthal: The Role of Technical Drawings in Archaeology." *Antiquity* 72 (275): 56–64. https://doi.org/10.1017/S0003598X00086270.

Verdesio, Gustavo. 2013. "Indigeneity and Time: Towards a Decolonization of Archaeological Temporal Categories and Tools." In *Reclaiming Archaeology: Beyond the Tropes of Modernity*, ed. Alfredo González-Ruibal, 168–80. New York: Routledge. https://doi.org/10.4324/9780203068632.ch13.

Wallace, Dwight T. 1957. "The Tiahuanaco Horizon Styles in the Peruvian and Bolivian Highlands." PhD dissertation. Department of Anthropology, University of California Berkeley.

Wallis, Neill J. 2015. "The Living Past: Itineraries of Swift Creek Images through Wood, Earthenware, and Ether." In *Things in Motion: Object Itineraries in Anthropological Practice*, ed. Rosemary A. Joyce and Susan D. Gillespie, 201–19. Santa Fe: SAR Press.

Webmoor, Timothy. 2005. "Mediational Techniques and Conceptual Frameworks in Archaeology: A Model in 'Mapwork' at Teotihuacan, Mexico." *Journal of Social Archaeology* 5 (1): 52–84. https://doi.org/10.1177/1469605305050143.

Weismantel, Mary. 2004. "Moche Sex Pots: Reproduction and Temporality in Ancient South America." *American Anthropologist* 106 (3): 495–505. https://doi.org/10.1525/aa.2004.106.3.495.

Van Dyke, Ruth M. 2006. "Seeing the Past: Visual Media in Archaeology." *American Anthropologist* 108 (2): 370–75. https://doi.org/10.1525/aa.2006.108.2.370.

FOUR

The Past as Kin

Materiality and Time in Inka Landscapes

DARRYL WILKINSON AND TERENCE D'ALTROY

INTRODUCTION

The marked lack of figurative imagery in Inka material culture is doubly anomalous. In the first instance, it appears peculiar when set within the comparative study of empires. Unfashionable as it is nowadays to make such generalizations, we might venture a claim that empires are unusually prolific in their production of images. Further, there is probably a very loose correlation between the degree of "complexity" (read: centralized and hierarchical politics) exhibited by any given society and its appetite for figuration (DeMarrais et. al 1996). The reasons for this are not necessarily all that obscure; the basic link between political complexity and monumental construction is widely accepted, and many a grand monument incorporates an equally grand set of iconic representations. Images are also valuable in their potential ability to transcend certain kinds of difference, something often of especial interest to imperial polities (D'Altroy and Schreiber 2004). A statue of Caesar Augustus, say, might be able to convey something to a group of Romano-British or Punic provincial elites that would be lost in a purely epigraphic Latin inscription—especially if the audience neither spoke nor read Latin. And as

DOI: 10.5876/9781607326427.c004

many scholars argue, images are in of themselves highly potent things. For example, in Paul Zanker's (1990) now seminal study of the *Power of Images in the Age of Augustus* we encounter a paradigmatic instance of the claim that visual culture (and its manipulation) was fundamental to realizing ancient imperial ambitions. As he puts it, the "power of visual imagery was by no means the least of the forces that propelled Rome into the world of Hellenistic culture and contributed to the dissolution of the old Republican order" (Zanker 1990: 335). In the case of Rome then, the empire did not just make images; it was also made by them. And of course few would restrict this idea of imagistic potency to the ancient Mediterranean alone.

And yet to wander the avenues of the former Inka capital of Cuzco provokes a certain feeling of strangeness. Put simply, it does not look much like the capitals of other great empires, adorned with their many statues, elaborate friezes, and monumental exhortations to believe in their founders' manifest destiny. Instead the walls of Cuzco's palaces and temples stand as seemingly anonymous witnesses, presenting only the unornamented (though no less impressive) ashlar masonry for which the Inkas are so well known. Nonetheless, it is worth mentioning that the Inka avoidance of figuration can be overstated, for it was not an absolute policy. Readers familiar with Cuzco will be aware of the serpentine carvings that slither across the city's ancient walls, for example. Still, it is the relative scarcity and modesty of these motifs that is perhaps the most noteworthy thing about them. As Tamara Bray (2009: 360–62) discusses, the only truly consistent exception to the Inka's tendency to aniconic expression is found in the small, typically metallic figurines associated with the Qhapaqucha ceremonial offerings. These miniatures either take the form of anthropomorphs or zoomorphs (camelids are the most common) and are particularly associated with highly structured, high-altitude ritual deposits that often include Spondylus shell, textiles, and in some cases, mummified human juveniles. But beyond these quite unusual contexts, figuration remains exceedingly rare within the Inka material culture that has survived to the present day.

Yet as we stated initially, Inka aniconism is *doubly* strange, for it cannot be easily explained in "local" terms either. The Inka's imperial predecessors in the Andean highlands (the Tiwanaku and Wari polities), and their contemporaries on the North Coast (e.g., the Chimú State), all made significant public use of anthropomorphic or zoomorphic representations (Arnold and Hastorf 2008; Cook 1983; Janusek 2006; Kolata 1993; Moore 1992). In other words, the contrast between the Inkas and the other Andean empires is, in this respect, rather stark. So while it is common to read how the Inkas built upon the technological and social foundations laid by their predecessors, the many techniques of statecraft they inherited did not apparently include the manipulation of figurative imagery. Ultimately, this absence is an empirical anomaly that cannot be left uninterrogated. If monumental imagery is constitutive of power, as is widely taken to be the case—*and* if pre-Inka Andean states made use of such methods to enhance their authority—then why would the Inkas have avoided such a tried and potent means of empire building? Whether we frame our problem as a comparative one, with respect to the character of the archaic state globally, or instead cast it as question of analyzing local historical change, it still demands some sort of explanation.

That said, accounting for the Inka's general avoidance of figurative imagery is a complex task, and obviously lies beyond the scope of a single essay. And in any case, there is hardly likely to be one catchall explanation for such a phenomenon. Our present goal is therefore somewhat less ambitious, insofar as we will attempt to answer a more restricted version of the question, namely, Why did the Inkas not produce figurative representations *of the past*? While there are the occasional, but still very rare, artifacts that depict a human or a zoomorph (such as those from the aforementioned Qhapaqucha offerings) there is not a single item of known Inka material culture (of precolonial date) that could be said to depict a historical event or narrative, at least in any iconic form recognizable to us. This may restrict us to only one genre of figurative representation, but by approaching the problem in a more targeted fashion we aim to avoid overly reductive or single-stranded explanations. Our conclusions here should thus be taken as an attempt to put together but a few pieces from a much larger puzzle, one whose solution will ultimately require much more work.

As will shortly become clear, we do not consider appeals to "aesthetics" to provide a sufficient account of the lack of figuration we observe in Inka material culture. Explaining Inka aniconism as a simple preference for certain kinds of abstract visual motif would require construing the aesthetic realm as an autonomous sphere of action, in a manner that would be highly anachronistic with regards to the precolonial Andes (Dean 2010: 1–16). Nor do we think it likely that the Inkas simply thought other forms of representing the past (e.g., oral narratives) were adequate and that additional figurative elaborations were thus unnecessary. Instead, we suggest that explaining the lack of figurative historical iconography in the Inka Empire requires deeper explanations. That is, they call for analyses that force us to grapple with the profound alterity of Inka ways of relating to time. In the final analysis, we suggest that the very idea of *representing* the past itself was a somewhat alien proposition in the Inka world—because for them, the past was far from being an absent or completed thing. It is not that they simply chose to avoid the depiction of past events, or that it never occurred to them to do so; rather for the Inkas, the very conceptions of the past upon which such practices rest were irrelevant.

BEYOND MEMORY, BEYOND REPRESENTATION

Monumental depictions of major historical events are far from uncommon in the archaeological record. Many classic examples will doubtless be familiar to most readers—the triumphal Arch of Titus in Rome depicting the first-century sacking of the Jewish Temple in Jerusalem, for example, or the victory stela of the Akkadian ruler Naram-Sim, showing his conquest of the Lullubi people of the Zagros Mountains, or the "success" of Ramses the Great at the Battle of Kadesh, as portrayed in the wall reliefs of the Ramesseum complex of New Kingdom Egypt. At one level, the interpretation of such artifacts needs no great excursus into high social theory. They can be quite straightforwardly understood as political propaganda, efforts to glorify the empire (and its rulers) by deploying historical narratives that corroborate their self-aggrandizing claims. In other words, they instantiate the old Orwellian maxim "He who controls the past controls the future. He who controls the present controls the past." The manipulation of posterity is always a matter of concern—for the powerful most of all. Or as Winston Churchill is reputed to have said, with characteristic wryness, "history will be kind to me, for I intend to write it."

In a more scholarly context, the monumental histories displayed by ancient empires are typical examples of what archaeologists often call "the past in the past." In an essay reflecting on the state of archaeological theory at the end of the 1990s, Chris Gosden and Gary Lock pointed out the irony in archaeologists' lack of interest in the historical consciousness of the peoples they studied (Gosden and Lock 1998: 2–3)—though even then there were some notable exceptions (e.g., Holtorf 1996). The intervening years have seen some considerable shifts in this respect, and now the "the past in the past" or "social memory" has emerged as one of the core themes of poststructuralist archaeology, with a proliferation of volumes directly addressing the subject (Harris 2009). These include general calls for attention to the importance of memory in all archaeological contexts (e.g., Van Dyke and Alcock 2003), as well as a range of monographs covering cases as diverse as Neolithic Britain, the Hellenistic Mediterranean, and the Mesoamerican Classic (e.g., Alcock 2002; Edmonds 2002; Joyce 2003). This is not the place to survey the already extensive archaeological literature on the topic, though more recent work has sought to emphasize memory as something that is grounded in materiality (if not necessarily monumental in scale) and something always under active construction (see Mills and Walker 2008). Ultimately, however, our aim here is to provide a counterpoint to the memory studies paradigm, by questioning the assumptions implicit within such approaches. We ask: What is assumed to be *universal* in memory-based frameworks? In this respect, the Inkas serve as our dissenting example, from which we argue that the very concept of "memory" itself does not adequately capture the logic of Inka temporal engagements with the landscape, and may in fact work to analytically efface them.

Put as succinctly as possible, the memory studies paradigm assumes that all societies, that all peoples, *represent* the past. It thereby construes the universal relationship with the past to be a mode of engagement centered on an *absent* object, one that, because it is not present, must therefore always be re-presented. At one level, our arguments here can be seen as relating to wider theoretical trends in the humanities and social sciences, typified by an increasing skepticism toward "representation" as a general interpretive paradigm. In a recent archaeological iteration of this shift, for example, Andrew Jones and Benjamin Alberti have called for moving away from the classically postprocessual notion that material culture is social because it acts as a vehicle for human-inscribed meanings (Jones and Alberti 2013). Memory studies are, we suggest, inextricable from a representationalist approach to material culture; they rely on the perception of the past as an absent object that must be engaged through its traces in the present. Our arguments here can therefore be seen as being in partial alignment with recent archaeological scholarship that critiques representation more broadly. As an alternative, we argue for a need to interpret Inka material practices in light of their temporal ontologies rather than ours, thereby questioning the implicit configurations of time that underlie the very idea of memory itself.

THE QUESTION OF TEMPORAL ALTERITY

In Western societies, relationships with the past are dominated by its absence. We might say that someone is "living in the past," but this is virtually always meant as a metaphor—they are of course living in the present just like everyone else. Living in the past, in a strictly literal sense, is impossible, unless one has access to some sort of time machine. Indeed, the phrase only makes sense as a critique if it is understood metaphorically. We can think about the past, describe the past, and engage with the past in any number of ways; but it is impossible to actually reside in it, for it is by definition no longer present. European linguistic terms are saturated with this temporal ontology. The word "present" from the Latin *praesent* (being at hand) implies a past and future that is, by contrast, *not* at hand. The prefix "re-," also of Latin derivation, appears in most verbs that deal with relating to the past. Thus when we engage with the past, it entails a process of *re*-collection; it is something that we *re*-member. Etymologically, "re-" refers either to a movement backward or a sense of an action occurring again.

Archaeologists are perhaps especially intimate with the past as an absent object, given that they typically *re*-construct it from the traces and clues it has left in its wake. Archaeologists might often say that they study the past, but in a more literal register they really mean that they study the traces of the past still available to us *in the present*. The logic of the "trace" is fundamental to archaeology, but also runs across a wide range of modernist fields of knowledge (forensics having a particularly close methodological kinship in this respect), and fundamentally deals in how we can know the absent or invisible thing (Borić 2010; Crossland 2009). Another particularly relevant example (especially with respect to memory) is Sigmund Freud's psychoanalytic method, which he explicitly understood via the metaphor of archaeological excavation, wherein a lost past is reconstituted from the study of its present fragments (see Thomas 2004: 161–69). Put simply, traces are indexical referents to something that is not (or no longer) there. Dušan Borić's (2010: 4) terms are quite apt in this respect, summarizing the modern idea of the trace (like memories) as "an imprint that *has been* left behind by a past agency and which remains in the present as its testimony." Footprints are an everyday sort of trace, but just as a footprint is not a foot, the traces of the past are not the past itself. And of course we can only see the footprint after the foot has *absented* itself, the trace not being revealed until the thing that made it is no longer there. These points are especially relevant to our arguments here because the logic of the trace is widely taken as a universal one in memory studies. Consider the following claim: "Like us, past peoples observed and interpreted *traces* of more distant pasts to serve the needs and interests of their present lives" (Van Dyke and Alcock 2003: 1, emphasis ours). Note here that past peoples' engagement with their pasts is explicitly likened to the reconstructive practices of the modern archaeologist (i.e., mediated through the interpretation of traces).

This absence of the past is thus a fairly fundamental proposition within the normative "temporal ontology" found in Western societies. And by the phrase "temporal ontology," we mean a theory, or less formally, a set of taken-for-granted propositions about time: what it is and how it works at a fundamental level. By extension, the trace logics that underpin archaeological, forensic, or psychoanalytical reconstructions could be thought of as a form of temporal epistemology—a framework by which the past can be *known* or at least *reliably inferred*. For most Westerners, time only moves in one direction (i.e., forward) and from the perspective of ego, one can only exist in the present. Both past and future are therefore necessarily absent, the one because it has already gone by, and the other because it has yet to be. Past and future still differ in several respects, however, despite both being absent. Most important, the past is singular (i.e., only one thing "really happened") while the future is infinitely plural (i.e., any number of things might yet happen). For example, the practice of a trace-oriented discipline such as forensics would be rendered a complete nonsense if a crime scene could be the product of several contradictory, and yet all equally real, events. Multiple interpretations of the past are fine, but only one actual past can exist, and this fixity is a core proposition in most Western temporalities.

Of course, we can only make such statements about temporal ontologies by allowing for a high degree of generalization—and so it is always good to acknowledge the existence of contradictions and diversity. For example, most individuals of a secular-scientific persuasion (i.e., those who reject the existence of supernatural phenomena) would affirm that only the past can leave traces in the present. However, believers in prophecy or fortune-telling would likely accept that the future, just like the past, also imprints traces upon the present. Divination is a trace-oriented discipline too, but obviously quite different to a modernist field such as forensics (if for no other reason that it does not normally hold up in court). And obviously those who believe in, say, a strictly Calvinist doctrine of Predestination would not accept the infinitely plural characterization of the future, since God has always preordained its eventual form. Elsewhere, some theoretical physicists have argued that in certain contexts (often regarding quantum-level phenomena) effects might actually precede their causes (reversing the normal direction of time and commonsense rules of causality). Such counterexamples are interesting because they show that even across scientific fields, quite different temporal ontologies may be deployed. Nonetheless, recognizing a lack of absolute coherency in Western temporal ontologies is not to say there is absolute *in*coherency.

Given the plurality of temporal ontologies evident within the Western world itself, arguably any study of temporality in the precolonial Americas would have to begin with a consideration of how indigenous modes of relating to time might well be very different from those assumed by archaeologists (i.e., people wedded through their disciplinary practice to trace logics). As is often the case, language is a useful place to begin, and indigenous Andean linguistic terms often suggest quite different understandings of time when compared to most European tongues. In particular, studies of both Quechua and Aymara have emphasized the importance of expressing temporal locations through spatial metaphors that deploy a logic distinct from that of their European counterparts.

In Aymara, the present and past are both associated with the subject's visual field, because they are both known, while the future remains unknown and is associated with the unseen (Miracle and Yapita 1981). The Aymara word *nayra* thus means "in front of," but is also linked etymologically to the eyes and the field of vision. Tellingly, Rafael Núñez and Eve Sweetser's studies of gestures produced by Aymara communities indicate that the recent past is associated with the space immediately before the speaker, while references to the distant past, are accompanied by gestures to the far-off horizon in front of the speaker (Núñez and Sweetser 2006). Conversely, the word indicating the future (*qhipa*) also means "behind," and so the future is located in gestures to the rear of the speaking subject. Most directly relevant to our arguments is the fact that only the future appears to be absent (from the vision), while both present and past are placed in the apprehensible visual field, albeit distinguished respectively as foreground and background. Similar patterns are noted in Quechua, where the term *ñawpa* refers to both the past and the space in front of the speaker. It is also closely related etymologically to the term *ñawi* ("eye" or "vision"), and so a phrase such as *ñawpa pacha* ("olden times") can be given a literal gloss as "time of the eye" or "time in the eye" (Bourdin 2014). It is also interesting to note here that the Quechua term for the "past" seems etymologically comparable to the English term for "present" (compare "in front of/in the eye" to "at hand/in view"). Finally, the unity of the past-and-present, in contradistinction to the absent future, also appears in the fact that a single tense covers both past and present, contrasting with a separate future tense.

At this point, linguistic evidence alone might lead us to be suspicious of any *a priori* assumption of an absent past in the Inka context. Moreover, the available ethnohistoric materials can offer considerable insight into Inka temporal ontology, especially when read with an awareness of the aforementioned linguistic data. A wide array of entities are described and discussed in the Spanish chronicles, all of which we might gloss as "historical actors." These include multiple categories of beings such as founding ancestors, culture-heroes, creator gods, enemy warriors, and so on. From a Western perspective, historical actors normally enter the narrative, play their part in the story, and then exit it again. Thus the finite lifespan of any individual is one of the fundamental logics upon which our historical narratives rest. In this respect, an influential view of history in Western thought is captured in the Latin phrase *res gestae*, literally meaning "the things done" (Collingwood 1993). In Roman contexts this was typically associated with a particular (usually important) individual, seen in texts such as *Res Gestae Divi Augusti* (*The Deeds of the Divine Augustus*). This text is an example of a particular genre of ancient funerary inscription, intended to sum up the lived life of a great man. In later ages, this mode of accounting the past was perhaps best exemplified in the "king and queens" genre of history, as was once inflicted on generations of British schoolchildren. In the Inka context, however, it seems that although historical actors enter the narrative, there is no guarantee that they will ever exit it. Rather than death as we understand it, there is often some form of transformation or metamorphosis, after which the historical actor continues his or her part in the story (albeit as a physically altered being). Moreover, the state of a person after death was often conceived of as a purified condition, in some senses superior to that of the fleshly living individual (Salomon 1995).

This perspective might be better illustrated by drawing a contrast with some ways of thinking more familiar to Western readers. Consider, for example, all the various important persons who populate the past within the different Abrahamic religions. They might include individuals thought to be divine (such as Christ) or prophetic leaders, who were human but nonetheless divinely chosen or inspired (Muhammad, say, or Moses). All those persons existed on earth as solid, flesh-and-blood humans, but only for a limited period of time. They entered history and eventually, they exited it too. Although they may be regarded as continuing to exist in some transcendent or heavenly form despite their physical bodies having long ago passed away, they are no longer present in the material world as ordinary historical actors. That said, they nonetheless remained extremely influential figures in the lives of many from Late Antiquity through to the present day. That influence was not direct however, but always *mediated*, through institutions (such as the Church) or texts (e.g., the New Testament and the Qur'an) or images (such as Byzantine iconography) or relics (such as the preserved body parts of saints)—and many other forms besides.

The situation in the precolonial Andes, and among indigenous peoples even after, appears to have been radically different. The Inkas too recognized a range of important "religious" figures who were key movers within their historical narratives (as assembled by the Spanish chroniclers). These included divine entities such as Wiraqocha, the creator-being and original maker of the world, or major culture-heroes such as Ayar Uchu (one of the first Inkas). But quite unlike the foundational figures of the Abrahamic traditions, the earthly careers of Inka divinities and heroes were not finite in duration. In a very real sense, they never actually died, and they remained fully instantiated in historical time. This presence often entailed a transformation into some kind of lithic being—usually called a *wak'a* (although not all wak'a were necessarily made of stone). The creator Wiraqocha, for example, was manifest in a number or rock formations located across the imperial heartland, while the culture-hero Ayar Uchu was transformed into an outcrop on the hill near Cuzco called Wanakawri, later venerated as one of the most important wak'a in the entire Inka Empire. It is important to understand that in no sense were these actors "dead" or merely "memorialized" through rocks. Although they had become lithic entities, they could still talk, think, eat and drink—and so continued to live rich social lives. Wak'a are often glossed in the archaeological literature as a type of shrine, which is true in a sense of course, but as Frank Salomon (1991: 19) puts it, these shrines, "are living beings, persons in fact." In other words the term "shrine" is based on a Christian analogue, and so implies an overly object-like entity when deployed in the Andes. And even though "person" is a better term than "shrine," as Tamara Bray (2015: 10) discusses, the kinds of personhood found in Western thought do not fully map onto the kind of "relational" persons expressed in Andean wak'a—and so the term must be used with some caution. Moreover, wak'a were not the only undying historical actors to populate the Inka world. The *mallki* (mummified bodies of the emperors) continued on as important political actors, despite their transformed corporeality, often attending affairs of state and engaging in all manner of palace intrigues.

It is worth taking a moment to reflect on just how profoundly alter this experience of the past is from a modern perspective. For us, historical actors come and go, ultimately being fully replaced every few generations or so. By contrast, for the Inkas the key historical actors just multiplied. Royal history for the Inkas was not a succession of kings and queens, but more like an *accumulation* of them.

Although this ever-accreting collection of historical figures was theoretically infinite, it is important to note that there were circumstances under which wak'a could be deactivated (van de Guchte 1999) and mallki destroyed or perhaps diminished in their vitality (D'Altroy 2015), meaning that such accumulation was always more limited in practice. Nonetheless, the possibility of historical personages accumulating over time without ever actually dying is obviously quite a different mode of figuring temporal relationships. Elsewhere, we have argued that the Inka equivalent of a "canon" (their version of the Confucian Analects, the Greek classics or the Abrahamic scriptures) was instantiated in the very landscape itself (D'Altroy and Wilkinson 2009). But following on from our present argument, this was much more than a mere inscription of narratives or information composed in nontextual medium. Today we can read the recorded words and deeds attributed to a prophetic being such as Moses and debate their meaning and import. The Inkas had no need of such secondary exegesis with respect to the inscribed landscape of their world. Their prophets (often literally so, since many wak'a possessed powers of divination) were vital beings still resident in the landscape around them. Thus there was simply no need to interpret their opinions from surviving records or narratives, for it was only a matter of visiting them in person and asking them directly.

These observations also have far-reaching implications for what we understand the political manipulation of the past to actually entail. The history of the early Christian Church could easily be written as a series of battles over the legitimacy of competing scriptures. The canonical gospels of Matthew, Mark, Luke, and John are just the victorious few, the survivors of a much wider corpus of gospel texts that once vied for authority among the first Christians (Ehrman 2003). Those struggles were fundamentally contests over *content*, in that the suppressed gospels contained accounts of Christ's life and sayings whose theological implications were rejected by the leaders of proto-orthodoxy. The word "gospel" literally means "the good news," but it took many centuries of struggle among competing factions to determine what exactly that news was. We say this to emphasize that this form of political contestation vis-à-vis the past was simply impossible in an Inka context. The Christian divinity departed from history, and so a contest ensued over what should be the "true" memory of what He had said and done. By contrast, the Inkas' divinities were still there, still performing deeds and still making pronouncements, and so the goal was not to shape the records of their lives, but to monopolize access to them—thereby serving as the gatekeepers between such beings and the wider world.

THE PAST AS KIN

If we accept that Inka engagements with the past were not predicated upon relations of absence, then the theoretical framework of memory studies becomes increasingly questionable—at least for the ancient Andes. The field of memory studies (in archaeology and elsewhere) fundamentally draws on the notion of a past that is accessed through representational practices. And as a matter of basic logic, one only has to *re*-present something if it is not present already. Social memories consist of a set of signs (whether iconic images, dramatic reenactments, textual narratives, or oral accounts) that work to signify the absent past for the purposes of present audiences. These signs are woven by conscious actors in the present, and so come under the influence of a wide variety of competing sectional interests (i.e., they are inherently political). But Inka relationships with the past were not built on relationships focused on an absence, so it is doubtful that we can adequately understand their temporal practices through such social memory approaches. What then is the alternative? How did the Inkas engage with the past, if not through memories? To begin to answer such questions with a finer grain, it is necessary to explore some specific ethnohistorical and archaeological cases in greater detail.

Spanish narratives of the ancient Andean past—based on information provided by indigenous informants during the 1500s and 1600s—make frequent reference to the powerful, godlike entity we mentioned earlier called Wiraqocha. The precise meaning of the name is debated, however, and its literal translation is usually given as something like "sea of grease" or "foam of the sea" (cf. Rowe 1960). The kind of being Wiraqocha represents is not exactly clear from the accounts given to us, but the standard gloss is that he served as the primary creator deity in Inka mythologies. That said, it is important to consider how the Spanish narratives were refracted through the lens of sixteenth-century Catholicism, and thus reflect much importation of European theologies (Pease 1970; Urton 1999: 37–38). As Arthur Demarest (1981) has argued, the term Wiraqocha can be applied to the deified Sun in certain seasonal contexts, but not others, and so may be better understood as an aspect or activity of the divine, rather than the proper name of a specific god. In any case, various Spanish chronicles make reference to a series of individual wak'a also called wiraqochas, residing at multiple locales within the Inka heartland (table 4.1). The relationship between this series of wiraqochas manifest as wak'a, and the singular Wiraqocha who takes on the role of the creator deity is not altogether clear. The wak'a may have been avatars or manifestations of *the* original Wiraqocha, or his personhood may have been distributed across multiple bodily expressions.

Clearly however, the wak'a-wiraqochas and Wiraqocha the creator god are directly linked. The table 4.1 lists the various wak'a of the imperial heartland that are directly named as wiraqochas in the Spanish chronicles, including whatever details are given for their respective geographical locations. If we look at these wak'a placed on a map of the heartland, as shown in figure 4.1, it becomes evident that they run in a single line from Raqchi in the southeast to the Amaybamba

TABLE 4.1. A list of the wak'as located near Cuzco associated with Wiraqocha, as described in the Spanish chronicles.

Name/description of wak'a	*Location*	*Reference(s)*
Stone quarry and wak'a named Wiraqocha	Fourth shrine on the fourth zeq'e of Antisuyu	(Cobo 1990: 55, 66)
Turuca, a wawqi (brother) of Tiki Wiraqocha	Second shrine on the first zeq'e of Antisuyu	(Cobo 1990: 63)
A wak'a; Chanka Wiraqochan	At Chuquichaca, on the eastern boundary of Vilcabamba	(Molina 2011: 43)
A wak'a; Atun Wiraqochan	At Urcos, on a nearby hill	(Betanzos [1557] 1996: 7–10; Molina 2011: 43; Sarmiento de Gamboa 2007: 52)
A wak'a; Apotin Wiraqochan	Amaybamba	(Molina 2011: 43)
A wak'a; Urusaywa Wiraqochan	Amaybamba	(Molina 2011: 43)
A wak'a; Chuqwichanka Wiraqochan	Huaypon, probably near Maras	(Molina 2011: 44)
A temple and idol of Teqzi Wiraqocha	Cacha, in Collao	(Cieza de León 2005: 305–7)

Valley in the northwest (figure 4.1). This path of movement follows the axis mundi of the imperial heartland—the Vilcanota River—which is also the route that Wiraqocha took from Lake Titicaca to the sea during his cosmogonic journey that brought the world into being (Urton 1999). At various points along Wiraqocha's mythical journey, there are wak'a that bear his name. Yet here it is important to take seriously the idea that for precolonial Andeans, the past was not defined as a necessarily absent entity. In other words, these wak'a were not *traces* of Wiraqocha, nor were they signs left to indicate his former passage across the landscape. Seeing them as a sort of divine footprint (i.e., an index of completed agency) is precisely the wrong way to interpret them. Consider the contrasting case of a medieval pilgrim going to Jerusalem to walk the Stations of the Cross along the Via Dolorosa. That mythic journey is today memorialized with markers at each significant point, and the pilgrims who *trace* the path are thus following in Christ's footsteps. Transplanting the rather different logic of Wiraqocha's cosmogonic journey, it would be more like following the Via Dolorosa and encountering Christ's living body at each of the twelve stations—fully present in multiple locations at once. And not only that, one could still interact with the deity at each site, being entirely instantiated in present historical time. Wiraqocha did not leave footprints; rather he left his multiplied lithic selves as a series of vital and sentient bodies, all residing at key junctures along the road.

The multiplication of certain persons across more than one physical body is not uncommon in Inka contexts. The bodies of Inka emperors were simultaneously

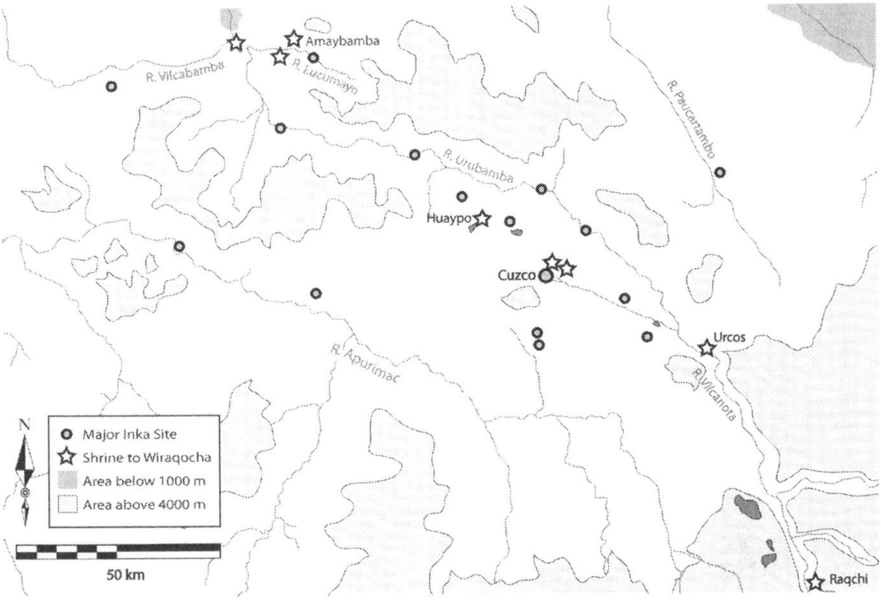

FIGURE 4.1. *A map of the area around Cuzco, showing sites associated with Wiraqocha. Note the southeast-to-northwest arrangement, following the path of Wiraqocha's cosmogonic journey.*

distributed across multiple material forms, including a series of lithic or metallic "brothers" (called *wawqi*) who were perceived as being fully identical in personhood with the emperor (i.e., they were not thought to be mere symbolic representations of him) (Dean 2010: 41–45). It is not surprising then that Inka gods should have been equally distributed across a range of bodies, and perhaps to an even greater number than were the emperors. Note from the table of recorded wiraqochas that at least one wak'a in question, called Turuca, was specifically described by Cobo as being a wawqi of Wiraqocha—precisely the same term used for the nonhuman bodies in which the Inka rulers were multiply manifest. Thus in a quite literal fashion, the fundamental process of cosmic creation was not something that was over and done with for the Inkas. From their perspective it was far more than an event that was only *remembered* in the present, for it was perpetually happening in the material landscape around them—always accessible and visible in the form of the god Wiraqocha's great journey. At all the key points in the creator's journey, one could still find him there in the form of his wak'a or wawqi. In that sense, we might say that the Inkas really retained no memories of Wiraqocha's journey at all, because you can only remember an event that actually ends—and which thereby passes into completed history.

A further observation is worth making regarding the wiraqochas described in the chronicles. A variety of material exchanges apparently took place between the wak'a and their human (i.e., Inka) caretakers. In one case particularly relevant to our concerns here, Cristóbal de Molina (2011: 43) describes how the wak'a of wiraqocha resident at Urcos would regularly be fed by *kamayuq* (specialists) when hungry, through the burning of appropriate offerings. According to Molina these specialists described themselves as kin to the wiraqocha—specifically as its sons and brothers. The instantiation of wak'a within human and nonhuman kin networks is a recurrent theme in Andean historical accounts (Bray 2015: 9). And the role of food (and food-sharing) in the constitution of kin relations is of especial significance in the Andean context also. As Bruce Manheim and Guillermo Salas Carreño put it, food "is the substance that relates and constitutes Quechua bodies" (Manheim and Salas Carreño 2015: 61). So for the Inkas, wak'a were not only persons, but often family as well, and as with Andean kinship in general, food-sharing was the necessary and material basis of such relatedness.

Some final points need to be emphasized here, concerning existence, continuity, and practice. The Inkas understood all things that had vitality—whether human or other—to be in constant need of replenishment (Salomon 1991). Human vitality was fed by favorable relations with the purified ancestors, who in turn needed to be taken care of with offerings of coca, food, and drink, so that they might continue to exist as hale beings. Thus, the people of the accumulating past and present were interlocked in a perpetual arrangement of mutual assistance that assured their continuity. This assumption of interdependence might be contrasted with a Western sensibility toward the relationship between past and present. We typically imagine the present to flow from, or emerge out of, the past; and so metaphorically, Western individuals often refer to the past as the upstream portion of a river, which feeds the downstream stretches (i.e., the present and future). Such idioms construe the past as having a determining effect on the present, but because the river flows in one direction only, this influence is not reciprocated. However, in the indigenous Andes the past and present seem to be mutually sustaining in such a fashion that neither has primacy, like two pillars leaning on each other for support. In other words, the past arises out of the present, no less than the present arises out of the past.

Peter Gose (2008) has analyzed the wiraqocha concept in some detail, arguing that in its original (i.e., precolonial) sense, it referred to the most primordial ancestral beings of the Andes. It is interesting then that the Inkas might have considered themselves to be the "brothers" of such a primeval ancestral entity, given that in Western terms it might seem contradictory for a sibling to be an ancestor, as these are normally mutually exclusive terms. But is this truly a contradiction in the original sources? The difficulty, as always, lies in the fact that our ontological assumptions may simply not translate into the Andean context. It may well have been the case that beings such as wak'a and wiraqochas existed in multiple times, always performing the primordial act of creation in the deep past, while continuing to constitute its enduring effects in the present. Indeed, if past and present are truly mutually sustained, then perhaps a wiraqocha needed to be fully alive and active in both at once. In that case, it might have been possible for a wiraqocha to be simultaneously an ancestor in the deep past, and also a brother in the present. Asking what a wiraqocha was may thus have no singular answer, considering we are talking about a being who could reside in multiple times and places simultaneously. For us, actually existent entities only reside in the present and so cannot occupy mutually exclusive statuses at the same time. Wiraqochas, however, may have had no such ontological limitations. Here we can only speculate, but in any event, it seems more likely that the "contradiction" comes from the incompatibility of our concepts with Andean ones, rather than any confusion on the part of the Inkas themselves.

ARCHAEOLOGY IN A LANDSCAPE WITHOUT MEMORY

At this point we turn to the archaeological record to complete our examination of how the past was engaged in the Inka world. Linguistic data, alongside ethnohistoric records, can certainly help to flesh out the broader framework, but to delve deeper into the precise material practices that sustained Inka engagements with historical beings such as wak'a, one must consider the evidence provided by archaeology. Wak'a, especially those that took on a lithic form, were subject to a wide range of archaeologically visible modifications under the Inkas. This is not the place to survey the full scope of variation in such practices, but Carolyn Dean (2010: 26–35) has suggested at least four basic modes of interaction: contouring, distancing, framing, and carving. For the purposes of our argument here, we wish to focus specifically on what she calls "framing"—not to imply, of course, that much could not also be said of the other three. Framing refers to the placement of lithic wak'a inside masonry structures, and also includes visual framing by means of doorways and windows (Dean 2010: 27–28). We will use the term "enclosing" since it implies a less purely vision-oriented practice than does framing, although it covers essentially the same phenomenon. Why then go to the trouble of enclosing wak'a? Considering that wak'a are living and sentient beings that reside in the landscape, there are clearly significant political consequences to their physical enclosure—just as there would be in enclosing a body made of flesh and blood. It facilitates a restriction of access, obviously, but to what end?

Here we wish to consider a specific example of archaeologically known wak'a, located in the Amaybamba Valley. As noted in the table above, the Amaybamba was the location of (at least) two wak'a associated with Wiraqocha—and represents the northwestern end of the horizontal axis formed by the creator god's cosmogonic journey across the Inka heartland. The Amaybamba should therefore be recognized as a region of considerable cosmological significance from the Inkas' point of view; the terminus of the great axis mundi that defined the heartland of their empire. It is also an ecologically transitional landscape, typical of the deep and narrow valleys that are found on the eastern slopes, connecting the colder Andean highlands to the vast forest expanses of Amazonia. Today, as well as in antiquity, it was the primary access point into the tropical lowlands adjacent to the Sacred Valley, and so also the gateway to the great forests of Antisuyu, the part of the world the Inkas identified with chaos and disorder (but also primal power). Along the valley bottom, ranging between 1,100 and 1,800 meters above sea level, the temperatures are warm and the vegetation dense—offering the perfect conditions for growing coca, which was the area's primary attraction in the eyes of the Inka. The entire landscape of the Amaybamba was therefore heavily restructured under the empire, and an extensive infrastructure was developed to connect the various plots associated with coca growing and harvesting, as well as to facilitate its redistribution to other locales within the heartland (Wilkinson 2013). At the core of this plantation system lay a palatial complex called Wamanmarka, which exhibits a number of typically "royal" architectural features (Wilkinson 2013). Much more might be said of the Inka occupation of this region, as well as its relationship to the other estate systems of the Cuzco region, but for our purposes it is the evidence for wak'a that is of greatest interest.

Two likely enclosed wak'a (in lithic form) were identified in the archaeological survey of the Amaybamba Valley. Although it is tempting to conclude that one or both of these may be the same wak'a of Wiraqocha that were said to have been located there in the documentary sources, we have no means at present to establish this with any certainty. Identifying archaeological sites with specific wak'a named in the Spanish chronicles has met with considerable success in the area immediately around Cuzco (e.g., Bauer 1998), but the farther we move away from the capital itself, the more fraught the task tends to become. This fact largely reflects the Cuzco-centric perspectives of the people who produced the chronicles—both their authors and their aristocratic Inka informants. In and around the capital, much more detail is provided in terms of toponyms and contextual geographical features, whereas wak'a found farther afield are more likely to be located using much vaguer terms. Nonetheless, the rock outcrops of the Amaybamba are treated in a fashion that is very typical of wak'a, even if a direct association with a documented counterpart is unlikely to ever be proved. And we might also note that some of the known wak'a, lying at other points along the same heartland cosmogonic axis, were treated similarly to the candidates from the Amaybamba discussed above. A good example of this is seen in one of the wak'a at Urcos, which also takes the form of a large rock outcrop enclosed in masonry (Bray 2015: 11).

In the Amaybamba, one of these probable wak'a is located at the site of Capillayoq, set on the north side of the primary rivercourse. The valley floor is at its widest and flattest around Capillayoq, offering the best agricultural fields in the region. Large quantities of tea and coffee are grown there today, but the density of the archaeological evidence suggests it once lay at the very heart of the coca plantation system established in the valley during the Late Horizon. There are several ruined structures at the site that are not especially remarkable in themselves, though they are clearly Inka in origin due to their rectilinear form and associated ceramic scatters dominated by imperial wares. One building does exhibit a higher quality of masonry than most of the others at the site, and occupies a larger footprint. The structure stands out, however, not only for its architecture but also for the native rock outcrop it encloses (see figure 4.2). The stone outcrop in question protrudes from the floor of the northeastern corner of the structure, is roughly trapezoidal in shape, and occupies approximately ten square meters of the interior floor space. It is otherwise unmodified (e.g., through carving), but evidently was intended as the focus for the architecture that surrounds it.

The structure has been modified through modern reconstruction efforts and virtually all of the surface artifacts cleared away. However, several ceramic fragments were still noted around the structure—including a number of sherds from a large Inka polychrome vessel. This is particularly interesting because such vessels are strongly associated with state hospitality, often used to serve maize beer (*chicha*) at commensal feasting events. There is however nothing around the structure that would indicate a space intended for the holding of feasts, such as a plaza or *kallanka* hall. It is a more-or-less isolated structure in the middle of what would have once been agricultural fields. Of course, with the Inkas we cannot ever assume that humans are the sole intended recipients of feasting—given that the chronicles make reference to feeding wak'a too. That being the case, we might very well expect the ceramic evidence in association with wak'a to point to their shared participation in state feasting ceremonies. Recent work by Tamara Bray (2012) supports this interpretation, identifying a number of archaeological cases where the ceramic evidence indicates commensal feasting events to which nonhumans (i.e., wak'a) were equally a party. Similarly, Steve Kosiba (2015: 189) points to evidence of wak'a feeding near Ollantaytambo, as indicated by Inka serving vessels in association with a carved and partially enclosed rock outcrop.

But what difference do the walls make? Here the "buildings" we find in association with rock outcrops emerge as quite important elements in the story too. It is possible that if wak'a are persons, then enclosing them in masonry might serve as a form of incarceration. The idea that wak'a might be imprisoned at the behest of the state would not be surprising given Inka accounts of how rocks could be subject to capture and various forms of corporal punishment. However, the kinship engendered between the Inka kamayuq and the wiraqochas they fed seems to discount such antagonistic interpretations, at least in all

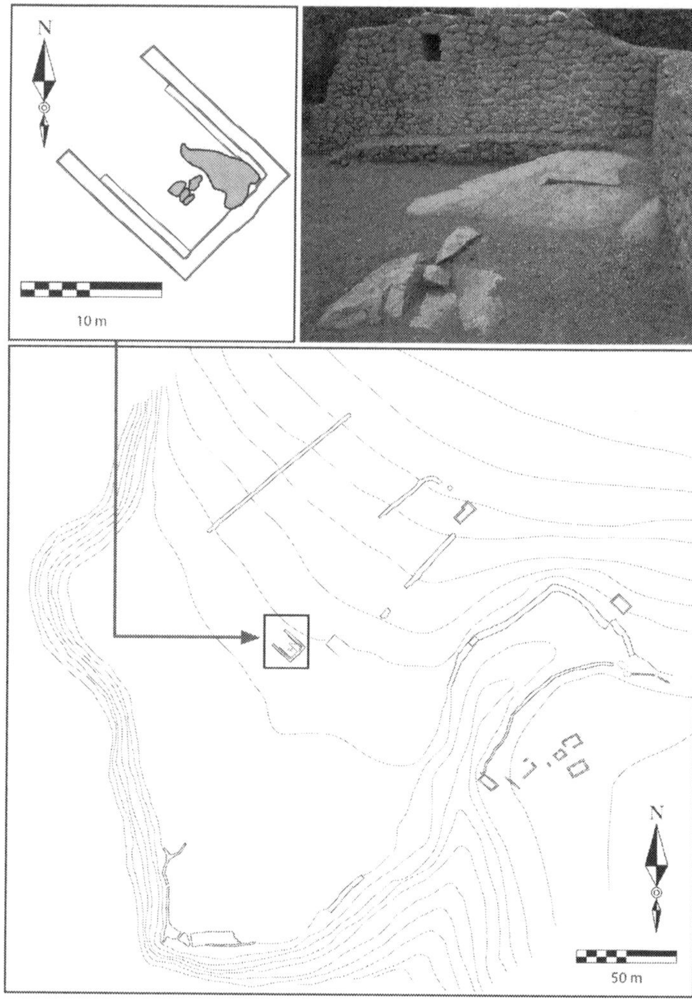

FIGURE 4.2. *A likely wak'a (top right) located at the site of Capillayoq in the Amaybamba Valley. Note the enclosure of the rock outcrop by masonry (top left).*

cases. Another possibility to consider then is that spatial positioning of subjects between material layers actually plays an important role in maintaining certain kinds of relationship, especially those that are especially close or *intimate*.

Susan Niles offers some interesting commentary on the Inka practice of enclosing rocks in masonry, saying that such stone outcrops "seem to be impediments to comfortable living. It may well be the case that the outcrops occasionally found in Inca domestic structures are not part of the house or its furnishings but are, rather, the denizens of the house" (Niles 1987: 28). In recent years, archaeologists have increasingly been concerned with accessing the embodied subjectivity of past actors, which draws in a wide range of concerns, including those dealing with gender, class, and ethnicity (e.g., Joyce 2005). And it is not such a great leap to extend such theoretical approaches to include the embodied subjectivity of nonhumans such as the wak'a also. That being said, archaeologists have given much less attention to the related phenomenon of *inter*subjectivity. As the ethnographer Michael Jackson (1998: 10) points out, the self "is everywhere reciprocal to other selves, both behaviorally and experientially." Intersubjectivity therefore places the emphasis on forms of subjective being that are necessarily dyadic. Intimacy is an excellent example of such, for it is a state that a subject can only share with a second, distinct subject; one always has to be intimate *with* another. The effects of physical layers are a fundamentally important element when it comes to maintaining intersubjective relationships. And clothing and buildings are perhaps the most basic kinds of layers that work to maintain and sustain intersubjectivity.

A very obvious example of how layers make intersubjectivity might be seen in the context of two lovers beneath their sheets. The fact that they share the same close quarters without layers around their bodies (i.e., they are naked) and they are both simultaneously enveloped in another layer (in the form of the bedclothes) does not simply signify their state of intimacy, but is actively constitutive of it. The spatial and material disposition of their interactions is essential and goes beyond mere symbolism. And this kind of "layer effect" extends into highly political arenas as well. There is a reason why the bathhouse was such a central institution across ancient Roman society, and offered the location par excellence for politicking. Being naked together, shorn of one's outer layers and markers of rank and status, allows for relationships of intimacy and trust to be engendered, which are of course vital in building political networks and alliances. While it is evident that clothing and other kinds of material regalia were central to the forming of embodied, subjective identities in the Roman world (as elsewhere), the absence of those wrappings was no less significant—especially in crafting shared intersubjective states such as intimacy, trust, and familiarity.

Moreover, there are good reasons for foregrounding the importance of layering practices with respect to the Andes specifically; for not only is food-sharing constitutive of kin relations, as we discussed earlier, but so too are the spatial contexts of such commensal practices. As Mary Weismantel (1995: 695) puts it, people "who eat together in the same household share the same flesh in a quite literal sense." So the sharing of food makes kin, but especially food-sharing that occurs within a series of place-making layers, such as those that form the walls of a house (see also Manheim and Salas Carreño 2015: 60–64). Returning to the Amaybamba and the enclosed rock outcrop at Capillayoq, then, it is worth noting that the masonry structure around the lithic inhabitant may have been equally central to the active creation of familial bonds. Dwellings constitute a layer around certain practices that take place between kin, while simultaneously excluding outsiders—both visually and spatially. In this regard, Susan Niles's view that masonry enclosures were in fact houses for wak'a may be right on point, because layers of this sort would have allowed for a kind of domestic intimacy to be formed between the Inka kamayuq and the wak'a they fed. The exchange of food behind shared walls in a close physical space forms bonds that are also "close" in the intersubjective sense of being familiar. That is to say, they are kin because they are house sharers as much as because they are food sharers—for the layer effects also constitute the conditions for the very intimacy upon which these relationships rest. In the case of the wak'a outcrop at Capillayoq, the process of kin-making was not fundamentally different to the constitution of family bonds between humans, and the fact that this particular context involves relations between humans and lithic beings is only remarkable to Western eyes.

Moreover, the power that is garnered through such processes of building familial bonds with wak'a is nothing particularly unusual from an anthropological perspective. Political strength that exists in the form of extensive networks of kin relations, which one can call upon when required, has been observed across many societies. In this case, it just so happens that many of the familial networks constructed by the Inkas included not only humans, but nonhumans too. And these nonhumans were in fact the major historical agents of the Inka world—a past that was not an abstract and objective list of things that occurred (i.e., res gestae), but a living congeries of sentient beings through whom history was created and by whom it was continuously shaped and reshaped. Maintaining positive, reciprocal ties with such beings—of which there were multitudes—was precisely what the political manipulation of the past looked like in the Inka world.

It is necessary to end the present section by emphasizing that the Inkas' manipulation of history being described here was in no sense metaphorical. Scholars talk of the manipulation of history all the time, but this is seldom meant literally, for it usually refers to representations and not the events themselves. Indeed such insistence is important, because we so often talk of controlling the past in a manner that is implicitly nonliteral. As a thought experiment, consider the following two contrasting examples of "changing history." It is a staple of science fiction stories to imagine traveling back in time and killing some historical figure (e.g., a dictator or tyrant) so that they and all their works are erased from time. This however would be rather different from destroying all the historical documentation and records of their regimes, so as to ensure they are entirely forgotten. In the science fiction case, the manipulation of the past is literal, while in the other, it is metaphorical; in the latter it is the representations that exist in the present that are being destroyed, while in the former, it is the events and past actors themselves. We suggest that the Inkas' interaction with the past is conceptually much closer to the nonmetaphorical case. They were not playing with memories for political purposes, but rather intervening in the very warp and weft of history itself, remaking it so that the past literally took on a different form. A common metaphor for conveying the immutability of the past is to describe history as something that is "carved in stone," referencing a substance we associate with durability and inflexibility. Yet for the Inkas the past seems to have been more set in intertwined threads (perhaps literally so, in the case of the *khipu*), that is, a set of material relationships that could always be unraveled, and then rewoven in new ways.

CONCLUSIONS

We began this essay with the question of why the Inkas did not produce figurative depictions of historical events, unlike so many of the other ancient empires familiar to us. We hope to have now offered a framework for beginning to answer that query. Various other societies might seek to manipulate the depiction of the past through the control of various representational media. But for the Inkas the past was not primarily engaged with through practices that are best understood as representational at all. Interaction with the past (in the form of wak'a) inhered in social relations, and indeed, more specifically, kinship relations. The past existed in the form of living and sentient nonhuman actors who were made the literal kin of the Inkas—through a range of material exchanges that were constitutive of kinship in the ancient Andes more generally. Quite likely the notion of "controlling" the past by making pictures of it would have seemed bizarrely ineffectual from the Inka point of view. Or perhaps it was even so foreign a notion as to simply have been unthinkable. It is difficult to tell for certain. Nonetheless, they clearly did not produce such depictions, and we consider this to be best explained by working through the material consequences of the radically different temporal ontologies found in the indigenous Andes. The Inkas sought to bend the past to their will, by tying it into kinship obligations, making it family in effect, so that it would owe them allegiance. The past will only give you its support if you feed it—and so the Inkas directed their political efforts accordingly. To the Western mind, images of the past are substitutions that make up for its absence. But if the past is not absent, and is instead united in continuity with the present, then such images have nothing to substitute for.

This is an essential reason why, we suggest, there is a complete lack of figurative depictions of historical narratives in the Inka world. The wak'a—the objects of political manipulation through practices such as enclosure—are the Inka equivalents of the monumentalized histories seen on, for example, Roman triumphal arches. However, they are only "equivalent" insofar as wak'a are indicative of an altogether different mode of relating to the past than what we find in such artifacts from the Old World. The Inka engagement with the wak'a had little to do with struggles over the control of content. The Inkas sought to be on good terms with the wak'a, by engendering relationships of affinity or even kinship between themselves and the lithic bodies of past actors who were still very much alive in the present. It is the difference between struggling over what gets depicted on a frieze or mosaic (a representational contest) and who gets to feed, fete, and flatter the object of social desire (a struggle over relationships). And herein lies the obvious value of enclosure as a material practice. By restricting access, one creates a potentially more exclusive and intimate relationship—locking others out and solidifying one's own privileged position.

All that said, it is worth underscoring that the deliberate evasion of figurative art is not in itself diagnostic of any particular temporal ontology. After all, it is not as if the Inkas' penchant for aniconism was unique or unheard of in other imperial polities. Probably the best-known example is the monotheistic prohibition against "graven images" that is seen in much Islamic and Jewish art, as well as certain strands of Christianity (Assmann 2011). This reflects a long-running theological anxiety about the pitfalls of idolatry and is therefore (so far as we know) a radically different logic to the one that drove Inka aniconism—the superficial similarities notwithstanding. By the same token, we cannot assume that an embrace of figurative iconography always implies a representationalist perspective with regards to images and the things they depict. With this in mind, the contrast we have drawn between the prolific historical representations produced by the Romans, Akkadians, and ancient Egyptians needs a few final words of comment and caution. For these seemingly "representational" depictions of history made by the other great empires of antiquity may not have been quite so representational as we have been led to believe. In particular, Zainab Bahrani's (2008) important work on anthropomorphism in ancient Mesopotamia gives us some pause here. In her analysis, such sculpted royal and divine figures are much more than mere depictions, for they also have real-world effects through (what to modern eyes looks like) magical means. In particular, the capture of statues of divinities in Mesopotamia could have seriously deleterious impact on entire populations, and so was both a method of, and cause for, war. Indeed, Bahrani's work indicates that representationalist models of historical memory may be no less problematic in ancient Mesopotamia than they are for the precolonial Andes. It might therefore be more accurate to suggest that the contrast we are drawing is less one between aniconic Inkas and iconophilic Mesopotamians or Egyptians, and more a contrast with how the monumental art of these various societies has traditionally been *interpreted*. In any event, we do not think that social memory theory accords well with the Inka case, and have sought to marshal our arguments to demonstrate this . . . and although we suspect that similar critiques could be produced for the Maya, the Mississippians, the ancient Egyptians, and so on, we will leave that debate to the respective experts in those fields.

In sum then, like all empire builders, the Inkas had competitors and dissenters who sought to contest their power, and so exclusivity in their relationship with the wak'a was a core concern. In such light, the practices of wak'a enclosure, evidenced in the Amaybamba and more widely, can be read as a sort of monopolization of their social lives. The contrast this leaves in archaeological terms could not be more stark. Contesting memory involves producing images, texts, and performances as the media through which the past is represented. And we apprehend these archaeologically in the production of iconographic and epigraphic inscriptions, and in their subsequent erasure, redaction, and replacement. The very different processes involved in the monopolization of relations with undying historical actors leaves masonry enclosures and ceramics—the remains of the attempts to house them and feed them. Above all, it speaks to a practice of making the nonhumans who constituted the past, and yet simultaneously resided in the present, into kin. Churchill thought history would be kind to him because he intended to write it. The Inkas imagined it would be kind to them because it was, after all, family.

REFERENCES

Alcock, Susan E. 2002. *Archaeologies of the Greek Past: Landscape, Monuments, and Memories*. Cambridge: Cambridge University Press.

Arnold, Denise Y., and Christine A. Hastorf. 2008. *Heads of State: Icons, Power, and Politics in the Ancient and Modern Andes*. Walnut Creek, CA: Left Coast Press.

Assmann, Jan. 2011. "What's Wrong with Images?" In *Idol Anxiety*, ed. Josh Ellenbogen and Aaron Tugendhaft, 19–31. Stanford: Stanford University Press.

Bahrani, Zainab. 2008. *Rituals of War: The Body and Violence in Mesopotamia*. New York: Zone Books.

Bauer, Brian S. 1998. *The Sacred Landscape of the Inca: The Cusco Ceque System*. Austin: University of Texas Press.

Betanzos, Juan de. [1557] 1996. *Narrative of the Incas*. Austin: University of Texas Press.

Borić, Dušan. 2010. "Introduction: Memory, Archaeology and the Historical Condition." In *Archaeology and Memory*, ed. Dušan Borić, 1–34. Oxford: Oxbow Books.

Bourdin, Gabriel Luis. 2014. "En los tiempos de ñaupa: El cuerpo y la deixis temporal en lenguas originarias de Sudamérica." *Península* 9 (1): 33–58. https://doi.org/10.1016/S1870-5766(14)70119-9.

Bray, Tamara L. 2009. "An Archaeological Perspective on the Andean Concept of Camaquen: Thinking through late Pre-Columbian Ofrendas and Huacas." *Cambridge Archaeological Journal* 19 (3): 357–66. https://doi.org/10.1017/S0959774309000547.

Bray, Tamara L. 2012. "'Ritual Commensality between Human and Non-Human Persons: Investigating Native Ontologies in the Late Pre-Columbian Andean World." eTopoi. *Journal of Ancient Studies* 2:197–212.

Bray, Tamara L. 2015. "Andean Wak'a and Alternative Configurations of Persons, Power and Things." In *Archaeology of Wak'a: Explorations of the Sacred in the Pre-Columbian Andes*, ed. Tamara L. Bray, 3–19. Boulder: University Press of Colorado. https://doi.org/10.5876/9781607323181.c001.

Cieza de León, Pedro. 2005. *Crónica del Perú: El señorío de los Incas*. Trans. Franklin Pease. Ayacucho: Fundación Biblioteca Ayacucho.

Cobo, Bernabé. [1653] 1990. *Inca Religion and Customs*. Austin: University of Texas Press.

Collingwood, R. George. 1993. *The Idea of History*. Oxford: Oxford University Press.

Cook, Anita G. 1983. "Aspects of State Ideology in Huari and Tiwanaku Iconography: The Central Deity and the Sacrificer." In *Investigations of the Andean Past*, ed. Daniel H. Sandweiss, 161–85. Ithaca, NY: Cornell University Press.

Crossland, Zoe. 2009. "Of Clues and Signs: The Dead Body and Its Evidential Traces." *American Anthropologist* 111 (1): 69–80. https://doi.org/10.1111/j.1548-1433.2009.01078.x.

D'Altroy, Terence N. 2015. "Killing Mummies: On Inka Epistemology and Imperial Power." In *Death Rituals and Social Order in the Ancient World: "Death Shall Have No Dominion,"* ed. Colin Renfrew, Michael Boyd, and Iain Morley, 404–22. Cambridge: Cambridge University Press.

D'Altroy, Terence N., and Katharina J. Schreiber. 2004. "Andean Empires." In *Andean Archaeology*, ed. Helaine Silverman, 255–79. New York: Blackwell.

D'Altroy, Terence N., and Darryl A. Wilkinson. 2009. "War, Peace, and the Landscape of the Inka Heartland." *Cardozo Public Law, Policy, and Ethics Journal* 7 (3): 663–76.

Dean, Carolyn. 2010. *A Culture of Stone: Inka Perspectives on Rock*. Durham, NC: Duke University Press. https://doi.org/10.1215/9780822393177.

Demarest, Arthur A. 1981. *Viracocha: The Nature and Antiquity of the Andean High God*. Cambridge: Peabody Museum of Archaeology and Ethnography, Harvard University.

DeMarrais, Elizabeth, Luis Jaime Castillo, and Timothy Earle. 1996. "Ideology, Materialization, and Power Strategies." *Current Anthropology* 37 (1): 15–31. https://doi.org/10.1086/204472.

Edmonds, Mark. 2002. *Ancestral Geographies of the Neolithic: Landscapes, Monuments and Memory*. London: Routledge.

Ehrman, Bart D. 2003. *Lost Scriptures: Books That Did Not Make It into the New Testament*. Oxford: Oxford University Press.

Gosden, Chris, and Gary Lock. 1998. "Prehistoric Histories." *World Archaeology* 30 (1): 2–12. https://doi.org/10.1080/00438243.1998.9980393.

Gose, Peter. 2008. *Invaders as Ancestors: On the Intercultural Making and Unmaking of Spanish Colonialism in the Andes*. Toronto: University of Toronto Press.

Harris, Oliver. 2009. "Review of *Memory Work: Archaeologies of Material Practices.*" *Cambridge Archaeological Journal* 19 (2): 265–66. https://doi.org/10.1017/S0959774309000353.

Holtorf, Cornelius J. 1996. "Towards a Chronology of Megaliths: Understanding Monumental Time and Cultural Memory." *Journal of European Archaeology* 4 (1): 119–52. https://doi.org/10.1179/096576696800688051.

Jackson, Michael. 1998. *Minima Ethnographica: Intersubjectivity and the Ethnographic Project.* Chicago: University of Chicago Press.

Janusek, John Wayne. 2006. "The Changing 'Nature' of Tiwanaku Religion and the Rise of an Andean State." *World Archaeology* 38 (3): 469–92. https://doi.org/10.1080/00438240600813541.

Joyce, Rosemary A. 2003. "Concrete Memories: Fragments of the Past in the Classic Maya Present (500–1000 AD)." In *Archaeologies of Memory*, ed. Ruth Van Dyke and Susan E. Alcock, 104–25. Oxford: John Wiley and Sons. https://doi.org/10.1002/9780470774304.ch6.

Joyce, Rosemary A. 2005. "Archaeology of the Body." *Annual Review of Anthropology* 34: 139–58.

Jones, Andrew, and Benjamin Alberti. 2013. "Archaeology after Interpretation." In *Archaeology after Interpretation: Returning Materials to Archaeological Theory*, ed. Benjamin Alberti, Andrew Meirion Jones, and Joshua Pollard, 15–35. Walnut Creek, CA: Left Coast Press.

Kolata, Alan L. 1993. *The Tiwanaku: Portrait of an Andean Civilization.* Oxford: Blackwell.

Kosiba, Steve. 2015. "Of Blood and Soil: Tombs, Wak'a, and the Naturalization of Social Difference in the Inka Heartland." In *Archaeology of Wak'a: Explorations of the Sacred in the Pre-Columbian Andes*, ed. Tamara L. Bray, 167–212. Boulder: University Press of Colorado. https://doi.org/10.5876/9781607323181.c006.

Manheim, Bruce, and Guillermo Salas Carreño. 2015. "Wak'a: Entifications of the Andean Sacred." In *Archaeology of Wak'a: Explorations of the Sacred in the Pre-Columbian Andes*, ed. Tamara L. Bray, 47–74. Boulder: University Press of Colorado.

Mills, Barbara J., and William H. Walker, eds. 2008. *Memory Work: Archaeologies of Material Practices.* Santa Fe: School for Advanced Research Press.

Miracle, Andrew, and Juan de Dios Yapita. 1981. "Time and Space in Aymara." In *The Aymara Language in Its Social and Cultural Context: A Collection of Essays on Aspects of Aymara Language and Culture*, ed. Martha James Hardman, 33–56. Gainesville: University Press of Florida.

Molina, Cristóbal de. 2011. *Account of the Fables and Rites of the Incas.* Austin: University of Texas Press.

Moore, Jerry D. 1992. "Pattern and Meaning in Prehistoric Peruvian Architecture: The Architecture of Social Control in the Chimu State." *Latin American Antiquity* 3 (02): 95–113. https://doi.org/10.2307/971938.

Niles, Susan A. 1987. *Callachaca: Style and Status in an Inca Community*. Iowa City: University of Iowa Press.
Núñez, Rafael E., and Eve Sweetser. 2006. "With the Future behind Them: Convergent Evidence from Aymara Language and Gesture in the Cross-Linguistic Comparison of Spatial Construals of Time." *Cognitive Science* 30 (3): 401–450. https://doi.org/10.1207/s15516709cog0000_62.
Pease, Franklin. 1970. "The Andean Creator God." *Numen* 17 (3): 161–75. https://doi.org/10.1163/156852770X00018.
Rowe, John H. 1960. "The Origins of Creator Worship among the Incas." In *Culture in History*, ed. Stanley Diamond, 408–29. New York: Columbia University Press.
Thomas, Julian. 2004. *Archaeology and Modernity*. London: Routledge.
Salomon, Frank L. 1991. "Introductory Essay: The Huarochirí Manuscript." In *The Huarochirí Manuscript: A Testament of Ancient and Colonial Andean Religion*, ed. Tom D. Dillehay, 315–53. Washington, DC: Dumbarton Oaks.
Salomon, Frank L. 1995. "The 'Beautiful Grandparents': Andean Ancestor Shrines and Mortuary Ritual as Seen through Colonial Records." In *Tombs for the Living: Andean Mortuary Practices*, ed. Frank L. Salomon and Jorge Urioste, 1–38. Austin: University of Texas.
Sarmiento de Gamboa, Pedro. [1557] 2007. The History of the Incas. Austin: University of Texas Press.
Urton, Gary. 1999. *Inca Myths*. Austin: University of Texas Press.
Van de Guchte, Maarten. 1999. "The Inca Cognition of Landscape: Archaeology, Ethnohistory, and the Aesthetic of Alterity." In *Archaeologies of Landscape: Contemporary Perspectives*, ed. Wendy Ashmore and A. Bernard Knapp, 149–68. Oxford: Blackwell.
Van Dyke, Ruth M., and Susan E. Alcock. 2003. "Archaeologies of Memory: An Introduction." In *Archaeologies of Memory*, ed. Ruth Van Dyke and Susan E. Alcock, 1–13. Oxford: John Wiley and Sons.
Weismantel, Mary. 1995. "Making Kin: Kinship Theory and Zumbagua Adoption." *American Ethnologist* 22 (4): 685–704. https://doi.org/10.1525/ae.1995.22.4.02a00010.
Wilkinson, Darryl A. 2013. "Politics, Infrastructure and Non-Human Subjects: The Inka Occupation of the Amaybamba Cloud Forests." PhD dissertation. Columbia University, New York.
Zanker, Paul. 1990. *The Power of Images in the Age of Augustus*. Ann Arbor: University of Michigan Press.

FIVE

Past-Forward Past Making

Late Pre-Hispanic and Early Colonial Andean Archaeology and History

ZACHARY J. CHASE

INTRODUCTION

Centuries ago, indigenous Andeans from Huarochirí in Peru's central highlands (figure 5.1) communicated the region's very deep past through mythical narratives about consorting, ancient, superhuman beings (*huacas*). One of these stories—recorded in Quechua in the early years of the seventeenth century and compiled into the renowned "Huarochirí manuscript"—tells of the male huaca Cuni Raya Viracocha and the female huaca Caui Llaca. The following comes from Frank Salomon and George Urioste's (Salomon and Urioste 1991: 46–50) translation.

"A long, long time ago," the narrative goes, the Huarochirano huaca Cuni Raya used to present himself to others as a poor and socially unconnected vagrant. Like other huacas and *villcas*,[1] Cuni Raya desired Caui Llaca, and through trickery was able to impregnate her. When the child was one year old, Caui Llaca called all huacas and villcas to Anchicocha (see figure 5.1), where they eagerly gathered, dressed to impress and anticipating the chance to pair with the beautiful huaca. Once assembled, Caui Llaca asked which of them had inseminated her,[2] addressing each of the males individually with the exception of one attendee: Cuni Raya

DOI: 10.5876/9781607326427.c005

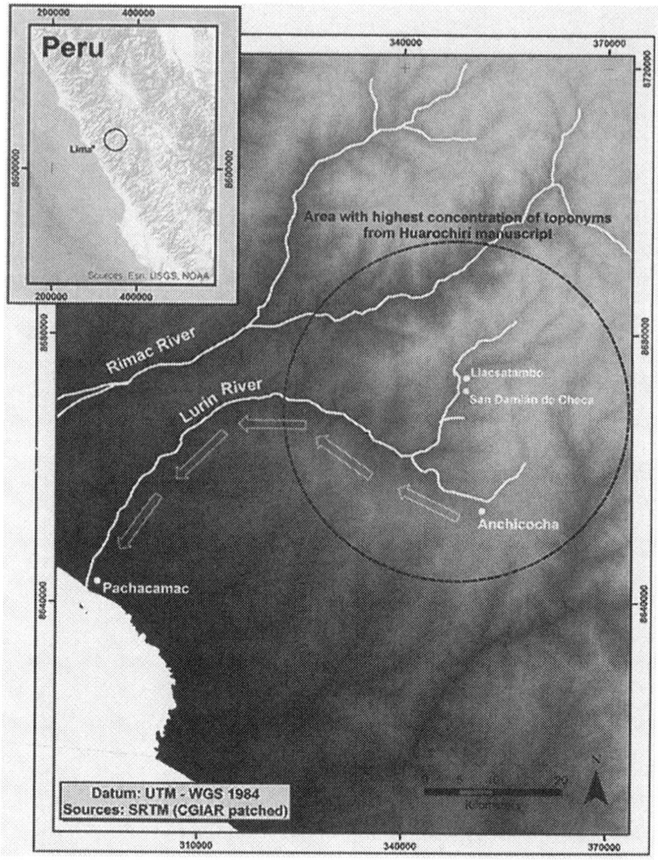

FIGURE 5.1. *Caui Llaca's flight and Cuni Raya's pursuit began in Anchi Cocha (currently known as Anchicocha) and ended at Pachacamac on the Pacific Coast, following the Lurín Valley and River. The circle demarcates the area with the highest concentration of toponyms mentioned in the Huarochirí manuscript.*

had shown up looking "like a friendless beggar" and was seated humbly at group's edge; such a figure could not possibly be the father. None of the huacas claimed parentage, so Caui Llaca changed tacks. She set her child down and said, "Go, identify your father yourself!" When the infant crawled to Cuni Raya and "instantly brightened up and climbed onto its father's knee," Caui Llaca was despondent:

> 'Atatay, what a disgrace! How could I have given birth to the child of a beggar like that?' she said. And taking along only her child, she headed straight for the ocean. And then . . . Cuni Raya Viracocha put on his golden garment. He started to chase her at once, thinking to himself, "She'll be overcome by sudden desire for me."

"Sister Caui Llaca!" he called after her. "Here, look at me! Now I'm really beautiful!" he said, and he stood there making his garment glitter. Caui Llaca didn't even turn her face back to him. "Because I've given birth to the child of such a ruffian, such a mangy beggar, I'll just disappear into the ocean," she said. She headed straight out into the deep sea near Pacha Camac, out there where even now two stones that clearly look like people stand. And when she arrived at what is today her dwelling, she turned to stone.

Yet Cuni Raya Vira Cocha thought, "She'll see me anyway, she'll come to look at me!" He followed her at a distance, shouting and calling out to her over and over.

Cuni Raya eventually reached the Pacific Coast, where he seduced the daughter of huaca Urpay Huachac, and, by scattering the contents of Urpay's small fish pond, populated the ocean with fish for the first time.

Figure 5.1 displays the myth's geographic context, based on matching the toponyms provided with place-names on current Peruvian military maps.[3] Note the correspondence between the mythical pursuit's path and the course of the Lurín Valley and River from Anchicocha to Pachacamac on the Pacific Coast. This "geomythology," wherein the female Caui Llaca is the Lurín Valley,[4] and the male Cuni Raya is the Lurín River, which enters and fertilizes the valley, is redundant with expressions of a specific cultural precept: fecundity (e.g., a child, a watered valley, a fish-stocked ocean) comes from antagonistic or competitive complementarity / duality (e.g., Caui Llaca was deceitfully impregnated and fleeing her pursuer; Cuni Raya, whose coastal "seduction" was similarly dubious and not sanctioned by Urpay). Such "insistence" (Ascher and Ascher [1981] 1997: 38–40), or meaningful repetition of important themes or principles, highlights a common characteristic of mythical narrative and time (Eliade 1984, 1954; Lévi-Strauss 1963: 211–12; Sahlins 1985). The mythical events' temporal remoteness ("a long, long time ago") is inversely correlated to the importance or value the myth's principles held for the seventeenth-century Huarochirí narrators.

Toward the chapter's end, we will revisit this myth to address matters of historical succession and other temporalities raised in this chapter. For now it serves as an introduction to the contents of the Huarochirí manuscript, their intimacy with locally known landscapes, and their potential to illuminate time-related matters. At the same time, my intention is to expose the historiographic limitations of these interpretive approaches. In addition to the workings of time in myth just discussed, the fact that reconfiguration of collective pasts was fundamental to formal interactions between Andean peoples at least as far back as Inka expansion (Chase 2015, 2016b, 2018; D'Altroy 2001; Kosiba 2015a, 2015b; Silverblatt 1988) requires not only assaying different theoretical frameworks, but the application of extratextual methods of data acquisition—in particular, archaeology. As I emphasize below, archaeological investigations, together with consideration of other theories, should improve interpretations of the histories of these Andean pasts. Similar to the way stereoscopic vision gives dimensional depth and resolution to photographs of geographic landscapes, a critical mass of recent archaeological data permits historical and historiographic revisions of ethnohistorical texts, which combine to provide new temporal resolution on Andean pasts, their production, and ways they were experienced (inter alia Bauer and Smit 2015; Bauer et al. 2013; Chase 2015; Covey 2006, 2008a, 2008b; Eeckhout 2008; Kosiba 2015a, 2015b; Makowski 2015). In the end, archaeological data that give us purchase on Huarochirí's chronological history become the very means to relativize, and thus better understand, indigenous Andean temporal and historical sensibilities as well as our own.

Over nearly a century, our understanding of the late pre-Hispanic and early Spanish colonial indigenous Andean past was shaped primarily by the contents of historical documents and their interpretations. This ethnohistory was oriented by specific historical and temporal models, the long debates over historicist versus structural Inka historiographies being the predominant and most explicit example (Covey 2006; Rowe 1945, 1967; Zuidema 1964, 1985, 1990; see introduction [chapter 1] to this volume). While we have seen exceptional erudition and creativity in recent efforts to reconcile these specific interpretative models (Gose 1996, 2008; Julien 2000; Pease 2010), archaeological research has not always prominently featured as the guiding interpretive apparatus and source of data in addressing historiography in the Andes.

Huarochirí presents an outstanding case in point. Scholarly understanding of the ancient Andes has been greatly influenced by the unique circa 1608 Quechua "Huarochirí manuscript" (Arguedas and Duviols 1966; Salomon and Urioste 1991; Taylor 1999). The manuscript was produced at the behest of parish priest Francisco de Avila, who used a Quechua-speaking assistant to collect local, native accounts of the past and present in the highlands east of Lima (Acosta Rodríguez 1987; Durston 2007; Salomon 1991). It constitutes the only book-length indigenous-language document about native religion and myth to come out of colonial South America. Huarochirí in its entirety was half of the Inka district of "Yauyos," the macroethnic label for the region's Indians (Dávila Brizeño [1586] 1965); the Quechua manuscript is primarily about one of those Indian groups. Its narrators and protagonists were the Indians of the *waranqa* (an Inka administrative demographic unit of 1,000 families) of Checa who identified their collective ceremonial center as Llacsatambo (figure 5.2), a pre-Hispanic hilltop settlement a few kilometers north of the Spanish colonial *reducción* (resettlement) of San Damián de Checa (figure 5.3). For many archaeologists and historians, the manuscript reveals an indigenous Andean cosmos otherwise hidden or lost. And indeed the text's manifest leitmotif is the superation of worlds past by worlds present—a historical etiology of its narrators' place in space and

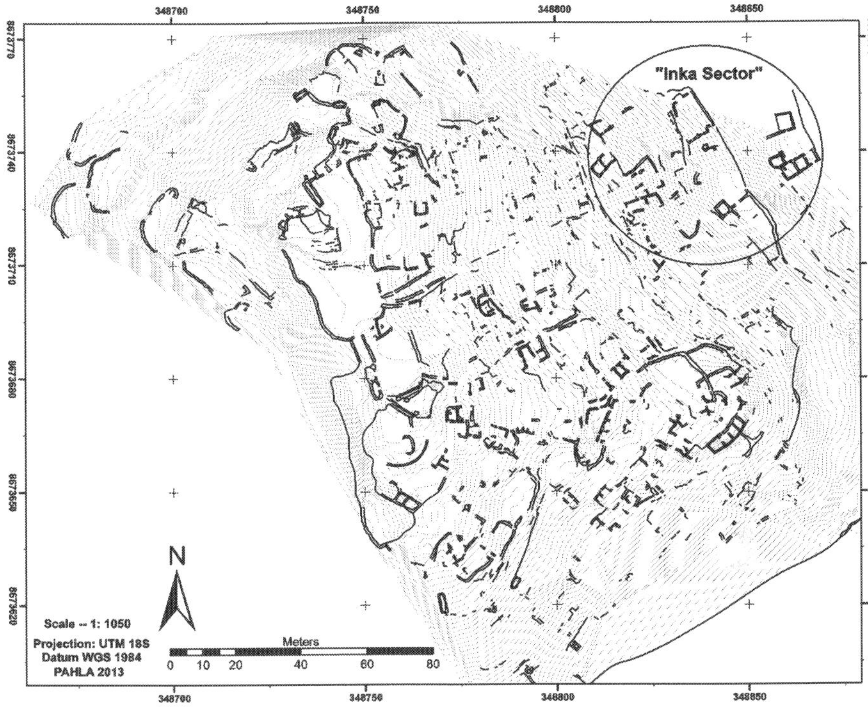

FIGURE 5.2. *Plan map of Llacsatambo, ceremonial center of the Huarochirí manuscript's narrators and their ancestors. (Map created as part of the Proyecto Arqueológico Huarochirí–Lurín Alto.)*

time. However, until recently, systematic archaeology of the manuscript's center has been virtually nonexistent despite multiple published requests from the region's most prominent ethnohistorians (Rostworowski 1988: 56; [1978] 2002a: 194, 201, 206, 211, 214; Salomon 1991: 24).

In this chapter I present results from the first systematic archaeology in the central area of the manuscript's production, and show how these require new theoretical orientations to clarify earlier historical reconstructions of Huarochirí's past, thereby providing deeper insight into the material and narrative construction of these Andean worlds, both in the pre-Hispanic and Spanish colonial eras. I argue that the collection and writing of the Huarochirí manuscript captured and homogenized an array of culturally distinct ways of creating and communicating the past within a specific structure of chronological historiography by which Andean realities were rendered legible to Spanish legal and religious bureaucrats. That is, the text not only worked to produce new understandings of collective origins, identities, and territoriality, but has also obscured the very means by which indigenous Andean groups had previously

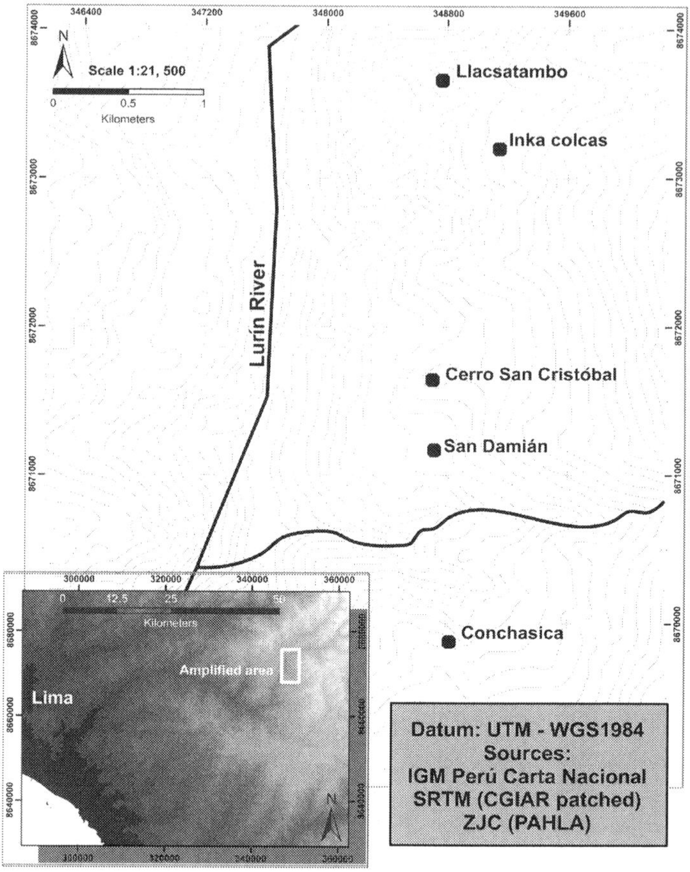

FIGURE 5.3. *Map of the central area ("Llacsatambo-San Damián axis") of the Huarochirí manuscript's narrators, including the Inka colcas associated with Llacsatambo.*

understood, constructed, codified, and communicated their past and situated existence in the world.

THE MYTH OF A PREHISTORY

To address these arguments let us first explore the diachronic production of a "regime of historicity" (Trouillot 2003: 38) that I call Huarochirí's "canonical (pre)history," and consider its historiographic, epistemological, and ontological underpinnings.[5] The "canonical prehistory" of Huarochirí is a specific sequential ordering of events and processes (drawn from and interpreting the contents of the Huarochirí manuscript), fitted into rough chronological order according to Andean archaeology's Horizon/Intermediate Period model and purporting

Late Middle Horizon (ca. **AD 900**)	Late Intermediate Period (ca. **AD 1000–1470**)	Late Horizon/Inka Period (ca. **AD 1470–1532**)	Early Spanish Colonial Period (**AD 1532–1608**)

Historical events ⇨ **Myths & Rituals** ⇨ **H. manuscript**

FIGURE 5.4. *Illustration of the chronology and causal underpinnings of the "canonical prehistory" of Huarochirí. The historical events and processes (waves of invasions from the highlands toward the coast) considered the foundation of Huarochirí's myths and rituals are projected to have begun in the late Middle Horizon. The myths and rituals narrated in the Huarochirí manuscript are treated as post hoc reflections of the series of late Middle Horizon through Late Intermediate Period invasions.*

to account for the goings-on in Huarochirí in the centuries preceding the manuscript's composition. It also comprises a more general model of history and historiography, underpinned by assumptions about chronological temporality and the causal relationship between myth, ritual performance, and historical events and processes. Specifically, the canonical prehistory of Huarochirí holds that the highlands Yauyos invasions that ostensibly drove previous populations toward the coast were carried on over several centuries, beginning around AD 900 (see figure 5.4). Further, the huacas, their myths and rituals recorded in the manuscript represent commemorative, *post hoc reflections* of the actual historical invasions and organization of social, political, agricultural, and religious order. This account has become the predominant, de facto explanation of Huarochirí's past, appearing in historical and archaeological publications, and popular literature alike.

The foundations of Huarochirí's canonical prehistory emerge first from the manuscript's Checa narrators themselves, who related their local ancestral past in sequential and successive terms within the recurrent pattern of highlands invasions and conquest/expulsion of previous inhabitants; these previous inhabitants were called Yuncas, a label that is important to unpack because it referred to aboriginal peoples, to original or prior temporal periods and conditions, and also to the lands, climate, peoples, and other biota of low, warm river valleys and the coast to the west of highlands Huarochirí.[6] The invasions were carried out by the narrators' forefathers and led by their tutelary huaca Tutayquiri (son of apical regional huaca Pariacaca): "This land was once all full of Yunca. As soon as Tutay Quiri's children had expelled those Yunca, they began to distribute among themselves, according to their own ayllus, the fields, the houses, and the ayllu designations" (Salomon and Urioste 1991: 119). Taking the form of torrential red and yellow rain, "Tutay Quiri [then] set out from Llacsa Tambo to conquer" residual Yuncas populations, washing them down the Lurín and Rímac River Valleys to the Pacific Coast. As we will later see, it is significant, that in the midst of this expulsion, Yunca peoples who revered and worshipped Tutayquiri were said to have been spared and permitted to remain in their villages, from that moment forward to be recognized as "brothers" to the Checa (Salomon and Urioste 1991: 79–80).

For decades Andeanists have recognized Huarochirí's multiple, repeated myths of highland invasion as an especially concentrated, and well-articulated expression of the widespread *huari-llacuaz* sociopolitical arrangement. As drawn from seventeenth-century documentation of the "extirpation of idolatries" in the highlands of Lima's archiepiscopate, and presented by Pierre Duviols (1986: lvi–lxv; 1973), the "dualist huari-llacuaz complex" (lxv)[7] was the basic principle of village-level social organization throughout the north-central highlands. Huari populations traced ancestry to primordial, autochthonous huaca-ancestors; they were sedentary agriculturalists, expert terrace and canal builders. *Llacuaces* were the descendants of highlands pastoralist invaders associated with high-mountain precipitation phenomena such as lightning/thunder, torrential rain, and hail. The hierarchical predicate of village social relations was right by invasion/conquest (llacuaz) over right by autochthony (huaris). The productive complementarity of this asymmetrical dualism (see Urton 2012) worked most obviously in agricultural cycles wherein huaris appealed to their ancestors' connection to crop fields while llacuaces could bring the seasonal rainwater from the heights above. Their respective and intertwined chthonic and invasive qualities evoke both the Cuni Raya/Caui Llaca myth and the way the Inka "envisioned agriculture as warfare—a victory claimed by disemboweling the earth" (D'Altroy 2015: 405; Bauer 1996).

The (Checa) Yauyos and Yuncas of Huarochirí were llacuaces and huaris, respectively (Gose 2008: 18–19; Salomon 1991). Questions of the temporal-historical nature of these relations will be discussed shortly. For now it is sufficient to note how within the Llacsatambo foundation narrative, the Checa narrators seem to frame their past in successive historical terms recognizable to the Spanish and modern historians alike.

Indeed, "historiography" in its most literal sense was the manuscript's express raison d'être, as its preface makes clear: because writing was unknown to "the ancestors of the people called Indians," their past had "faded from view." The remedy was the written manuscript documenting "the lives of the ancestors of the Huaro Cheri people" (Salomon and Urioste 1991: 41). Still, with their multiplication, even in written form, the narratives' general historical order confounded the cleric behind their compilation. In his *Tratado* (the 1608 beginnings of a synthesis and commentary on the contents of the Huarochirí manuscript through its first seven chapters), Avila admitted the chronological puzzle before him and his determination to figure it out. He wrote, "In all of these fables and history I have not been able to find out their order and succession, which was first and which came after, because all of these things are extremely ancient; still, by the time it is all rendered into writing it is possible that I may have figured it out, or at least as accurately as possible" (in Arguedas and Duviols 1966: 206).[8]

Although not the only effective criterion in sixteenth- and seventeenth-century Spanish historiography (see e.g., MacCormack 1991, 2007; Ossio 2008: 187; Pease 1997, 2010), biblical history was the primary contributor to Avila's struggle with the chronology of the manuscript's contents; it also underpinned the logic informing his hope of achieving a historically accurate reconstruction of Huarochirí's past. As Salomon (1991: 3) notes, "The tendency to force non-Christian testimony into patterns congruent with 'universal history' and a unified Bible-based chronology is conspicuous in many Peruvian chronicles, both indigenous and Spanish, and Avila's *Tratado* shows that he partook of it." Thus in the manuscript's early chapters we see Genesis-like periods of darkness and telluric creation; explicit explanations of peoples' relationship to their "gods"; and the origins of the natural order, cultural practices, and traditions. Later material is associated with known historical phenomena such as Inka presence; the advent of the Spanish; and Spanish colonial institutions, events, and people. Between these two temporal poles are events generally pertaining to the Checa and their ancestors but lacking absolute historical referents more precise than, for example, local events having taken place "before the Inca's birth" (Taylor 1999: 222–25).[9] These were the historiographic elements of the Huarochirí manuscript that were "packed away" in the Madrid archive over the centuries that followed.

The Huarochirí narratives' next major appearance came with Clements Markham's 1873 English translation of Avila's (Markham [1873] 1964: 121–47) 1608 *Tratado*. Markham saw this as a source on the pre-Inkaic Andes, and argued for historically literal interpretations of primordial myths (Markham [1873] 1964: xviii). Markham's (1906) continued reliance on written records to reconstruct the Andean past partook of and perpetuated the hermeneutic standard of Huarochirí's canonical prehistory, found earlier in Avila's interpretation of Huarochirí's narratives as stylized renderings of chronologically historical occurrences. An archaeological echo of the hermeneutic was sounded in Pedro Villar Córdova's ([1935] 1982: 346–347) *Arqueología del departamento de Lima*, as the account of the foundational mythical battle wherein Pariacaca vanquished the ancient and anthropophagus huaca Guallallo was attributed to Huarochiranos' historical origins on the coast and subsequent migration to the highlands.

Markham's approach was characteristic of the European-American practice of "ancient history" (see Prescott [1847] 2002). But through the turn of the century, as Max Uhle forged Andean archaeology's "chronological horizons" and "periods" (Rowe 1998: 17; Uhle 1902; 1912: 23–24; [1903] 2003;), he endeavored also to dismantle both the specific conclusions and methodological tendencies of Markham's ancient historical approach.[10] Uhle held that in the absence of any scientific archaeological basis, Markham resorted to colonial texts recording "obscure traditions" (Uhle [1910] 1970: 379–80).[11] Not that Uhle rejected entirely the use of historical documents, as long as the information applied was appropriate to the problem at hand. In his writings relating to the Huarochirí manuscript's region, Uhle used the *Relaciones geográficas* and other "fact-finding" reports (1912) to conclude that the "Yauyos and the inhabitants of the valleys of Lima and of Pachacamac [i.e., the Rímac and the Lurín] were all Aymaras . . . [A]t the apex of Aymara civilization, tribes of this race expanded throughout Peru as Quechua tribes did during the Inka period" (Uhle [1910] 1970: 382).[12] With this Uhle provided the chronological basis for situating the commencement of the Huarochirí manuscript's invasion narratives in the Middle Horizon (ca. AD 500–1000), a standard of the canonical prehistory (see Uhle 1912: 14–17, 23–29). Moreover, despite fundamental differences, invasion was the mechanism at the base of the hypotheses of both Markham and Uhle.

Other key elements of Huarochirí's canonical prehistory can be detected in Markham's disposition toward Spanish colonial texts as culturally encoded Andean history as opposed to historically specious (Chase 2016b, 2018; Cerrón Palomino 2015: 50; cf. Bauer and Smit 2015; Bauer et al. 2013). At the same time, a muted form of Uhle's positivistic tone is later echoed in the canonical prehistory. For instance, as will be seen, even if narrative content and form were culturally variable, they were held to reflect economically and politically strategic activities and experiences within a universal empirical rationalism.

The twentieth century then saw the development of what could, for the sake of brevity, be summarily called "integrative Andeanism." This interpretive approach desegregated geographic and temporal divisions in order to define a unified Andean culture integral to autochthonous, ancient, and ongoing Andean civilization.[13] Huarochirí's canonical prehistory was certainly informed by these notions and their promotion of multidisciplinary approaches, which led to the real watershed for the widespread scholarly use of the Huarochirí manuscript in the interpretation of the Andean past: its 1966 Spanish translation by Peruvian luminary of letters, José María Arguedas (Arguedas and Duviols 1966). Vocationally devoted to conveying the authentic Andean experience, the gifted author approached this mission through ethnography and folklore studies, linguistics, poetry, and fiction. For Arguedas, the Huarochirí manuscript was "the near uncontaminated message of [Andean] antiquity," imparted by "quite common Indians . . . still immersed in the ancient religion in an embodied way, [they were] agents of pre-Hispanic life" (Arguedas and Duviols 1966: 10).[14]

Scholars Karen Spalding (1967, 1974, 1984, 2012) and María Rostworowski (1988, [1978] 2002a, [1999] 2002b, [1972] 2004) shared this esteem for the manuscript's contents and pioneered the modern historical study of Huarochirí. Spalding, who provided the paleographic transcription for Arguedas's translation (Arguedas and Duviols 1966: 14), saw in the manuscript's contents expression of Andean social, political, and economic institutions. Her early work included the most systematic reproduction of Huarochirí's history since Avila.[15] For Rostworowski, too, one of the Huarochirí manuscript's virtues was that it largely bypassed the imperial Inka bias of many Spanish chronicles as well as the "distorted understanding of pre-Columbian reality" occasioned by the Spanish tendency to interpret and document the Andean world in European terms ([1978] 2002a: 193–94).[16]

Rostworowski, who has made perhaps the greatest contribution to the canonical prehistory, treated the manuscript's myths as recollections of the pre-Inkaic Yauyos' invasions and occupation of Huarochirí's fertile river valleys in repeated "waves" over centuries prior to Inka and Spanish colonization of the area (Rostworowski [1978] 2002a: 205, 211–12). Rostworowski found Uhle's periodization for the settlement of the Lurín and Rímac River Valleys reinforced by an item from a 1611 letter sent from Huarochirí to Archbishop Lobo Guerrero. Avila's anti-idolatry campaigns had located in the area of Santiago de Anchucaya the mummified body of a man named Tutayquiri, who was said to have died 600 years earlier (Rostworowski [1978] 2002a: 207; see also Chase 2016a: 98–112).[17] It was possible, then, that the Checa invasions and conquest of aboriginal inhabitants of Huarochirí began in the late Middle Horizon (ca. AD 900–1000) and continued throughout the Late Intermediate Period (LIP [AD 1000–1400]) (Rostworowski [1978] 2002a: 211).

A clear summary of the influence of the chronological aspect of the canonical prehistory is provided by historian Waldemar Espinoza, who describes a Huarochirí that had been successfully and productively invaded and reorganized centuries prior to the Inka advent. Descendants of the Yauyos invaders "ruled over the *yungas* populations" (though note that the Yauyos also "defeated and had run the ancient *yungas* inhabitants out of the area") and together had mastered agriculture, craftsmanship, and good government to such a degree that the expansive Inka entirely refrained from interfering (Espinoza Soriano 1992: 120–21).

Reinforced by the specific chronological anchors mentioned above, Rostworowski's historical interpretation of the Huarochirí narratives is eventful and "adventful," assuming the causal primacy of historical events (e.g., invasions and conquests) that have no cultural or interpretive motif or motivation. At the same time, they did not happen in a vacuum but appear to have been informed by "universal" rational-economic behaviorism (e.g., the Yauyos sought to expand their agricultural territory and diversity, or they "coveted" Yunca coca lands). These prime movers were subsequently reenacted commemoratively in rituals, recalled and communicated through myths.

Beyond her acceptance of the antiquity of the Tutayquiri mummy, Rostworowski's euhemeristic literalism extends to her historical interpretations of a structurally organized ethical charter for the narrators' generation (i.e., "we" of the present as opposed to those of a former *pacha* or "world-age"). This charter comes in the recurrent episode in the Huarochirí narratives involving a highlands huaca testing the generosity and hospitality of a people by sitting on the outskirts of a collective feast dressed as a beggar (cf. Cuni Raya). Aside from single, exceptional individuals who are spared for their attention to the huaca (see Salomon and Urioste 1991: 61–62, 127), the people ignore the disguised huaca, fating themselves to defeat and consignment to an inactive, inert (Salomon 1998: 9), and one-dimensional past, as they are washed from their settlements at the return of the wrathful, stormy huaca. Rostworowski suggests that these myths may be vestigial recollections of the historical episodes of reconnaissance; the "huacas" were highlands scouts or spies collecting intelligence prior to attacks on Yunca settlements (Rostworowski [1978] 2002a: 208–9). To my view such a reading cannot do justice to the semiosis that made meaningful collective pasts and presents in late pre-Hispanic and early Spanish colonial Huarochirí. Yet within the canonical prehistory, a strict chronological historicism restricts ritual, myth, even the huacas, to being post hoc reflections of "real" historical events and people, thereby acting to foreclose on perceiving and reimagining other ways of past making.

By connecting the manuscript narratives to details from other Spanish colonial documents and ethnographic data, Rostworowski made provocative suggestions aimed to rouse other Andean scholars—archaeologists in particular—to further, more concerted research into Huarochirí's past. Unfortunately the results have been to the contrary. The dearth of systematic archaeology in Huarochirí may owe to the assumption that we already know what happened there (hence, the "myth of a prehistory") because of the existence of the manuscript and the prominent ethnohistorical work based on it. The cursory archaeology that has been done (Bueno Mendoza 1992; Coello Rodríguez 2000; Coello Rodríguez and Díaz Arce 1995; INC 2008; Patrocinios and Tapia 2002) has tended to fit field observations a priori within the periodization laid out by the ethnohistory.

In a recent, thought-provoking article, Jane Feltham (2005; see also Feltham 1983, 1984, 2009) summarizes "what we know from ethnohistory about the Yungas and the Yauyos together with . . . what we might expect to find in the archaeological record that would verify these statements" (Feltham 2005: 133). Feltham's ethnohistorical summary matches Huarochirí's canonical prehistory (though with more muted statements of periodization); her list of archaeological expectations is shaped by the same conceptual or interpretive framework and is therefore internally reasonable and logically consistent. But she notes that it is rarely met in the archaeological record. For instance, despite assertions of (repeated) invasion and conquest, "there is little archaeological evidence for warfare" on the coast and in the middle Lurín Valley (Feltham 2005: 135–36); Inka rule in the Lurín "from an ethnohistorical point of view . . . seems to have been unimportant" though in the archaeological record it "dominates" the region (Feltham 2005: 141). In the face of these discrepancies Feltham suggests rereadings of the colonial documents, but stops short of a wholesale reconsideration of the "canonical prehistory." She concludes by acknowledging the productivity inherent in potential discrepancies between the archaeological and historical records, an observation resonant with this chapter.

PAST MAKING: HISTORICAL SENSIBILITIES, TIME, ARCHAEOLOGY

What, then, can be offered in contrast to Huarochirí's canonical prehistory? The first and most basic answer, I think, is more intensive, rigorous, and systematically collected archaeological data from the region (just what Rostworowski and Salomon have long suggested). But as the extant ethnohistory and archaeological research has illustrated, a fundamental shift in our understanding of the semiosis of indigenous Andean temporality / time, the past, and "history"—how and what they meant and worked—seems equally important.[18]

"Chronicles of the Impossible": This is Frank Salomon's (1982) term for the results of native Andean scribal efforts to reconcile the fundamental incongruities between indigenous Andean and Western (i.e., Spanish colonial and present-day) historical sensibilities. As we have seen, writings on the Huarochirí manuscript narratives from Avila to the present show that nonindigenous Andean authors also wrestled with "combining two drastically differing viewpoints about the nature and usefulness of the past" (Salomon 1982: 9). In a move beyond treating the Spanish drive to orient and situate specific Andean events within biblical history, Alan Durston's (2007) recent work on the Huarochirí manuscript includes the suggestion that these were due not just to incongruities between two particular histories, *both chronologically structured*. Rather, the Huarochirí manuscript's "narratives were part of a commonly held mythology that was probably familiar to the writer himself, who was attempting to establish 'correct' versions, often guided by concepts that were alien to this mythology, *such as chronological sequence*." Thus, more than just an attempt to fix Andean narratives within a biblical-classical framework, "the [Huarochirí manuscript] reflects an effort to produce a synthesis that was consistent with European *notions of temporal sequence and causation*" (Durston 2007: 232, emphasis mine; see Seoane and Culquichicón this volume [chapter 8]).[19]

These observations (from Salomon and Durston) implicate entirely different epistemological and ontological criteria than those that underpin the practice of history as we commonly understand it (cf. Curatola Petrocchi 2012; D'Altroy 2015: 119–73) and are indispensable to addressing an aporia as old as the first contact between Spaniards and Andeans. In both formal and semiotic terms, the ways Andeans registered, performed, and notated the past have been taken as beyond consequential comparison with (say) European historical writing (but see Salomon 2004; Salomon and Niño-Murcia 2011; Urton 2002, 2003, 2015), an obstacle that may well be responsible for the Huarochirí manuscript being one of a kind (Cummins 1994: 189–90). Thus, scholarly work on Andean historical consciousness and its representational modes is a staple of modern Andean studies, extensively and deftly treated by scholars such as Thomas Abercrombie (1998), Rolena Adorno (1982, 2000), Peter Gose (1996, 2008), Juan Ossio (2008), Tom Cummins (1994, 1998), and Joanne Rappaport (1994, 1998; Rappaport and Cummins 2012), in addition to those already cited and many others (see also introduction [chapter 1] to this volume).

Similarly, a recent surge of critical studies combining archaeology and previously accepted readings of Andean historical documents (Astuhuamán Gonzáles 2008; Bauer and Smit 2015; Chase 2015; Covey 2006, 2015; Eeckhout 2008; Kosiba 2015a, 2015b; Makowski 2015; Topic 1998; Topic et al. 2002; Wernke 2013) addresses these themes from different but related angles. In what follows, I propose a theoretical approach to the history and prehistory of Huarochirí and then show its fit with recently gathered archaeological data from Huarochirí and how it allows us to reconsider our ethnohistorical readings. I conclude by zooming out to consider these issues in other parts of the Andes and comparatively with Mesoamerica.

In order to avoid the disciplinary sense of "history" (or "structure" for that matter), and its ostensibly irreducible epistemological and ontological connotations (Durston's "chronological sequence" and "notions of temporal sequence and causation"), I instead use the phrases "historical sensibility" or "past making" to denote different forms of meaningful engagement with the legitimate past (Hughes and Trautmann 1995; Lucas 2005; Palmié 2002, 2014).[20] In doing so, my approach primarily follows leads provided by Michel-Rolph Trouillot (1995) and Neil Whitehead (2003) (see also introduction [chapter 1] to this volume). "Historical sensibilities" refers to the cultural variability of historical consciousness and of ways of conceiving, codifying, and communicating the past; "past making" is intended to evoke the concept of place making and refers to the practices involved in the construction of collectively meaningful and legitimate pasts, both generally and in particular instances (cf. Abercrombie 1998; Boone and Urton 2011; Covey 2006; Rappaport and Cummins 2012; Salomon 1982, 1999, 2002, 2004; Urton 1984, 1988, 1990). Within this framework, the phrase "past-forward" from the chapter's title refers to an Andean historical sensibility in which desirable or politically exigent presents and futures depended on the establishment of corresponding pasts, and vice-versa.[21] It also has to do with the way certain versions of the Andean past became fixed—"concretized," to use Gary Urton's phrase (1990: 125) from his exposition of the late sixteenth-century geographical placing of Inka Pacariqtambo—and then repeated in histories of the Andes for centuries thereafter (see Bauer and Smit 2015; Chase 2015; Covey 2006, 2015; D'Altroy 1987, 1992, 2005; Eeckhout 2008; Makowski 2015; Pease 1997, 2010; Salomon 1999; Topic 1998).

As Trouillot (1995) cogently argued, "history" (the term and our commonly understood concept) partakes both of "the sociohistorical process" and "our knowledge of that process." But in order for "that which happened" to become "that which we know of what happened" a host of intervening steps must obtain and coalesce. Most basically, for "x" to have occurred historically—Trouillot uses the famously unfamiliar Haitian Revolution and Republic of 1803—it must first in its own time be conceivable and recognized as an event, as something happening. Once recognized, some sort of record must be made that future interested parties can access, and these parties must find significance in the event recorded. Finally, as knowledge of the significant event is produced (in popular retellings, holidays, ceremonies, commemorations, scholarly publications, national monuments, etc.), it must resonate powerfully enough within current contexts and sensibilities to perdure in collective understanding. Without this confluence of factors, Trouillot's two senses of history do not correspond (see also Sayre [chapter 2], this volume; Rappaport and Cummins 2012: 153–54).

Ultimately, what Trouillot's sequence tells us is that the historical nature of any event or process in the past—history's existence itself—is contingent on the values of a series of presents. My intention is for this interplay of past and present to be reminiscent of anthropological definitions of myth, and particularly treatments of myth and history (Dundes 1984; Hill 1988; Rumsey and Weiner 2001; Sahlins 1985, 2004), because at the heart of both lies cultural significance. We can think of Trouillot's "sociohistorical" processes as both the "concrete circumstances [that are] essential to the very possibility of signification" (Keane 2005: 186) and as the emergence of meaning over time. I am not arguing that myth and history are the same, only that they share similar constitutive frames from which both derive their currency. The complex endeavor to understand and account for different historical sensibilities involves tracking multiple historical phenomena that interact in "complex referential loops"[22]—cycles and epicycles whereby the gross contours and specific crucial details of collectively understood pasts and desired and inhabited presents are continually coconstituted. The cogency of Trouillot's historiography comes from the way it turns the suppositions of historian Otto von Ranke's *"wie es eigentlich gewesen"* on themselves by revealing that "the historical" has, as it were, a *chaîne opératoire*.[23] Telling things "as they really happened" requires uncovering the epistemological, conceptual, and political conditions that create or silence particular pasts over time.

To advance on Trouillot's insightful, power-oriented twist on R. G. Collingwood's (1946) historiography, archaeologists, and historians must also focus on the social, semiotic, and cultural construction of the criteria involved in "past making."[24] In this vein, Neil Whitehead (2003: ix–xi) announced his preference for and use of the concept of "historicities" over history or historiography (see note 20, above). After pointing out that "working with incomplete as well as socially and culturally biased materials" is a basic "condition of historiography" far and wide (Whitehead 2003: ix), Whitehead names the constitutive foci of historicity oriented research as "the investigation of the cultural schema and subjective attitudes that make the past meaningful." Thus such research "reflects both the historical experiences of a given group and the cultural significance of recalling the past." Of course, "multiple histories may occur from multiple historicities" (cf. Abercrombie 1998: 196 on Spanish scribes' vexation at the "multiplicity" of the Andean past). But as I argue has been the case with the Huarochirí manuscript, "our record of histories has expanded much farther than our understanding of the historicities [or past makings] that create them, and this disjuncture in our understanding has produced a rather defective framework of analysis in anthropology" (Whitehead 2003: xi).

To be clear, I advocate for real, positive knowledge of the indigenous Andean past—I could hardly do otherwise while bringing archaeological data to bear on our readings of historical texts and expounding the virtues of doing so—and this is just to the point. First, the materials that come to constitute archaeological data could also be culturally and politically instrumental in the production of presents and pasts in the past. Second, the ways they did or could do so, that is, the cultural logics of effective, meaningful social and cosmological action—what Swenson (2003) has called "emic functionalism"—are amenable to anthropological efforts to understand them. Indeed, greater comprehension of the choices and activities of past individuals and societies depends on our respect for and active efforts to figure out how different semiosis worked, a precept that would seem to go without saying in sociocultural ethnography among living populations.

PERFORMATIVE PAST MAKING

As I have argued elsewhere (Chase 2015, 2016a, 2016b, 2018), collective pasts associated with ceremonial centers, ancestral landscapes, and huacas (with their related narratives and rituals) played a critical role in mediating colonial encounters between indigenous Andeans, like those of Huarochirí, and the expansive Inka and Spanish polities. These dynamics drew from and contributed to "performative past making" (cf. Sahlins 1985: xii–xiii). By performative I mean both formally presented and received acts that can be considered to follow some sort of "script" (i.e., performances), and the class of semiotic acts (i.e., performatives) that do or can make a change in the world by their execution and reception (Austin 1962; Inomata and Coben 2006). A focus on semiotic performatives is so apt for studying culture and history because the power of performatives is to make new collective realities, but only to the degree that their form and substance are cogent within extant, shared cultural contexts. The performative past making on display in late pre-Hispanic and early Spanish colonial Huarochirí consisted of the conjunction of these two senses of performativity (performances and performatives) and is eloquently articulated in the same narratives of the Checa conquest of Llacsatambo discussed above that formed the foundation of the canonical prehistory. Recall that the narrators' ancestors, led by huaca Tutayquiri, invaded the Llacsatambo territory, driving out the Yunca population and taking possession of their houses, fields, and huacas. Then, "as soon as Tutay Quiri finished his conquests, his children came here [Llacsatambo] and danced their dance of origin, just as they'd once danced it in Vichi Cancha,"[25] their place of origin (*pacarina*) (Salomon and Urioste 1991: 120). The standard huari-llacuaz components—aboriginal agriculturalists, invading highlanders, and a conquest event—are complemented here by an explicit performative act (the *taki* or dance), and not one that merely *marked* the conquest event, but that created new Checa origins at Llacsatambo through formal performance. The conquest was also celebrated annually in the Chanco pilgrimage, which started and ended at Llacsatambo; participants treaded Tutayquiri's path in conquering this territory with and for the Checa ancestors while reciting "'We go in Tutay Quiri's steps; We go in the path of his power'" (Salomon and Urioste 1991.: 79–81; cf. Taylor 1999: 165).[26]

In the canonical prehistory's chronology, this invasion was one of many that likely took place sometime between the late Middle Horizon and Late Intermediate Period, and prior to the Late Horizon; the associated myths and rituals recorded in the Huarochirí manuscript are considered reenactments that guarded the memory of these events, which ignores the performative, generative aspect of the *taki* ritual explicitly described by the narrators. Also bypassed is the temporal connotations of the description of the Chanco pilgrimage from the Huarochirí manuscript's Quechua text itself. The root of the Chanco's ritual recitation was the term *callpa* (Taylor 1999: 164; see note 26, above), usually translated as power or strength. But as Sabine MacCormack has pointed out, *callpa* was also associated with rituals to divine the future. In this sense *callpa* "involved gaining access to a deity's power or to the forces inherent in nature and *making them manifest in society*" (MacCormack 1998: 339; emphasis mine), which suggests that whatever the Chanco pilgrimage's historically reflective quality, it was in equal measure generatively past making. That is, its annual performance was a *projection* of Checa territory and status as conquering *llacuaces*. To the manuscript's narrators, the ritual pronouncement was integral to the pilgrimage's regiment, and MacCormack's insight regarding *callpa* recalls Webb Keane (1997: 50–51) on the "metapragmatic" content of utterances such as prayers or spells, through whose recitation "the speaker may obtain those powers [recited] in turn." He continues: "This appears to work by recontextualizing narratives as metapragmatic statements: Their linguistic form remains the same, but their function shifts. Rather than being construed as accounts of actions that were carried out in the past, the words are taken as reports on and directives for the action they themselves carry out in the moment of speaking."

FIGURE 5.5. *The Inka colcas located a few hundred meters southeast of Llacsatambo. The colcas likely provisioned those assigned to tasks related to Llacsatambo, including constructing the site, caring for huacas and mummified ancestors, and other rotating labors. Excavations within the colcas and radiocarbon dates place their construction and use firmly in the Late Horizon. (Photograph by author.)*

ARCHAEOLOGY OF HUAROCHIRÍ

In order to recover some of the first systematically collected archaeological data from the area that manuscript narrators called home, and to address issues raised by the canonical prehistory, the Proyecto Arqueológico Huarochirí–Lurín Alto (PAHLA; see Chase et al. 2011) began intensive archaeological fieldwork and analysis in the San Damián–Llacsatambo axis, starting in 2010 (figure 5.3). Based on architectural survey and research from adjacent areas, eight test excavations were distributed between the hilltop site of Llacsatambo (figure 5.2) and Inka *colcas* southeast of Llacsatambo (figures 5.3 and 5.5). The distribution strategy was to sample spaces connected to obvious Inka architectural styles (i.e., in the Inka storehouses and Llacsatambo's supposed "Inka sector") and spaces associated with structures of more regional architectural style, to gain chronological understanding of site occupations.

Llacsatambo's architectural spacing and construction combine elements of regional LIP and Inka styles. The site's standing architecture consists largely of single or multichambered tombs joined to walled patios or court spaces, often with domestic or work structures conjoined or situated nearby, though these are found most frequently within the walled patio space. The highest areas of the site are occupied by multichambered tombs of two to four levels. The most

FIGURE 5.6. *A concentration of architecture and architectural features such as orthogonal walls and double-jamb windows and niches led to the designation of an "Inka sector" at Llacsatambo prior to test excavations. Excavations and radiocarbon dates from within structures of obviously Late Horizon style and structures conforming more to regional Late Intermediate Period styles indicate that the construction and use of these two structure varieties was much closer chronologically than previously assumed in the "canonical prehistory." (Photographs by author.)*

pronounced Inka architectural influence is found in double-jamb windows and niches in structures with orthogonal walls (figure 5.6), concentrated especially in the site's "Inka sector." There are also trapezoidal niches and tall narrow doors distributed throughout the site (though clustering more on the outer eastern perimeter of the densest architecture).

Hundreds of ceramic sherds representing medium- to large-sized food and liquid storage and serving vessels were recovered through surface collection and excavation. The ceramic assemblage includes several pieces of typical Inka form and style (e.g., neck fragments and *asas falsas* [i.e., ornamental handles] from the Inka *urpu* [aríbalo] vessels used for serving *aqha*, or corn beer). Handles associated with food storage, serving, or preparation vessels make up more than 20 percent of the assemblage, and carbonized coca remains were recovered from three of the test units. Thus patterns consistent with the collective feasting of shared ancestors are apparent. Both ethnohistoric and archaeological evidence indicate that the site contained a concentration of ancestral remains and huacas corresponding to (or being the necessary constituents of) various social and political relations (Chase et al. 2011; Taylor 1999: 247–65).[27]

The excavations undertaken by PAHLA have not revealed evidence of a series of occupations at Llacsatambo previous to Inka arrival, as the Checa narrators claimed, nor occupation dating back to the late Middle Horizon, as projected in Huarochirí's canonical prehistory. Rather, all excavation units had

FIGURE 5.7. *Examples of Inka fine-ware from surface collections and excavations at Llacsatambo. (Photographs by author)*

comparable stratigraphy indicating a single, relatively brief and late occupation of Llacsatambo (see Chase 2015: 99–100).[28] Inka fine-ware (figure 5.7) was recovered from the lower levels of five of the six Llacsatambo units.

Four reliable C14 dates have been obtained from carbon samples excavated from secure contexts. All have been calibrated to two sigma. The dates from the lowest cultural levels in the excavations in the Llacsatambo structure with double-jamb windows and niches place its construction and use between AD 1448 and AD 1623. Those obtained from the lowest level of cultural activity in the Inka storehouses date them to between AD 1455 and AD 1518. The dates from the lowest levels of cultural activity in two domestic or work structures at Llacsatambo without architectural features or forms clearly identifying them as Inka influenced are AD 1318–AD 1428 and AD 1403–AD 1448.[29] In all cases, the carbon samples were obtained from below occupation levels in contexts likely comprising architectural fill randomly gathered to prepare level building and living surfaces atop the declining and rocky surface of the hills upon which Llacsatambo and the colcas were constructed.

In summary, the archaeological data recovered thus far do not support the chronology or causal/representational "order" of the canonical prehistory. But they *do* make sense when viewed from the vantage point of a different, performative historicity. That is, if among the pre-Hispanic and early colonial Checa the accepted, "emically-functional" relationship between the present and the past was one in which any desired present was intimately dependent on a vital, active, ongoing past and vice-versa, then interaction with the past would be dialogical and emergent, in the manner of a conversation. If the archaeological evidence is correct, it is likely that the ancestors of the Huarochirí manuscript built the ceremonial center of Llacsatambo, rather than usurping it from preexisting Yuncas in what we would call a historical event. But if this is so, why narrate the llacta's past in terms of a llacuaz invasion of huari settlements? I am suggesting that this owes to the fact that the founders were not interested in nor attempting to "do history," but were instead performing a successful, fertile llacta, which, as we have seen, required the competitive, complementary pairing of residents who could draw on chthonic, agricultural powers and residents who could bring the vital waters from above.

To be sure, this is not an argument for synchronic structure over diachronic history: Andean archaeology's Early, Middle, and Late Horizons "all involved the expansion of highland systems into lowland regions" (Quilter 2014: 224; Carneiro 1970). More specifically, in contrast to PAHLA's findings to date, Tom Dillehay's archaeological research in the Chillón Valley confirmed the historical nature of highlands to coast invasions (Dillehay 1976, 1977). But these are no less participants in the huari-llacuaz narrative for their chronological, event-based, historical reality. Rather, they promote a better understanding of the culturally informed nature of eventful actions. Among all of the highlands villages with population segments tracing ancestry to autochthonous or invading ancestors (huari or llacuaz, respectively), the historical, compositional admixtures of performed and lived pasts are surely unimaginably variegated. Concerning the temporality and historicity of huaris and llacuaces, Yauyos and Yuncas, the claim to different origins could be the expression and form of eventful political ranking (e.g., invasion/conquest, diplomatic subjugation) *and/or* the expression of long-term modus vivendi (i.e., the competitive complementary of sedentary and seminomadic peoples, the productive opposition of earthy, fixed agricultural plots and the seasonal, forceful advent of water from above) (cf. Gose 2008), as well as temporal routines/rhythms.

The precise degree to which lived historical situations influenced the structure of the performative past and vice-versa for any given highlands village will be difficult to know with (especially where no known narrative accounts exist). However, as the data from Llacsatambo in Huarochirí seem to illustrate, archaeology can greatly aid in separating out the elements and processes whose concretion is the content of colonial historical documents. Techniques such as stable isotope analysis can provide data from which to make more sure inferences about populations' geographic origins and migrations.[30]

In summary then, despite archaeology's potential to provide more precise chronological resolution to canonical history, precise historical certitude may be a social scientific fetish whose demands we cannot meet. This becomes apparent given the nature of performative historicity, kinship and identity, and political allegiance—especially in times pregnant with the potential for rapid and momentous change such as Inka expansion and Spanish invasion. Still, archaeological research is precisely suited to help determine whether particular known cases of late pre-Hispanic Andean tracing of ancestry to ancient or recent dwellers or invaders reinforced or precipitated specific historical-political conjunctures. We may also be able at least to gather the data necessary to address the possibility of real, long-term filiation between populations with different geographical origins. But the light this will shed on the Andean past will be limited if we do not allow the data to force us to rethink our anthropological theories and models.

DISCUSSION

The historical sensibility I have presented for the late pre-Hispanic and early colonial Andes, though proper to Andeans of this period in a cultural sense, is not irreducible, exotic beyond comparison with temporal sensibilities and historicities in other cultures and time periods. For example, Alexandre Kojève's characterization of "Hegel's historical Time . . . as a movement that starts with the future and moves through the past into the present" (qtd. in Fabian [1983] 2014: 200n20) is predicated on teleological primacy—something like an Aristotelian "final cause" being the already extant fact "driving" the transformational process of an object through causal stages which chronologically precede the finished, fully formed object. "Whereas biological Time would be characterized by the primacy of the past," Kojève continues, "it may be that the Time in which the Present takes primacy is cosmic or physical Time" (Fabian [1983] 2014: 200n20). Again, here "primacy" is *causal* or *influential*, not chronological.[31] This is not the same as *ahistorical*, but rather akin to the performative history Marshall Sahlins attributes to eighteenth-century Hawaiians (Sahlins 1985: xii–xiii). Kojève's general concept of "cosmic" time and Sahlins' specific exposition find correspondence with what I characterize as the historicity of late pre-Hispanic and early Spanish colonial Huarochiranos. Different expressions of comparable historical sensibilities and past making in the Andes have been illustrated in Gary Urton's (1990) seminal study of the early colonial construction of pre-Hispanic Inka "reality," and more recently by Peter Gose (1996, 2008; see below), who made a convincing argument modeling the intersections of sociopolitical rankings, new historical events, cosmological principles, and spatial arrangements in Inkaic Cuzco.

Anthropological research suggests that we ought probably to think about a plurality of pasts in the Andes. Indigenous Andeans surely experienced what I would call a mundane slippage of time from present to past, visible in aging and other entropic processes. This also appears in Andean expression in a chronologically comparative (but cyclically tinged) way; for example, last year's crops did not come in as well as this year's. A chronologically cumulative past is also detectable in the Inka's historical structures/structured history mentioned above (Gose 1996; see also introduction [chapter 1], this volume).

A different sort of Andean past, perhaps the one Salomon (1999: 20) describes as a reality parallel to the present and accessible through special ritual means (Salomon 1998: 11–12), was at work in more specifically marked, significance-rich activities and contexts, such as those involving and invoking ancestors (which, of course, overlap with the agricultural activities of the chronologically comparative and in the structured historicities mentioned above) (see Wilkinson and D'Altroy, this volume [chapter 4]). As Salomon (citing Meyer Fortes in Salomon 1995: 315) reminds us in his article on the interactions between living ancestors and the merely living in Tawantinsuyu's capital and provinces, not all of one's dead relatives become ancestors. Or as Claude Lévi-Strauss wrote of our sense of the historical past: "History . . . does not account for the present, but it makes a selection between its elements, according only some of them the privilege of having a past" (Lévi-Strauss 1966: 231). The dynamic of ancestor making and the "selectivity" of history are akin to performative past making both in terms of processes of selection and recognition of present situation through established categories of legitimate pastness. This is the past I am addressing in writing of the performative historicity of late pre-Hispanic and early colonial Huarochirí above.

Scholars of ancient and Spanish colonial Mesoamerica have long noted (and debated) the cyclical and linear conceptions of time and history as recorded in the region's pictorial and alphabetical documents (inter alia Boone 2000, 2007; Carrasco 1982; Clendinnen 1987; Farriss 1995; Hassig 2001; Krech 2006) A brief comparison with one recent theorization of social memory and "time leaps" in the expression of Mesoamerican temporal sensibilities (Megged 2010: 26–34) will clarify the distinction between history and the historicity folded within and between Huarochirí's pages and stratigraphy. Time leaps have to do with the structural homology between Mesoamerican historical narratives and cosmic time, specifically the repeating structure of cataclysm and the reestablishment of social, political, spatial, and anthropogenic order. Historically this was achieved through performance of primordial, foundational actions, the recollection and depiction of which were inflected by historical, on-the-ground specifics (Megged 2010: 27–31). In these historical narratives, chronological sequence is seemingly disrupted at transitional moments by "leaps back" to representations of the gods' mythical, primordial deeds (those apposite to the historical crises at hand), followed by a return to the historical events (Megged 2010: 29–30; figures 5.2 and 5.3).

As we have seen in the Andean historicity presented above, the sensibilities thus depicted (comparable to what I have elsewhere called "temporal collapse" [Chase 2015]) are only apparently disjunctive from within a monolithic concept of history. For example, on an AD 1382 inauguration depiction, Joyce Marcus comments that "the scene refutes the notion that scribes were writing history, since at least three of the people shown . . . could not have been in attendance, [as] they lived 100 years or more before" the central figure; or again, on a Copán inauguration scene: "this . . . cannot be considered 'history' because it includes rulers whose reigns were separated by centuries" (qtd. in Megged 2010: 31). These characterizations are fair enough if we are intent on restricting the definition of history to the dogmatic disciplinary, institutional practice solidified in Europe and the United States from the nineteenth century to the present.[32] And, though to my view there is no question that the documents in question should be considered historical, these are the distinctions that lead me to advance the concept of past making to explain the Huarochirí data. (That is, in the Mesoamerican past making in these depictions the presence of mythical beings does not compromise, but rather constitutes their legitimate historicity.)

Marcus's suggestion of historical discrepancies in these Mesoamerican documents (viz. the impossibility of the interactive presence of beings whose lives could not have been coeval) brings to mind a well-known but underemphasized detail from Spanish histories of the Inka. One of the earliest of these histories is by Juan Diez de Betanzos, a Quechua-speaking Spaniard with affinal connections to Andean royalty. Chapters 6–8 of Betanzos's ([1557] 2010) *Suma y narración de los Incas* deal with the pivotal events by which the Inkas won primacy over competing polities to become *the* imperial power of Cuzco and then the Central Andes, namely, confronting and defeating in battle their chief rival, the Chancas, led by their lord, Uscovilca.

As written by Betanzos ([1557] 1996 : 19–30), Uscovilca "ruled," "found out," "decided," "consulted," "ordered"; Uscovilca "gave" a unit of soldiers to the two captains he "himself sent" toward Antisuyo because he "wanted" to take on Cuzco himself. The reigning Inka desired "to negotiate with" the bellicose Uscovilca, but in the end retreated while his son Inka Yupanqui continued to marshal fighting men when he saw "Uscovilca was coming."[33] Ultimately, Uscovilca was captured during the battle and killed. "And when his people saw him dead and also the great number of their own slaughtered," they fled (Betanzos [1557] 2010: 74–75).[34]

Fifteen years after Betanzos's history, Pedro Sarmiento de Gamboa completed his commission to gather the official history of the Inkas as part of the social, geographic, and bureaucratic reorganization efforts of the Viceroy Toledo

(Sarmiento de Gamboa [1572] 2001; [1572] 2007). Sarmiento interviewed Cuzco Inkas and *kurakas* in Jauja and Huamanga, and his manuscript was approved by dozens of Inka kin. Uscovilca is first mentioned in chapter 26 of Sarmiento's *Historia de los Incas*. He and his younger brother, "thieves and cruel tyrants" both, are named as the founders of the Chanca upper and lower moieties. Immediately following, Sarmiento explains that the brothers had died long before the fabled confrontation with the Inkas. Thereafter the Chancas carried Uscovilca's statue into battles and this was why the Chancas' feats were always attributed to Uscovilca. (Sarmiento de Gamboa [1572] 2001: 84–85). The same metonymic dynamic can be seen in the slippages in the Huarochirí manuscript's narratives about the conquests of Tutayquiri and the Checa.

Brian Bauer and colleagues (2013: 28, 58–59; see also Bauer and Smit 2015) reconcile the discrepancy by following Sarmiento in treating Uscovilca as a mythical ancestor and a metonym of the Chanca—a logical and useful interpretive move that aligns the narrative with modern historical sensibilities. However, my tedious list of verbs communicating Uscovilca's intentions and actions militates against dismissal of Betanzos's Uscovilca as historiographic oversight or shorthand and points instead to the need to take seriously the different historical sensibility and past making expressed and practiced by Betanzos's Inka informants in their account of Uscovilca (see Eeckhout and Owens 2015).

Death and the dead here are excellent ways to think about different temporal registers and different cultures of past making. We live a tradition with the temporal assumptions that whatever else death may be, it is a universal and inevitable consequence of time passing (or living beings passing through time); it is by this same virtue a marker of the passage of time (for us). But as Gose (1996) observed of the Inka, death was also fundamentally about (relative) sociopolitical rank, vocation, even spatial/geographic organization (cf. Salomon 1998 on huacas). In Huarochirí and neighboring areas, death was something more than a characteristic of the passage of time. It could be an achievement, and not just one of individual accomplishment, but one so charged with bounty and potency as to be beneficial to whole social collectivities.

This evokes a play on William Faulkner's oft-cited declaration "The past is never dead. It's not even past." In late pre-Hispanic and early Spanish colonial Andean past making (at least among the Inka and in Huarochirí), a certain kind of past was *only* past because it was dead. But being dead and past did not mean the cessation of existence.

CONCLUSION

Let us now revisit the narrative of Cuni Raya and Caui Llaca from the chapter's beginning (figure 5.1). As discussed, the myth's appearance in the manuscript's second chapter is significant, as the chapters were organized to approximate the structure of the Hebrew Bible: accounts relating the most ancient, foundational acts of creation and ordering of the world (i.e., those presumably of greatest antiquity and thus most identified with the Book of Genesis) were placed at the manuscript's beginning (Salomon 1991; Arguedas and Duviols 1966: 206–8). Recall the sequential structure that helped us to place and interpret the myth geographically. In terms of interpreting the historicity of the Huarochiri manuscript narratives, does the form of this sequential structure mean that we ought to take the Lurín Valley as having "occurred" or formed first, only then to be followed and filled by the Lurín River? We know geologically that this is not how valleys and rivers work. Their formation and course over time are the product of their ongoing, mutual influence: river water goes where the land offers least resistance, eroding valley channels that further guide the water in the same process. In like manner Caui Llaca and Cuni Raya intimately and perpetually intertwine in myth, in the past and in the present. These geological facts also prompt us to think of the seasonal, cyclical occurrence of the mythical "event" (i.e., the annual swelling of the Lurín River during the rainy season), at once primordial and annually present, which is perhaps to say eternal (see also Seoane and Culquichicón this volume [chapter 8]). What the data from Huarochirí seem to indicate about chronology and past making in this late pre-Hispanic and early colonial society is that Avila and subsequent archaeologists and ethnohistorians erroneously conflated *narrative* sequences with *chronologically historical* sequences.

Trouillot's analytic of the production of the past is quite appropriate for the Andes. From the robust anthropology of history in Andean scholarship cited above, let us conclude by revisiting just a handful of works. Decades ago, Salomon (1982: 11) informed us that "for Andeans, the sources of diachronic knowledge were completely different [from those of Spanish historicity] and had never been organized on the principles of absolute chronology, cause-and-effect, or eschatology." Instead "the useful past was embodied" in multiple three-dimensional media. At nearly the same time, Urton (1984, 1988) showed how history and social structure were codified and communicated in agricultural land divisions or in a twentieth-century adobe wall in Pacariqtambo. Thomas Abercrombie (1998) and Gose (2008, 1996), working primarily (but not exclusively) with texts, elucidate the coconstitution of past and present in a plurality of practices and materials that attend Andean past making. To this three (four, or five?)-dimensional past making, this chapter adds archaeologically recovered materials. For Huarochirí, these data provide an additional temporal dimension to the text-based canonical prehistory, and combine with the Huarochirí manuscript narratives to produce a more detailed vision of the performance of the past. What emerges supports through counterposition Gose's (2008) recent exposition of the mechanism that underwrote productive coexistence in the central Peruvian highlands (of which huari-llacuaz and the reception of Spaniards as Viracochas were expressions), which he terms "invaders as ancestors." The emerging picture of the Checa at Llacsatambo is one of "ancestors as invaders."

NOTES

Unless otherwise noted, the translations are my own.

1. Of the difficulty defining *villca*, Salomon (Salomon and Urioste 1991: 46n44) suggests that the manuscript and other sources indicate "that a *villca* is a human being who partakes of a *huaca*'s status." The Colonial lexicographer of Aymara, Ludovico Bertonio ([1612] 2006: 742), provides the following definitions: "The Sun as they called it anciently and now call Inti" (El Sol como antiguamente dezian y ahora dizen Inti) and "Shrine dedicated to the Sun or other Idols" (Adoratorio dedicado al Sol u otros Idolos). The use of *huaca* and *villca* together seems to be a product of the linguistic strata of the manuscript's production: the manuscript was written in Quechua, which Huarochiranos spoke together with "an ethnic tongue of the Jaqaru (Aymara-family) group" (Salomon and Niño-Murcia 2011: 22). The author may have been covering all bases, both in terms of communicating the variety of beings referred to and in linguistic terms. (Cf. Mannheim and Salas Carreño 2015: 55–59.)

2. Salomon (Salomon and Urioste 1991: 47n49) notes that the root of the Quechua term here for "made me pregnant" (*yumahuarcanquichic*) is sperm (*yumay*).

3. Geographic localization and sociopolitical affiliation of dozens of the manuscript's toponyms by Karen Spalding (1984: overleaf), María Rostworowski (1988: 56, 2002 [1978]), and appearing on the map in Salomon and Urioste's (1991) edition of the manuscript (based on the work of John Treacy) are of immense value to Huarochirí research.

4. This interpretation is corroborated in another myth in the manuscript's chapter 5, where a human woman is turned upside down and lithified (turned into a huaca) with her legs and genitalia in a position to be seen and to receive offerings of coca leaves (Salomon and Urioste 1991: 59).

5. It is important to mention that the label "canonical prehistory" is not pejorative, nor is it the work of any one scholar.

6. *Yuncas* (or yungas) referred also to the low, eastward descending Andes, but here I concentrate on the westward element. Also, as in Salomon and Urioste's (1991) translation of the Huarochirí manuscript, the capitalized Yunca refers to peoples so-called, whereas yunca refers to the geographic and environmental characteristics mentioned above.

7. "Complejo dualista huari-llacuaz" (Duviols 1986: lxv).

8. "En toda esta historia y fábulas no he podido averiguar el orden y sucesión de ellas, cual fue el primero y cual después, porque son todas cosas antiquísimas, mas podrá ser que, para cuando esto se vuelva a escribir, lo tenga sabido o a lo menos lo más verosímil."

9. "Antes del nacimiento del inga" (Taylor 1999: 222). In a footnote Taylor offers: "Or perhaps: 'before the lineage of the Inca was established'" ("O, tal vez: 'antes que se estableciera el linaje de los incas.")

10. In addition to Uhle's specific rejoinder (Uhle [1910] 1970) to Markham's (1906) textually based "Megalithic Age" hypothesis, Uhle's familiarity with Markham's work—including Markham's sketchy English translation of the *Tratado* (Markham [1873] 1964: 121–47)—is on display in Uhle's *Pachacamac* monograph (Uhle [1903] 2003), wherein he drew from translations of four different Spanish colonial texts (from Avila, H. Pizarro, Juan Polo de Ondegardo, Cristóbal Molina) contained in two of Markham's (1873, 1872) publications.

11. The particular "obscure traditions" cited by Markham in the critiqued article are not those of Huarochirí but instead Inka narratives of invasions from the south, and Markham used these stories to theorize Tiwanaku's history. But these invasion myths have a great deal in common with those from Huarochirí (as well as Huamachuco and other central Andean areas). See Chase 2018.

12. "Los Yauyos y los habitantes de los valles de Lima y de Pachacamac eran todos Aymaraes. . . . [E]n el gran momento de la civilización aimará, tribus de esta raza se extendieron por el Perú como lo hicieron las tribus quichuas en el período de los Incas" (Uhle [1910] 1970: 382).

In terms of and historical linguistics, the origins and spread of Aymara-related languages are far from decided to general consensus (see Beresford-Jones and Heggarty 2010; Cerrón-Palomino 1998; 2000a, 2000b, 2004, 2013, 2015; Heggarty and Beresford-Jones 2009; Urton 2012). At any rate, here it is Uhle's method of inference, not the accuracy of his assessment that is relevant to the point.

13. I must acknowledge the irony of my simplistic ("overintegrating") lumping together the works of these different scholars. Their similarities and distinctions are treated in greater detail in Zachary Chase (2016a). "Integrative Andeanism" comes out of works by Julio C. Tello, Luis Valcárcel, R. Tom Zuidema, and John V. Murra, for example.

14. "El mensaje casi incontaminada de la antigüedad"; "indios bastantes comunes . . . sumergidos aún y de manera muy encarnizada en la antigua religión, actores de la vida prehispánica" (Arguedas and Duviols 1966: 9–10).

15. Spalding's archaeological research (with Thomas Patterson) at middle Rímac Valley architectural complexes devoted to Huarochirí huaca Chaupi Ñamca stands out for its demonstration and discussions of Inka state involvement with these cult centers (Spalding 1984: 98–101; cf. 89; Thatcher and Hellmuth 1968–71).

16. "Sus informaciones hayan distorsionado el entendimiento de la realidad precolombina" (Rostworowski 2002a [1978]: 194).

17. In a separate document (Polia Meconi 1999: 306) relating the same events, a different priest estimated the mummy's age to be 800 years.

18. From another angle, the distinction I am drawing can be expressed in relation of George Lau's recent, well-taken argument that "the danger of taking Moche imagery literally lies . . . in assuming that Moche images provide real, impartial, or complete texts of the past" (Lau 2004: 176). I agree wholeheartedly with Lau and further suggest that we also ought to be aware that the very grounds of what different peoples consider

"real, impartial, or complete" (or "literal," for that matter) are not universal (much less expressed in universal forms), but are culturally and historically contextual.

19. In contrast, though indigenous author Felipe Guaman Poma's *Primer nueva corónica y buen gobierno* (Guaman Poma de Ayala [1613] 1980) does include a host of overtly biblical categories, objects, and characters, its contents are *built* more on Andean temporal/historical structures than on a "biblical frame" (Adorno 2000; MacCormack 1998; Ossio 2008; Salomon 1982).

20. In recent literature on the anthropology of history and time "historicity" has been the term commonly used for my "historical sensibility" (see Hirsch and Stewart 2005 [and contributors]; Whitehead 2003). I am grateful for the insight and thoughtful suggestion of one of the volume's reviewers, which led me to use the latter term instead of "historicity" to avoid confusion. As the reviewer correctly pointed out, "historicity" has long referred to the concept that history or sequentiality through time encapsulates and pervades everything. Or, as Jan Assmann (2003: viii) citing Reinhardt Koselleck puts it, "the semantic paradigm . . . of historicism" means, in essence, "the total historification of reality."

21. See also Federico Kauffmann Doig's (1978: 545) translation of *ñaupa pacha* as *tiempo adelante*.

22. This phrase comes from Michael Dietler's (2005: 34) description of the chronologically circular construction of Hellenic colonization, both in the ancient Mediterranean and in modern British adoption and reimagining of the ancient Greeks.

23. Von Ranke's programmatic phrase for the essential approach and aim of an objective, modern history has been recognized by Moses Finley, for example, as "probably the most famous of all pronouncements about the nature of history" (in Stroud 1987: 379). It is roughly translated as (re)producing history "as it really happened," that is, facts that are free of overdetermined interpretations or theories. The phrase appears in the preface to his *Geschichte der romanischen und germanischen Völker von 1494 bis 1514* (von Ranke 1824) Note, however, that the traditional, scientific historical interpretation of this work has been challenged by Peter Novick (1988: 21–31), who argues Von Ranke's intentions were in a more interpretive vein.

24. Indeed, the division of culture and politics is artificial, as recent expositions of Gramscian hegemony illustrate (Crehan 2002; Kolata 2013: 1–27; Sahlins 2004).

25. On Vichi Cancha, note again the fractal (Abbott 2001), asymmetrically dual (Urton 2012), huari-llacuaz "insistence" expressed in origins from the earth's produce and from vital fluid from on high: "Some people say that there was a wild *quinua* plant in the vicinity of Upper Paria Caca . . . and that humans emerged from its fruit there. Other people say that blood once fell from the high heavens. It hit the ground at the spot called Vichi Cancha, in the area of that same *quinua*" (Salomon and Urioste 1991: 117, italics in original). Vichi Cancha was the *pacarina* (place of origin) of Llacsatambo's aboriginal Yuncas and invading Checa Yauyos both.

26. The entire passage in Quechua reads: "Caytam canan cay checacunaca cay tutayquirip puriscanta paypac *callpa*nta purini ñispa huatacunapi cay llactamanta Tucoy

ynantin caricuna chacocta ruraypac llocsic carca cay nobiembre quellapi" (Taylor 1999: 164; emphasis added).

27. Regarding ethnographic evidence, in the Huarochirí manuscript, Llacsatambo is associated with at least five different huacas: Pariacaca, Tutayquiri, Maca Uisa, Llocllay Huancupa, and Cati Quillay (Salomon and Urioste 1991: 74, 79, 99, 101, 103).

28. Obviously, this stratigraphy was also found within the Inka colcas.

29. The distinct split (i.e., at AD 1448) between the former and latter sets of dates must be noted as possibly (perhaps even probably) indicating pre-Inka and Inka-era occupations of Llacsatambo. The squaring of these dates with the stratigraphy and stratigraphic location of the ceramics of Inka decoration is ongoing. The primary point here is that these dates militate against the centuries-long occupation of Llacsatambo as expressed in the ethnohistory.

The carbon samples were processed by the NSF-Arizona AMS Laboratory and DirectAMS Radiocarbon Dating Service.

30. A field study and museums collection search to gather skeletal data from Huarochirí for stable isotope analysis is currently being arranged between PAHLA and colleagues from Vanderbilt University and Harvard University / Peabody Museum.

31. This argument is perhaps made clearer with consideration of synonyms or near synonyms of "influential," which include "powerful," "important," "dominant," and "significant."

32. Of course, such a pure, secular history would not only ignore concepts and agents that do not fit within this specific framework, but which nonetheless are active forces in the lives of peoples and nations, but also ignore that it is itself a culturally and historically dependent frame of reference (at the very least in regards to the values espoused and promoted in it).

33. Here I combine my own translation with those of Hamilton and Buchanan (in Betanzos [1557] 1996:19–30). The corresponding Spanish words and phrases are (Betanzos [1557] 2010: 65–74): "era señor," "tuviese noticia," "paraciéndole que era bien ver," "entró en consulta," "mandó / les fue mandado por Uscovilca," "les diese," "envió ansimismo," "quería," "con él capitular," "viniese," "venía."

34. "E como los suyos le viesen muerto y viesen la gran matanza que en ellos se hacían . . . huyeron" (Betanzos [1557] 2010: 74–75).

REFERENCES

Abercrombie, Thomas A. 1998. *Pathways of Memory and Power: Ethnography and History among an Andean People*. Madison: University of Wisconsin Press.

Abbott, Andrew. 2001. *Chaos of Disciplines*. Chicago: University of Chicago Press.

Acosta Rodríguez, Antonio. 1987. "Francisco de Avila. Cusco 1573 (?)—Lima 1647." In *Ritos y tradiciones de Huarochirí del siglo XVII*, ed. and trans. Gerald Taylor. *Historia Andina* 12:551–616. Lima: Instituto de Estudios Peruanos.

Adorno, Rolena, ed. 1982. *From Oral to Written Expression: Native Andean Chronicles of the Early Colonial Period*. Syracuse, NY: Maxwell School of Citizenship and Public Affairs, Syracuse University.

Adorno, Rolena. 2000. *Guaman Poma: Writing and Resistance in Colonial Peru*. Austin: University of Texas Press.``

Arguedas, José María, trans., and Pierre Duviols, ed. 1966. *Dioses y hombres de Huarochirí: Narración quechua recogida por Francisco de Avila [¿1598?]*. Lima: Instituto Francés de Estudios Andinos and Instituto de Estudios Peruanos.

Ascher, Marcia, and Robert Ascher. [1981] 1997. *Code of the Quipu*. New York: Dover.

Assmann, Jan. 2003. *The Mind of Egypt: History and Meaning in the Time of the Pharaohs*. Cambridge, MA: Harvard University Press.

Astuhuamán Gonzáles, César W. 2008. "Los otros Pariacaca: Oráculos, montañas, y parentelas sagradas." In *Adivinación y oráculos en el mundo andino antiguo*, ed. Marco Curatola Petrocchi and Mariusz S. Ziółkowski, 97–119. Lima: Fondo Editorial, Pontificia Universidad Católica del Perú / Instituto Francés de Estudios Andinos.

Austin, John L. 1962. *How to Do Things with Words*. Cambridge, MA: Harvard University Press.

Bauer, Brian S. 1996. "Legitimization of the State in Inca Myth and Ritual." *American Anthropologist* 98 (2): 327–37. https://doi.org/10.1525/aa.1996.98.2.02a00090.

Bauer, Brian S., Miriam Aráoz Silva, and Lucas C. Kellett. 2013. *Los chancas: Investigaciones arqueológicas en Andahuaylas (Apurímac, Perú)*. Lima: Instituto Francés de Estudios Andinos, UIC College of Liberal Arts and Sciences, Institute for New World Archaeology.

Bauer, Brian S., and Douglas K. Smit. 2015. "Separating the Wheat from the Chaff: Inka Myths, Inka Legends, and the Archaeological Evidence for State Development." In *The Inka Empire: A Multidisciplinary Approach*, ed. Izumi Shimada, 67–80. Austin: University of Texas Press.

Beresford-Jones, David G., and Paul Heggarty. 2010. "Broadening Our Horizons: Towards an Interdisciplinary Prehistory of the Andes." *Boletín de Arqueología PUCP* 14:61–84.

Bertonio, Ludovico. [1612] 2006. *Vocabulario de la lengua aymara Transcripción del texto original de 1612*. Arequipa, Peru: Ediciones El Lector.

Boone, Elizabeth Hill. 2000. *Stories in Red and Black: Pictorial Histories of the Aztecs and Mixtecs*. 1st ed. Austin: University of Texas Press.

Boone, Elizabeth Hill. 2007. *Cycles of Time and Meaning in the Mexican Books of Fate*. Austin: University of Texas Press.

Boone, Elizabeth Hill, and Gary Urton, eds. 2011. *Their Way of Writing: Scripts, Signs and Pictographies in Pre-Columbian America*. Washington, DC: Dumbarton Oaks.

Bueno Mendoza, Alberto. 1992. "Arqueología de Huarochirí." In *Huarochirí: Ocho mil años de historia*, vol. 1: 13–66. Lima: Municipalidad de Santa Eulalia de Acopaya.

Carneiro, Robert L. 1970. "A Theory of the Origin of the State: Traditional Theories of State Origins Are Considered and Rejected in Favor of a New Ecological Hypothesis." (New Series) *Science* 169 (3947): 733–38. https://doi.org/10.1126/science.169.3947.733.

Carrasco, David. 1982. *Quetzacoatl and the Irony of Empire: Myths and Prophecies in the Aztec Tradition*. Chicago: University of Chicago Press.

Cerrón-Palomino, Rodolfo. 1998. "Examen de la teoría aimarista de Uhle." In *Max Uhle y el Perú antiguo*, ed. Peter Kaulicke, 85–120. Lima: Pontificia Universidad Católica del Perú, Fondo Editorial.

Cerrón-Palomino, Rodolfo. 2000a. *Linqüistica aimara*. Biblioteca de la tradición andina 21. Cuzco: Centro de Estudios Andinos Regionales "Bartolomé de las Casas."

Cerrón-Palomino, Rodolfo. 2000b. "El origen centroandino del aimara." *Boletín de Arqueología PUCP* 4:131–42.

Cerrón-Palomino, Rodolfo. 2004. "El aimara como lengua oficial de los Incas." *Boletín de Arqueología PUCP* 8:9–21.

Cerrón-Palomino, Rodolfo. 2013. *Las lenguas de los incas: El puquina, el aimara y el quechua*. Frankfurt am Main: Peter Lang. https://doi.org/10.3726/978-3-653-02485-2.

Cerrón-Palomino, Rodolfo. 2015. "The Languages of the Inkas." In *The Inka Empire: A Multidisciplinary Approach*, ed. Izumi Shimada, 39–53. Austin: University of Texas Press.

Chase, Zachary J. 2015. "What is a Wak'a? When is a Wak'a?" In *The Archaeology of Wak'as: Explorations of the Sacred in the Pre-Columbian Andes*, ed. Tamara L. Bray, 75–126. Boulder: University Press of Colorado. https://doi.org/10.5876/9781607323181.c004.

Chase, Zachary J. 2016a. "Performing the Past in the Historical, Ritual, and Mythological Landscapes of Huarochirí, Peru (ca. AD 1400–1700)." PhD dissertation, University of Chicago.

Chase, Zachary J. 2016b. "Pacha y presentación: La conversión colonial de Huarochirí prehistórico." *Boletín de Arqueología PUCP* 21: 13–37.

Chase, Zachary J. 2018. "Local Ritual Landscapes: Inca Appropriation, Reconstruction and Incorporation." In *The Oxford Handbook of the Inca*, ed. R. Alan Covey and Sonia Alconini, 219–40. Oxford: Oxford University Press.

Chase, Zachary J., Imbertis Herrera Abraham Magno, and Paredes Sánchez Luis Enrique. 2011. *Informe técnico: Proyecto arqueológico huarochirí–lurín alto*. Lima, Peru: Archived at the Ministry of Culture.

Clendinnen, Inga. 1987. *Ambivalent Conquests: Maya and Spaniard in Yucatan, 1517–1570*. Cambridge: Cambridge University Press.

Coello Rodríguez, Antonio. 2000. "El Camino inca en el distrito de San Damián (provincia de Huarochirí) Perú." In *Caminos precolombinos: Las vías, los ingenieros y los viajeros*, ed. Leonor Herrera y Marianne Cardale de Schrimpff, 167–93. Bogotá: Instituto Colombiano de Antropología e Historia.

Coello Rodríguez, Antonio, and Ernesto Díaz Arce. 1995. "Un tampu inka en San Damián (Huarochirí-Perú)." *Sequilao: Revista de Historia, Arte y Sociedad* 8:125–40.

Collingwood, R. G. 1946. *The Idea of History*. Ed. Jan Van Der Dussen. New York: Oxford University Press.

Covey, R. Alan. 2006. "Chronology, Succession, and Sovereignty: The Politics of Inka Historiography and Its Modern Interpretation." *Comparative Studies in Society and History* 48 (1): 169–99. https://doi.org/10.1017/S0010417506000077.

Covey, R. Alan. 2015. "Inka Imperial Intentions and Archaeological Realities in the Peruvian Highlands." In *The Inka Empire: A Multidisciplinary Approach*, 83–95. Austin: University of Texas Press.

Covey, R. Alan. 2008a. "The Inca Empire." In *The Handbook of South American Archaeology*, ed. Helaine Silvermann and William H. Isbell, 809–30. New York: Springer. https://doi.org/10.1007/978-0-387-74907-5_40.

Covey, R. Alan. 2008b. "Multiregional Perspectives on the Archaeology of the Andes during the Late Intermediate Period (ca. A.D. 1000–1400)." *Journal of Archaeological Research* 16 (3): 287–338. https://doi.org/10.1007/s10814-008-9021-7.

Crehan, Kate A. F. 2002. *Gramsci, Culture and Anthropology*. Berkeley: University of California Press.

Cummins, Tom. 1994. "Representation in the Sixteenth Century and the Colonial Image of the Inca." In *Writing without Words: Alternative Literacies in Mesoamerica and the Andes*, ed. Elizabeth Hill Boone and Walter D. Mignolo, 188–219. Durham, NC: Duke University Press.

Cummins, Tom. 1998. "Let Me See! Reading Is for Them: Colonial Andean Images and Objects 'como es costumbre tener los caciques Señores.'" In *Native Traditions in the Postconquest World*, ed. Elizabeth Hill Boone and Tom Cummins, 91–148. Washington, DC: Dumbarton Oaks.

Curatola Petrocchi, Marco. 2012. "Los cinco sentidos de la etnohistoria." *Memoria Americana* 20 (1): 61–78.

D'Altroy, Terence. 1987. "Transitions in Power: Centralization of Wanka Political Organization under Inka Rule." *Ethnohistory (Columbus, Ohio)* 34 (1): 78–102. https://doi.org/10.2307/482267.

D'Altroy, Terence. 1992. *Provincial Power in the Inka Empire*. Washington, DC: Smithsonian Institution Press.

D'Altroy, Terence. 2001. "Politics, Resources, and Blood in the Inka Empire." In *Empires: Perspectives from Archaeology and History*, ed. Susan Alcock, Terence D'Altroy, Kathleen Morrison, and Carla Sinopoli, 201–26. Cambridge: Cambridge University Press.

D'Altroy, Terence. 2005. "Remaking the Social Landscape: Colonization in the Inka Empire." In *The Archaeology of Colonial Encounters*, ed. Gil J. Stein, 263–95. Santa Fe: School of American Research Press.

D'Altroy, Terence. 2015. *The Incas*. 2nd ed. Malden, MA: Wiley-Blackwell.

Dávila Brizeño, Diego. [1586] 1965. "Descripción y relación de la provincial de los yauyos." In *Relaciones geográficas de Indias*, vol. 1, ed. Marcos Jiménez de la Espada, 155–65. Madrid: Ediciones Atlas.

de Betanzos, Juan. [1557] 1996. *Narrative of the Incas*. Austin: University of Texas Press.

de Betanzos, Juan. [1557] 2010. *Suma y narración de los Incas*. Lima: Universidad Nacional Mayor de San Marcos.

Dietler, Michael. 2005. "The Archaeology of Colonization and the Colonization of Archaeology: Theoretical Challenges from and Ancient Mediterranean Colonial Encounter." In *The Archaeology of Colonial Encounters: Comparative Perspectives*, ed. Gil Stein, 33–68. Santa Fe, NM: SAR Press.

Dillehay, Tom D. 1976. "Competition and Cooperation in a Prehispanic Multi-Ethnic System in the Central Andes." PhD dissertation. University of Chicago.

Dillehay, Tom D. 1977. "Tawantinsuyu Integration of the Chillón Valley, Perú: A Case of Inca Geo-Political Mastery." *Journal of Field Archaeology* 4:397–405.

Dundes, Alan, ed. 1984. *Sacred Narratives: Readings in the Theory of Myth*. Berkeley: University of California Press.

Durston, Alan. 2007. "Notes on Authorship of the Huarochirí Manuscript." *Colonial Latin American Review* 16 (2): 227–41. https://doi.org/10.1080/10609160701644516.

Duviols, Pierre. 1986. *Cultura andina y represión: Procesos y visitas de idolatrías y hechicerías Cajatambo, Siglo XVII*. Cuzco: Centro de Estudios Rurales Andinos Bartolomé de Las Casas.

Duviols, Pierre. 1973. "Huari y llacuaz: Agricultores y pastores, un dualismo pre-hispánico de oposición y complementariedad." *Revista del Museo Nacional (Lima)* 39:153–91.

Eeckhout, Peter. 2008. "El santuario del Oráculo de Pachacamac y los peregrinajes a larga escala en los Andes prehispánicos." In *Adivinación y oráculos en el mundo andino antiguo*, ed. Marco Curatola Petrocchi and Mariusz Ziółkowski, 161–80. Lima: Fondo Editorial, Pontificia Universidad Católica del Perú/Instituto Francés de Estudios Andinos.

Eeckhout, Peter, and Lawrence S. Owens. 2015. "The Impossibility of Death: Introduction to Funerary Practices and Models in the Ancient Andes." In *Funerary Practices and Models in the Ancient Andes: The Return of the Living Dead*, ed. Peter Eeckhout and Lawrence S. Owens, 1–11. New York: Cambridge University Press. https://doi.org/10.1017/CBO9781107444928.002.

Eliade, Mircea. 1954. *The Myth of the Eternal Return. Or, Cosmos and History*. Princeton: Princeton University Press.

Eliade, Mircea. 1984. "Cosmogonic Myth and 'Sacred History'." In *Sacred Narratives: Readings in the Theory of Myth*, ed. Alan Dundes, 137–51. Berkeley: University of California Press.

Espinoza Soriano, Waldemar. 1992. "Huarochirí y el estado inca." In Huarochirí: Ocho mil años de historia, vol. 1: 117–94. Lima: Santa Eulalia de Acopaya.

Fabian, Johannes. [1983] 2014. *Time and the Other: How Anthropology Makes Its Object*. New York: Columbia University Press. https://doi.org/10.7312/fabi16926.

Farriss, Nancy. 1995. "Remembering the Future, Anticipating the Past: History, Time, and Cosmology among the Maya of Yucatan." In *Time: Histories and Ethnologies*, ed. Diane Owen Hughes and Thomas R. Trautman, 107–38. Ann Arbor: University of Michigan Press.

Feltham, Jane. 1983. "The Lurín Valley, Peru, AD 1000–1532." PhD dissertation. University of London.

Feltham, Jane. 1984. "The Lurín Valley Project: Some Results for the Late Intermediate Period and the Late Horizon." In *Current Archaeological Projects in the Central Andes: Some Approaches and Results*, ed. Ann Kendal, 45–72. BAR International Series 210. Oxford: British Archaeological Reports.

Feltham, Jane. 2005. "Yungas and Yauyos: The Interface between Archaeology and Ethnohistory as Seen from the Lurín Valley." In *Wars and Conflicts in Prehispanic Mesoamerica and the Andes*, ed. Peter Eeckhout and Geneviève Le Fort, 128–45. BAR International Series 1385. Oxford: British Archaeological Reports.

Feltham, Jane. 2009. "La arqueología de sisicaya." In *La revisita de Sisicaya, 1588: Huarochirí veinte años antes de dioses y hombres*, ed. Frank Salomon, Jane Feltham, and Sue Grosboll, 57–101. Lima: Fondo Editorial de la Pontificia Universidad Católica del Perú.

Gose, Peter. 1996. "The Past Is a Lower Moiety: Diarchy, History, and Divine Kingship in the Inka Empire." *History and Anthropology* 9 (4): 383–414. https://doi.org/10.1080/02757206.1996.9960887.

Gose, Peter. 2008. *Invaders as Ancestors: On the Intercultural Making and Unmaking of Spanish Colonialism in the Andes*. Toronto: University of Toronto Press.

Guaman Poma de Ayala, Felipe. [1613] 1980. *Primer nueva corónica y buen gobierno*. Mexico City: Siglo Veintiuno Editores.

Hassig, Ross. 2001. *Time, History, and Belief in Aztec and Colonial Mexico*. Austin: University of Texas Press.

Heggarty, Paul, and David Beresford-Jones. 2009. "Not the Incas? Weaving Archaeology and Language into a Single New Prehistory." *British Academy Review* 12:11–15.

Hill, Jonathon D. 1988. *Rethinking History and Myth: Indigenous South American Perspectives on the Past*. Urbana: University of Illinois Press.

Hirsch, Eric, and Charles Stewart, eds. 2005. *Ethnographies of Historicity, Special Issue of History and Anthropology* 16 (3).

Hughes, Diane Owen, and Thomas R. Trautmann, eds. 1995. *Time: Histories and Ethnologies*. Ann Arbor: University of Michigan Press.

Inomata, Takeshi, and Lawrence A. Coben. 2006. *Archaeology of Performance: Theaters of Power, Community, and Politics*. Lanham, MD: AltaMira Press.

Instituto Nacional de Cultura (INC). 2008. *Proyecto de inventario y registro del patrimonio cultural arqueológico de la nación: Programa Qhapaq Ñan. Macro región centro*. Lima: Instituto Nacional de Cultura, Lima.

Julien, Catherine J. 2000. *Reading Inca History*. Iowa City: University of Iowa Press.

Kauffmann Doig, Federico. 1978. *Manual de arqueología peruana*. Lima: Piesa.

Keane, Webb. 1997. "Religious Language." *Annual Review of Anthropology* 26 (1): 47–71. https://doi.org/10.1146/annurev.anthro.26.1.47.

Keane, Webb. 2005. "Signs Are Not the Garb of Meaning: On the Social Analysis of Material Things." In *Materiality*, ed. Daniel Miller, 182–205. Durham, NC: Duke University Press. https://doi.org/10.1215/9780822386711-008.

Kolata, Alan L. 2013. *Ancient Inca*. Cambridge: Cambridge University Press.

Kosiba, Steven B. 2015a. "Of Blood and Soil: Tombs, Wak'as, and the Naturalization of Social Difference in the Inka Heartland." In *The Archaeology of Wak'as: Explorations of the Sacred in the Pre-Columbian Andes*, ed. Tamara L. Bray, 167–212. Boulder: University Press of Colorado. https://doi.org/10.5876/9781607323181.c006.

Kosiba, Steven B. 2015b. "Tracing the Inca Past: Ritual Movement and Social Memory in the Inca Imperial Capital. Navigating the Inca Landscape. Roads and Ritual Routes in Ancient Cuzco." In *Tribus: Jarbuch des Linden-Museums, special edition of Perspectives on the Inca*, ed. Monica Barnes, Inés de Castro, Javier Flores Espinoza, Doris Kurella, Karoline Noack, 178–205. Germany: Stuttgart.

Krech, Shepard. 2006. "Bringing Linear Time Back In." *Ethnohistory (Columbus, Ohio)* 53 (3): 567–593. https://doi.org/10.1215/00141801-2006-005.

Lau, George. 2004. "Object of Contention: An Examination of Recuay-Moche Combat Imagery." *Cambridge Archaeological Journal* 14 (2): 163–84. https://doi.org/10.1017/S0959774304000113.

Lévi-Strauss, Claude. 1963. *Structural Anthropology*. New York: Basic Books.

Lévi-Strauss, Claude. 1966. *The Savage Mind*. Chicago: University of Chicago Press.

Lucas, Gavin. 2005. *The Archaeology of Time*. London: Routledge.

MacCormack, Sabine. 1991. *Religion in the Andes: Vision and Imagination in Early Colonial Peru*. Princeton: Princeton University Press.

MacCormack, Sabine. 1998. "Time, Space, and Ritual Action: The Inka and Christian Calendars in Early Colonial Peru." In *Native Traditions in the Postconquest World*, ed. Elizabeth Hill Boone and Tom Cummins, 295–343. Washington, DC: Dumbarton Oaks.

MacCormack, Sabine. 2007. *On the Wings of Time: Rome, the Incas, Spain, and Peru*. Princeton, NJ: Princeton University Press.

Makowski, Krzysztof. 2015. "Pachacamac: Old Wak'a or Inka Syncretic Deity? Imperial Transformation of the Sacred Landscape in the Lower Ychsma (Lurín) Valley." In *The Archaeology of Wak'as: Explorations of the Sacred in the Pre-Columbian Andes*, ed. Tamara L. Bray, 127–66. Boulder: University Press of Colorado. https://doi.org/10.5876/9781607323181.c005.

Mannheim, Bruce, and Guillermo Salas Carreño. 2015. "Wak'as: Entifications of the Andean Sacred." In *The Archaeology of Wak'as: Explorations of the Sacred in the Pre-Columbian Andes*, ed. Tamara L. Bray, 47–72. Boulder: University Press of Colorado.

Markham, Clements R. 1872. *Reports on the Discovery of Peru*. London: Hakluyt Society.

Markham, Clements. 1906. "The Megalithic Age in Peru." Stuttgart. *Internationaler Amerikanisten-Kongress* 1904 (part 2): 521–29.

Markham, Clements R. [1873] 1964. *Narratives of the Rites and Laws of the Yncas*. New York: Burt Franklin.

Megged, Amos. 2010. *Social Memory in Ancient and Colonial Mesoamerica*. Cambridge: Cambridge University Press.

Novick, Peter. 1988. *That Noble Dream: The "Objectivity Question" and the American Historical Profession*. Cambridge: Cambridge University Press. https://doi.org/10.1017/CBO9780511816345.

Ossio, Juan M. 2008. *En busca del orden perdido: Le idea de la historia en Felipe Guaman Poma de Ayala*. Lima, Perú: Fondo Editorial de la Pontificia Universidad Católica del Perú.

Palmié, Stephan. 2002. *Wizards and Scientists: Explorations in Afro-Cuban Modernity and Tradition*. Durham, NC: Duke University Press. https://doi.org/10.1215/9780822383642.

Palmié, Stephan. 2014. "Historicist Knowledge and Its Conditions of Possibility." In *The Social Lives of Spirits*, ed. Ruy Blanes and Diana Espírito Santo, 218–39. Chicago: University of Chicago Press.

Patrocinios, Pedro, and Liliana Tapia. 2002. "Proyecto: Determinación de sectores y tipología de arquitectura funeraria en Llaquistampu (San Damián)." Technical report archived with Ministry of Culture, Lima, Peru.

Pease, G. Y. Franklin. 1997. "Los cronistas y la escritura de la historia incaica." In *Arqueología, antropología, e historia: Homenaje a María Rostworowski*, ed. Rafael Varón Gabai and Javier Flores Espinoza, 115–26. Lima: Instituto de Estudios Peruanos: Banco Central de Reserva del Perú.

Pease, G. Y. Franklin. 2010. *Las crónicas y los Andes*, 2nd ed. Lima: Fondo de Cultura Económica.

Polia Meconi, Mario, ed. 1999. *La cosmovisión religiosa andina en los documentos inéditos del Archivo Romano de la Compañía de Jesús, 1581–1752*. Lima: Pontificia Universidad Católica del Perú.

Prescott, William H. [1847] 2002. *History of the Conquest of Peru*. New York: Phoenix Press.
Quilter, Jeffrey. 2014. *The Ancient Central Andes*. New York: Routledge.
Rappaport, Joanne. 1994. *Cumbe Reborn: An Andean Ethnography of History*. Chicago: University of Chicago Press.
Rappaport, Joanne. 1998. *The Politics of Memory: Native Historical Interpretation in the Colombian Andes*. Durham, NC: Duke University Press.
Rappaport, Joanne, and Tom Cummins. 2012. *Beyond the Lettered City: Indigenous Literacies in the Andes*. Durham, NC: Duke University Press.
Rostworowski, María. 1988. *Conflicts over Coca Fields in XVIth-Century Peru*. Ann Arbor: University of Michigan, Museum of Anthropology.
Rostworowski, María. 1978] 2002a. "Señoríos indígenas de Lima y Canta." In *Pachacamac: Obras Completas II*, ed. María Rostworowski, 193–373. Lima: Instituto de Estudios Peruanos.
Rostworowski, María. [1999] 2002b. "Estudio introductorio al informe de Rodrigo Cantos de Andrade." In *Pachacamac: Obras completas II*, ed. Maria Rostworowski, 173–87. Lima: Instituto de Estudios Peruanos.
Rostworowski, María. [1972] 2004. "Las etnías del valle del Chillón." In *Costa Peruana Prehispánica: Obras completas III*, ed. María Rostworowski, 27–81. Lima: Instituto de Estudios Peruanos.
Rowe, John H. 1945. "Absolute Chronology in the Andean Area." *American Antiquity* 10 (3): 265–84. https://doi.org/10.2307/275130.
Rowe, John H. 1967. "Form and Meaning in Chavin Art." In *Peruvian Archaeology: Selected Readings*, ed. John H. Rowe and Dorothy Menzel, 72–103. Palo Alto, CA: Peek Publications.
Rowe, John H. 1998. "Max Uhle y la idea del tiempo en la arqueología americana." In *Max Uhle y el Perú antiguo*, ed. Peter Kaulicke, 5–24. Lima: PUCP.
Rumsey, Alan, and James F. Weiner, eds. 2001. *Emplaced Myth: Space, Narrative, and Knowledge in Aboriginal Australia and Papua New Guinea*. Honolulu: University of Hawaii Press.
Sahlins, Marshall. 1985. *Islands of History*. Chicago: University of Chicago Press.
Sahlins, Marshall. 2004. *Apologies to Thucydides: Understanding History as Culture and Vice Versa*. Chicago: University of Chicago Press.
Salomon, Frank. 1982. "Chronicles of the Impossible: Notes on Three Peruvian Indigenous Historians." In *From Oral to Written Expression: Native Andean Chronicles of the Early Colonial Period*, ed. Rolena Adorno, 9–39. Syracuse: Syracuse University.
Salomon, Frank. 1991. "Introductory Essay: The Huarochirí Manuscript." In *The Huarochirí Manuscript: A Testament of Ancient and Colonial Andean Religion*, ed. Frank Salomon and George L. Urioste, trans. Frank Salomon and George L. Urioste, 1–38. Austin: University of Texas Press.

Salomon, Frank. 1995. "'The Beautiful Grandparents': Andean Ancestor Shrines and Mortuary Ritual as Seen Through Colonial Records." In *Tombs for the Living: Andean Mortuary Practices*, ed. Tom D. Dillehay,315–53. Washington, DC: Dumbarton Oaks.

Salomon, Frank. 1998. "How the Huacas Were: The Language of Substance and Transformation in the Huarochirí Quechua Manuscript." *RES* 33:7–17.

Salomon, Frank. 1999. "Testimonies: The Making and Reading of Native South American Historical Sources." In *The Cambridge History of the Native Peoples of the Americas, South America*, edited by Frank Salomon and Stuart Schwartz, vol. 3: 19–95. Cambridge: Cambridge University Press. https://doi.org/10.1017/CHOL97805 21630757.003.

Salomon, Frank. 2002. "Unethnic Ethnohistory: On Peruvian Peasant Historiography and Ideas of Autochthony." *Ethnohistory* 49 (3): 475–506.

Salomon, Frank. 2004. *The Cord Keepers: Khipus and Cultural Life in a Peruvian Village*. Durham, NC: Duke University Press. https://doi.org/10.1215/9780822386179.

Salomon, Frank, and Mercedes Niño-Murcia. 2011. *The Lettered Mountain: A Peruvian Village's Way with Writing*. Durham, NC: Duke University Press. https://doi.org/10 .1215/9780822394341.

Salomon, Frank, and George L. Urioste, eds. and trans. 1991. *The Cord Keepers: Khipus and Cultural Life in a Peruvian Village*. Austin: University of Texas Press.

Sarmiento de Gamboa, Pedro. [1572] 2001. Historia de los incas. Madrid: Miraguano Ediciones.

Sarmiento de Gamboa, Pedro. [1572]2007. History of the Incas, trans. and ed. Brian Bauer and Vania Smith. Austin: University of Texas Press.

Silverblatt, Irene. 1988. "Imperial Dilemmas, the Politics of Kinship, and Inca Reconstructions of History." *Comparative Studies in Society and History* 30 (1): 83–102. https://doi.org/10.1017/S001041750001505X.

Spalding, Karen. 1967. "Indian Rural Society in Colonial Peru: the Example of Huarochirí." PhD dissertation. University of California, Berkeley.

Spalding, Karen. 1974. *De indio a campesino: Cambios en la estructura social del Perú colonial*. Lima: Instituto de Estudios Andinos.

Spalding, Karen. 1984. *Huarochirí: An Andean Society under Inca and Spanish Rule*. Stanford, CA: Stanford University Press.

Spalding, Karen. 2012. *El diario histórico de Sebastián Franco de Melo: El levantamiento de Huarochirí de 1750*. Lima: Centro Peruano de Estudios Culturales.

Stroud, Ronald S. 1987. "'Wie es eigentlich gewesen' and Thucydides." *Hermes* 115, Bd. H. 3 (3rd Qtr.): 379–82.

Swenson, Edward R. 2003. "Cities of Violence: Sacrifice, Power and Urbanization in the Andes." *Journal of Social Archaeology* 3 (2): 256–96. https://doi.org/10.1177/1469605303 003002006.

Taylor, Gerald. 1999. *Ritos y tradiciones de Huarochirí*. Lima: Instituto Francés de Estudios Andinos.

Thatcher, John, and Nicholas Hellmuth. 1968–71. Field notes and "A Seriation of the Ceramics of Huarochirí." In author's possession, courtesy of Dr. Thomas Patterson.

Topic, John R. 1998. "Ethnogenesis in Huamachuco." *Andean Past* 5:109–27.

Topic, John R., Theresa Lange Topic, and Alfredo Melly Cava. 2002. "Catequil: The Archaeology, Ethnohistory and Ethnography of a Major Provincial Huaca." In *Andean Archaeology I: Variations in Sociopolitical Organization*, ed. William H. Isbell and Helaine Silverman, 303–36. New York: Kluwer Academic–Plenum Press. https://doi.org/10.1007/978-1-4615-0639-3_11.

Trouillot, Michel-Rolph. 1995. *Silencing the Past: Power and the Production of History*. Boston: Beacon Press.

Trouillot, Michel-Rolph. 2003. *Global Transformations: Anthropology and the Modern World*. New York: Palgrave. https://doi.org/10.1007/978-1-137-04144-9.

Uhle, Max. 1902. "Types of Culture in Peru." *American Anthropologist* 4 (4): 753–59. https://doi.org/10.1525/aa.1902.4.4.02a00080.

Uhle, Max. [1910] 1970. "Las civilizaciones primitivas en los alrededores de Lima." In *100 años de arqueología en el Perú*, ed. Rogger Ravines, 379–391. Lima: Instituto de Estudios Peruanos.

Uhle, Max. 1912. *Los orígenes de los Incas*. Buenos Aires: Coni Hermanos.

Uhle, Max. [1903] 2003. *Pachacamac: Informe de la expedición peruana William Pepper de 1896*. Lima: Universidad Nacional Mayor de San Marcos.

Urton, Gary. 1984. "Chuta: El espacio de la práctica social en Pacariqtambo, Perú." *Revista Andina (Cusco)* 2 (1): 7–56.

Urton, Gary. 1988. "La arquitectura pública como texto social: La historia de un muro de adobe en Pacariqtambo, Perú (1915–1985)." *Revista Andina (Cusco)* 6 (1): 225–63.

Urton, Gary. 1990. *The History of a Myth: Pacariqtambo and the Origin of the Inkas*. Austin: University of Texas Press.

Urton, Gary. 2002. "Recording Signs in Narrative-Accounting *Khipus*." In *Narrative Threads: Accounting and Recounting in Andean Khipu*, ed. J. Quilter and G. Urton, 171–96. Austin: University of Texas Press.

Urton, Gary. 2003. *Signs of the Inka Khipu: Binary Coding in the Andean Knotted-String Records*. Austin: University of Texas Press.

Urton, Gary. 2012. "The Herder-Cultivator Relationship as a Paradigm for Archaeological Origins, Linguistic Dispersals, and the Evolution of Record-Keeping in the Andes." *Proceedings of the British Academy* 173:321–43.

Urton, Gary. 2015. "The State of Strings: Khipu Administration in the Inka Empire." In *The Inka Empire: A Multidisciplinary Approach*, ed. Izumi Shimada, 149–64. Austin: University of Texas Press.

Villar Córdova, Pedro. [1935] 1982. *Las culturas prehispánicas del departamento de Lima.* Lima: Ediciones Atusparia.

von Ranke, Leopold. 1824. *Geschichte der romanischen und germanischen Völker von 1494 bis 1514*. Wiesbaden: E. Vollmer.

Wernke, Steven A. 2013. *Negotiated Settlements: Andean Communities and Landscapes under Inka and Spanish Colonialism*. Gainesville: University Press of Florida. https://doi.org/10.5744/florida/9780813042497.001.0001.

Whitehead, Neil L. 2003. "Introduction." In *Histories and Historicities in Amazonia*, ed. Neil L. Whitehead, vii–xx. Lincoln: University of Nebraska Press.

Zuidema, R. Tom. 1964. *The Ceque System of Cuzco: The Social Organization of the Capital of the Inca*. Leiden: Brill.

Zuidema, R. Tom. 1985. "The Lion in the City: Royal Symbols of Transition in Cuzco." In *Animal Myths and Metaphors in South America*, ed. Gary Urton, 183–250. Austin: University of Texas Press.

Zuidema, R. Tom. 1990. *Inca Civilization in Cuzco*. Austin: University of Texas Press.

SIX

Topologies of Time and History in Jequetepeque, Peru

EDWARD SWENSON

INTRODUCTION

Philosophers have long recognized that space and time are difficult to conceptualize apart. Time's passage and ineffable qualities are described in spatial terms, whereas space can only be inhabited and experienced in temporally embodied movements (Bergson 2001; Heidegger 1985; Merleau-Ponty 1962: 411–15; Munn 1992). By extension, technologies of quantitative time (charting the phases of the moon, counting the ticks of a clock's hand, etc.) measure the indivisible flow of experience into static, spatial units. In stressing Spinoza's view that time is fundamentally about duration, Henri Bergson (2001: 108) critiques the reduction of time to space, arguing that "as soon as we try to measure it, we unwittingly replace it by space" (see also Harootunian 2005: 25, 38–40). Nevertheless, an analysis of how space has been variably constructed to fix time in place can tell us a great deal about culturally distinct ideologies of change, continuity, and history. However, in devising chronologies and explaining historical transformation, Andean archaeologists have largely ignored indigenous temporalities and cultural variations in the "spatial mediations" of time (see introduction [chapter

DOI: 10.5876/9781607326427.c006

1], this volume). Distinct materialities/spatialities could very well have structured unique temporalities, and archaeologists have much to gain in developing new methods to interpret the literal *constructions* of alternate temporal worlds (see Roddick this volume [chapter 3]; Swenson 2015).

Archaeologists hold an especially privileged place to interpret the periodicities of routinized practices including seasonal occupation of sites or recurrent depositions in formal middens—the accumulated "marks of repetition" visible in the material record (Olivier 2004: 210; thus contra Connerton 1989: 102; see González-Ruibal 2014: 38). Tim Ingold's (2000) call that archaeological research should be designed to investigate the "temporalities of the landscape" stresses this point precisely. Ultimately, transformations in the "sedimentation" of habitual practices in the archaeological record can be fruitfully compared to archaeological signatures of commemorative rites that were complicit in the creation of explicit mythohistories and social memories (Connerton 1989; Gosden 1994: 35; Mills and Walker 2008; Olivier 2004; 2011: 10; Swenson 2017; Van Dyke and Alcock 2003). As discussed in the introduction to this volume (chapter 1), the temporal and historical are difficult to tease apart, and rituals of place making and formal time-reckoning have played a decisive role in temporalizing social behavior and in setting the parameters for political action. By extension, architectural planning often relates to the "management of change," explaining the consonance of building with rituals of time and social memory (Hodder and Pels 2010; Olivier 2004; Swenson 2011). Therefore, archaeologists are well equipped to examine how the temporalities of quotidian life complemented or contradicted the political production of space that intended to naturalize people's placement in history (see introduction, Swenson 2017). In the end, alterations in the periodicities of practice offer a stronger basis from which to interpret the range and depth of sociopolitical transformations than do changes in monumental architecture alone (Barrett 2004; Roddick 2013).

The main objective of this chapter is to demonstrate that shifts in political and religious ideologies in the Jequetepeque region of northern Peru were underwritten by transformations in social constructions of place, landscape, and ultimately time. The chapter will highlight considerable diversity in both temporal routines and their institutional regulation (Trouillot's "historicity II"; see Introduction [chapter 1], Sayre [chapter 2], and Chase [chapter 5], all in this volume) in one particular region of the North Coast of Peru. The analysis will focus principally on a comparison of the Late Formative site of Jatanca (500–100 BC) with the neighboring Late Moche ceremonial center of Huaca Colorada (AD 650–850), both located in the southern bank of the valley. However, settlement and landscape data from the Middle Moche Period (AD 400–650) will also be briefly considered in the analysis. The distinctive temporal orders (topologies of space-time) in Jequetepeque are reflected in varied techniques of architectural construction and in the contrasting use of public and residential space. Moreover, settlement data from rural sites throughout the valley and from the domestic zones of larger centers point to significant shifts in the temporal rhythms of both quotidian practices and episodic ritual events, especially between the Middle and Late Moche Periods. Surprisingly, the data suggest that despite pronounced differences in "affective time" and ideologies of history between Jatanca and Huaca Colorada, Late Moche temporalities more closely resembled those of the Late Formative Period than the intervening Middle Moche phase. Ultimately, the pronounced alterations in ritual calendars and the timing of everyday life in the Late Moche Period appear to have been the result of changing religious observances and related shifts in economic and political practices.

TOPOLOGIES OF TIME AND HISTORY

In the introduction to a recent edited volume (*Chrono-topologies: Hybrid Spatialities and Multiple Temporalities*), Leslie Kavanaugh (2010a) remarks that the unification of space and time as ontological categories represents one of the greatest revolutions in twentieth-century theoretical physics, as exemplified in particular by Einstein's famed theory of relativity. Of course, philosophers have long debated the nature of this fundamental binary of existence, and the concept of "topology" has recently been employed to capture the complex and shifting intersections of space and time in matter, persons, and environments (Boyer 2010; Deleuze and Guattari 1987; Serres and Latour 1995; Witmore 2007). Archaeologists also increasingly recognize the polychronic constitution of landscapes as composed of temporally distinct but variably enfolded and potentially contiguous peoples, places, and things (Witmore 2007: 195). Therefore, the space-time continuum within the domain of historical interpretation can be productively apprehended as a multilayered material medium. This layered field of action refers not to stratigraphic levels of succession but to the coexistence of timescapes comprised of multiple temporal cycles and unique places of memory (Alcock 2002; Bradley 2002; Lucas 2005; Olivier 2004; Koselleck 2004; Witmore 2007: 201). If there are simultaneously many different places, then the topological metaphor would suggest the simultaneity of distinct times, something Ernst Bloch (1991) has characterized as "contemporaneous non-contemporaneity" (*Ungleichzeitigkeit des Gleichzeitigen*), "where fragments of the past unexpectedly and suddenly rise to impinge upon the present . . . where past and present are not necessarily successive but simultaneously produced, or coexist as uneven temporalities" (Harootunian 2005: 42, 47). In other words, unique places, structures of practice, and ideologies are characterized by distinct rates of flux and endurance (Lucas 2005: 17). Stressing the material contiguity of space and time, Reinhart Koselleck (2004) further contends that the triad of past-present-future should be replaced by memory-experience-expectation (see also Lefebvre 1991: 37). Memories of different pasts are invoked by ongoing experiences with temporally disparate material worlds in the present, an engagement that sets the parameters for both novel apprehensions of the past and of expected futures to come.

In his survey of archaeological landscapes in the Southern Argolid, Greece, Christopher Witmore (2007: 195) draws from Michel Serres to describe such impingements of the material past upon the present as "percolating time" (*couler*—to flow, but as through a filter; see Serres and Latour 1995: 58). Witmore (2007: 196, 206) argues: "Rather than delineating stable and well-defined temporal distances, a topology [the "science of nearness and rifts"; Serres and Latour 1995: 60] plots the relations, the nodes of interaction and the passages between various pasts." He continues (2007: 196): "While linear time is conventionally oriented by unidirectional historical narrative, "percolating" time may best be synthesized topologically." Witmore holds that topology best describes archaeological reconstructions of history by revealing that the present is composed of "aggregate mixes of multiple material pasts" (Witmore 2007: 196).

Many Andean peoples similarly conceptualized time as physical and materially metamorphic, and ritual performance was commonly predicated on controlling, channeling, or fusing temporally varied life-forces immanent in the landscape (Allen 1988; Abercrombie 1998; see D'Altroy and Wilkinson [chapter 4], Spence Morrow [chapter 7], this volume). In fact, alternative materializations of time shaped historically specific constructions of personhood, community, and place in the Andes and elsewhere in the Americas. As discussed in the introduction (chapter 1), Andean people often did not make a radical break between the past and present, and the Quechua term *pacha* connotes the generative enfolding of time and space. This term also affirmed that authority accrued to those with the power to physically manufacture, accelerate, or turn back time (*wak'as*, elites, demiurges, etc.) (Ossio 2002; Randall 1982). Therefore, in the Andean world, it appears that "time [was] not something undergone but creatively produced" (Gurvitch 1964: 30; see also Kavanaugh 2010b, 104).

Topology thus serves as a useful heuristic because it resonates with certain Andean conceptions of space-time while providing a framework from which to interpret transformations in the political *construction* of both everyday temporal rhythms and more formal discourses on myth and history. A focus on the topological dimensions of landscape further reveals that history and temporality cannot be understood in isolation; both were conditioned by variations in the physical endurance of things, buildings, and environments. Ultimately, the following case study intends to demonstrate that an analysis of past political topologies can provide a more nuanced interpretation of social change than would traditional chronological frameworks (see also Roddick [chapter 3], Sayre [chapter 2], this volume).

It deserves mention that my application of topology as an interpretive device does not imply that diverse Andean peoples thought like Bergson, Serres, or

Deleuze. In fact, an important point of the following comparison is to demonstrate pronounced variation in the *topographic* construction of place and time by elites and other community members. Although aspects of topology in recent philosophical works might compare to the ideologies underwriting the building projects of the Moche site of Huaca Colorada, the inhabitants of the earlier center of Jatanca seemed to have built their landscape as a means to deny or arrest the dynamic enfolding of "multiple material and co-existing pasts." For my purposes, then, describing Jatanca as a topology intends to stress this active, spatial intervention in regulating time and creating history. Moreover, the term "topology" as an etic category allows for an assessment of other modalities of time as engineered in space, including especially embodied or "affective" time. The latter refers to how specific places can afford intense feelings and apprehensions of time's ruptures or durations (see Locke 2010).

DIVERGENT MATERIALIZATIONS OF TIME IN JEQUETEPEQUE

Fixing Time in Place: The Durable Topology of Jatanca

The Jequetepeque Valley of northern Peru, situated roughly 400 kilometers north of Lima, was characterized by intensive irrigation agriculture and relatively high population densities from the Cupisnique Period to the Late Horizon (800 BC–AD 1532). The valley is well known for its numerous and dispersed ceremonial centers, especially dating to the Moche Period (AD 350–800), and at the time of the conquest, Jequetepeque was described as housing one of Peru's most "noteworthy" oracles (Anonymous [1594] 2008; Hyland 2011: 69). In fact, the concentration and variety of ritual architecture dating to different periods in the valley strongly suggest the deep history of its pan-regional religious significance. The most comprehensive archaeological work in the region has concentrated on the Moche (AD 400–850) and later traditions (Lambayeque and Chimú; 850–1430), eras defined by social stratification, economic specialization, and varying degrees of political centralization (Castillo 2010; Cutright 2009; Donnan 2007; Donnan and Cock 1997; Mackey 2009; Swenson 2006, 2007, 2008; Zori and Johnson 2011). The Moche are well known for their vibrant and naturalistic art, elaborate architecture, and sacrificial political theology, and recent work has demonstrated considerable variety in the promulgation of Moche religious values and material culture on the North Coast (Castillo and Uceda Castillo 2008; Chapdelaine 2011; Quilter and Castillo 2010; Quilter and Koons 2012). In fact, the following analysis will demonstrate that despite continuities in certain Moche aesthetic and ideological traditions in Jequetepeque, variation in settlement, in architectural construction, and in the temporalities of practices highlights important sociopolitical transformations within the larger Moche period. Therefore, the study works both within and beyond the standard chronology of the Jequetepeque Valley to arrive at a richer understanding of historical developments in the region. Ultimately, the comparison reveals temporal routines changed significantly in the Moche era, thus exposing the limitations of chronological phases as stable and well-defined units.

Although both Jatanca and Huaca Colorada appear to have anchored a cult related to the worship of the nearby mountain of Cerro Cañoncillo in the Pampa de Mojucape (see Swenson and Warner 2016), the convergence of time, landscape, and architecture differed remarkably between the two centers (figure 6.1). The construction techniques, aesthetic presentation, and rectilinear configuration of the labyrinthine compounds of Jatanca depart dramatically from Huaca Colorada's vertical design and history of constant renovation. Although both centers constituted ritually charged environments, with prominent platforms built in a breathtaking locales, their contrasting architectonics point to differing "constructions" of community, authority, and ultimately time itself.

The dissertation research of John Warner (2010) and our later collaboration demonstrated that Jatanca constituted the largest Late Formative site (500 BC–100 BC) in the Jequetepeque Valley (Swenson 2011, 2015; Swenson and Warner 2016). The ceremonial center occupies an area of three square kilometers and is dominated by extensive compounds that together contain over 200 rooms. (Warner 2010) (figure 6.2). Canals run between and around the precincts, which are surrounded by dense pot sherd scatters, the only visible traces of a once-expansive domestic area built of perishable cane structures.

The five main compounds at Jatanca consist of expansive north plazas enclosed by high walls built of poured adobe (*tapia*, or rammed earth) that conjoin multichambered precincts to the south. Despite discrepancies in size and internal configuration, adherence to established spatial codes dictated the construction and use of the five major ceremonial structures at Jatanca (Swenson 2011: 292–98; Warner 2010: 339–42, 429–34). For instance, the expansive north plazas characterize the raised "Acropolis" to the east and the four other compounds of the monumental sector to the west (figure 6.2). The plazas could have accommodated hundreds of spectators, and these spacious courtyards were anchored at their southern extreme by an elevated stage or proscenium (Swenson 2011, 2015; Swenson and Warner 2016; Warner 2010). This raised terrace separated the plaza from the labyrinth of internal rooms, corridors, and smaller patios that constituted a considerable volume of each complex (figure 6.3). The prominent stages housed two platforms with ramps in mirror opposition, and they clearly formed the ceremonial nerve-center of Jatanca's separate precincts (Swenson 2011: 294; Swenson and Warner 2016; Warner 2010: 455–63). The ramped platforms appear to have staged rites of presentation, exchange, and supervision, and ceremonies conducted on the daises would have been visible to large gatherings in the northern plazas. The duplication of smaller, interior plazas housing miniature stages and opposed ramped platforms within two of the western compounds

FIGURE 6.1. *Map of the southern Jequetepeque Valley illustrating the location of Cerro Cañoncillo, the Late Formative site of Jatanca, and the Moche center of Huaca Colorada.*

suggests that rites of procession related to initiation or pilgrimage were orchestrated within the monumental complexes (Swenson 2011, 2015) (figure 6.3). The dyadic configuration of the platforms may have ritually materialized dualistic energies or cosmic powers. They could also have served as formal spaces for religious leaders or huacas who represented two halves of a united moiety or religious sodality (Swenson 2015: 695; Swenson and Warner 2016: 10; Warner

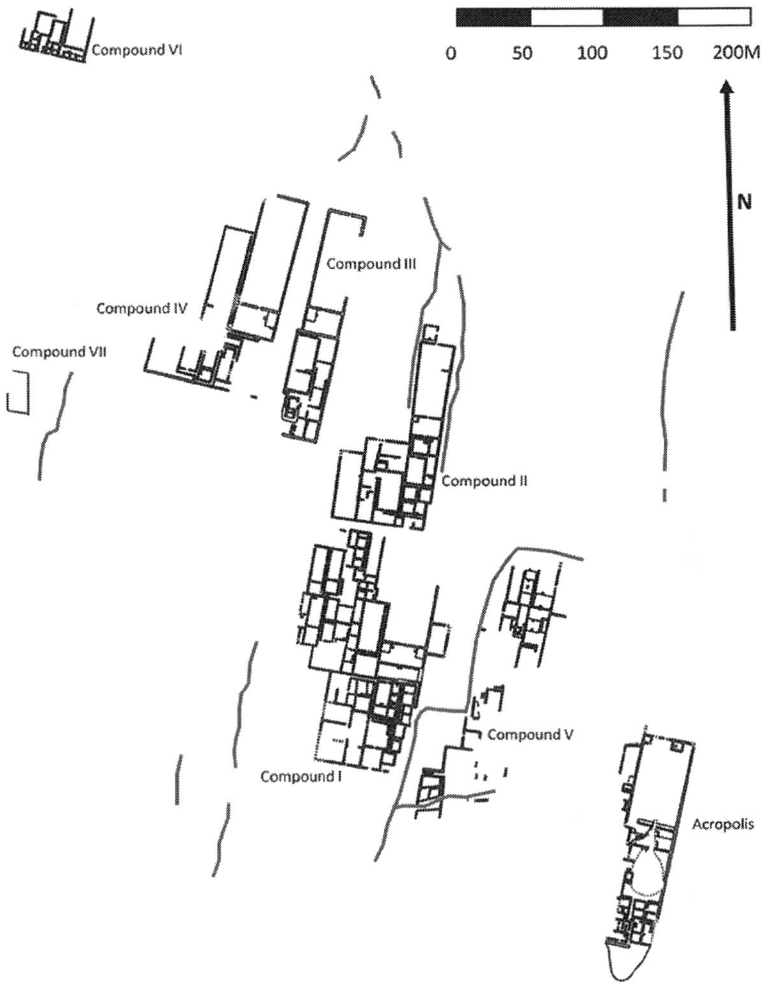

FIGURE 6.2. *Map of the monumental core of Jatanca (adapted from Warner 2010: 118).*

2010: 522–36). The evidence of replicated ceremonial space at Jatanca implies a "competition between stages" and the existence of an ethos of ritual and political pluralism at the site. Nevertheless, this pluralism was obviously grounded in a shared ideology predicated on prescribed modes of ritual action. As argued elsewhere, Jatanca's built environment points to semiautonomous and lineage-based *parcialidades* confederated into a larger political and religious community (Swenson 2011, 2015: 701; Swenson and Warner 2016; Warner 2010).

Six years of excavations have identified a general pattern underlying the organization of ritual performances and social practices at Jatanca. Centrally placed

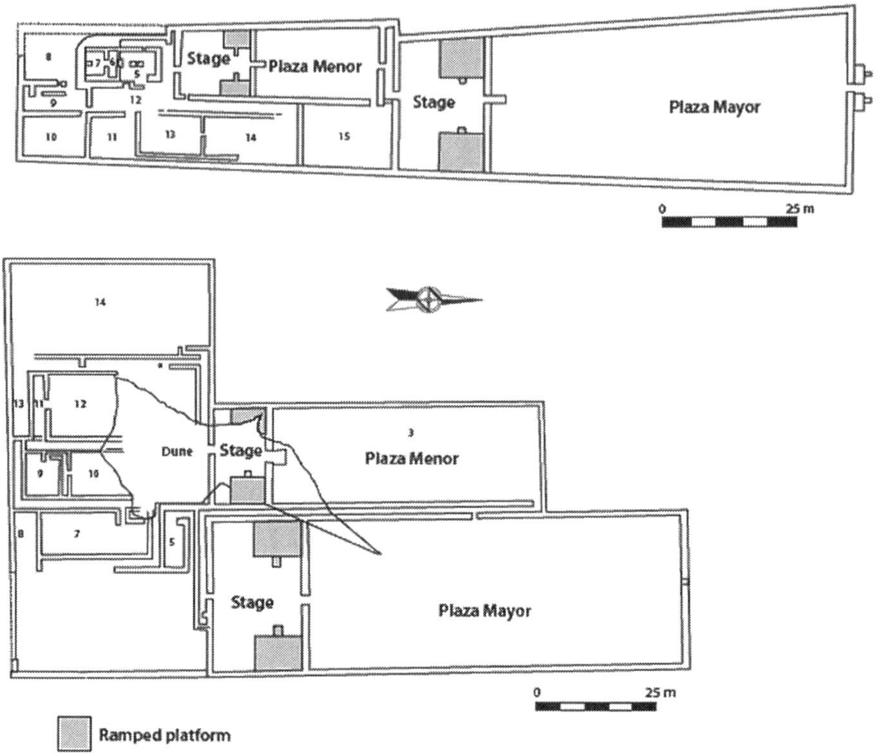

FIGURE 6.3. *Plans of Compounds 3 and 4 at Jatanca. Note the nested plazas and ceremonial proscenia supporting two ramped platforms in mirror opposition (adapted from Warner 2010: 356, 364).*

chambers aligned to the south of the ceremonial stage were associated with formal furnishings, ritual caches, large post emplacements with spondylus bead offerings, and aligned concavities to hold ceramics (Swenson 2011: 298; Warner 2010: 463–64) (figure 6.4). More exclusive rituals were no doubt performed here and these well-constructed precincts may also have served as the residences of ritual specialists or political leaders. In contrast to the more elaborate and centrally aligned series of rooms, peripheral chambers located beyond the main axis of ceremonial architecture are usually characterized by irregular hearths, shallow deposits of garbage, and internal partitions built of cane (Swenson et al. 2008, 2009). Therefore, the side chambers within the compounds appear to have served as the literal "backstage production" area supporting the ritual economy of the settlement (Swenson 2015). However, these spacious quadrilateral chambers were also delineated by solid Tapia walls and formed an integral part of the individual monumental compounds.

The peculiar layout and construction history of Jatanca permit interpretations of how time was experienced within the compounds ("affective time") while providing a foundation from which to approximate official constructions of memory at the ceremonial center. The center's horizontal and labyrinthine configuration indicates that ritual acts were rigidly sequenced in space and time. As discussed above, Jatanca is marked by the intense nesting of ceremonial architecture, and the internal replication of smaller plazas with ramped structures suggests that the monuments were designed for long processions perhaps related to initiation or pilgrimage (Swenson 2011, 2015) (figure 6.3). Therefore, the duplication of iconic ceremonial space and the phased, reiteration of key rites characterized the orchestration of ritual events at Jatanca. The architectural design of the precincts would have enhanced the emotional and psychological effects of encountering secluded huacas, an encounter that may have been at once prolonged and momentous (figure 6.3). For instance, the deepest chamber in Compound 3 was distinguished by a formal platform fronted by a patio containing two rectangular daises or altars (Swenson et al. 2009: 91–99) (figure 6.4). A powerful huaca or official was likely stationed in this exclusive room, a chamber that could only have been accessed by first passing through the two separate but nested plazas with opposed, dual platforms. In a similar manner, one of the deepest and most secluded chambers in the south-central end of the Acropolis was adorned with a beautifully plastered bench associated with fine artifacts including an exquisite spear point (Swenson et al. 2008: 40–36). A series of private rooms in Compound 1, similarly aligned with the ceremonial architecture of the plaza, were also characterized by an elaborate bench complex and a cache of fine-line ceramics (Swenson et al. 2009: 45–61; Warner 2010) (figure 6.4).

Therefore, the evidence suggests that Jatanca's compounds afforded an experience of protraction through time, an extended and dilatory temporality; for the select few that could pass through its imposing labyrinthine walls, time felt drawn out and decelerated, an alternative temporality removed from the regular rhythms of daily life. Fasting and other rites of self-abnegation, endured over possibly extended periods, may have permitted initiates to enter the deeper chambers and confer with powerful huacas or political leaders, as was documented for later Andean religious complexes including the famed pilgrimage center of Pachacamac (Eeckhout 1999; Patterson 1985; Swenson 2015: 701). The replicated monumental architecture of Pachacamac also consisted of nested plazas, and it could take months for pilgrims to pass from one courtyard to another as they purified themselves physically and spiritually for their anticipated audience with the great oracle. Time was not simply retarded within the ritual field, but its duration and feel appear to have been topologically inseparable from one's place within the disorienting maze of the compounds. In other words, the center's architecture engendered a distinct if heightened sense of anticipation.

FIGURE 6.4. *Bench-like structure and a cache of fine-line ceramics located in a deep and secluded sector of Compound 1 at Jatanca.*

In fact, the movement through Jatanca's labyrinthine corridors and concatenation of chambers likely induced an affective time (see Locke 2010) of suspense, anxiety, and expectation. This apprehension of time was likely intensified by the alternation of periods of prolonged, even tedious, waiting with peak moments of ritual theater involving dance, music, and memorable encounters with ritually charged peoples and things.

Interestingly, the materiality of Jatanca's horizontally configured architecture also points to a temporal ideology of permanence, transcendence, and prolonged duration. Construction of the solid tapia perimeters demanded the coordinated and synchronized labor of a large number of workers to construct the oversized wall molds, mix the clay and water, and pour the vast quantities of liquid adobe into the wooden frames (Campana 2000: 130–31; Swenson 2015; Warner

2010: 338–39). The canals passing through the site and dating to the Formative Period no doubt provided the water for the construction of the tapia perimeters (Warner 2010). Although the building of the walls conforms to McFayden's notion of "quick architecture" (McFadyen 2006), a process that involves the rapid and coordinated movement of laborers, the entire construction event must have been a time-consuming and protracted affair (Swenson 2015: 699). In fact, building large tapia walls segments leaves little room for improvisation, and a preconceived architectonic plan, entailing precise scheduling and long-term planning, underwrote architectural construction at Jatanca. The solid tapia walls are remarkable for their preservation and durability. Over 2,000 years old, they materialized a perduring temporality of hard stone as opposed to degradable adobe (Swenson 2015: 700). Indeed, the compounds were built to last, and their fixity and "powerful persistence" attest to their "negentropic" spatial and temporal effects (see Witmore 2007: 217).

Significantly, architectural remodeling was rare at Jatanca. Chambers, corridors, and daises were fastidiously curated spaces, and buildings were rarely demolished or rebuilt. In most precincts, one massive tapia floor was laid down between the walls, and when reflooring episodes occurred (rarely more than three to five floors), they almost always respected and maintained the original wall layouts of the precincts in question (Swenson 2015: 700; Swenson and Warner 2016: 23). The painstaking conservation of the buildings thus appears to have denied or resisted a conception of change based on cycles of growth, decay, death, and regeneration. The few examples of remodeling documented at the site are equally significant in that the new constructions deviated completely from the original plan, unlike the multiple renovations at Huaca Colorada (Swenson et al. 2008: 68–69; 2009: 103–4; see below). The reconstruction events at Jatanca razed earlier walls (they were not carefully buried as at Huaca Colorada), and these renovations seem to have intended to standardize the replication of ceremonial space between the different compounds (Warner 2010: 385–97, 429–34). Maintaining the integrity and physical constancy of the monument is further reflected in the lack of burials at the settlement. The people of Jatanca clearly interred their dead elsewhere (most likely at the mortuary site of Puémape; see below), and the few humans remains discovered at the center appear to have been intrusive and anomalous (Swenson 2015: 700; see below). Taboos may even have been in place that prohibited the regular interment of the dead within the confines of the center. In contrast to the later funerary sties of San José de Moro and Pacatnamú, Jatanca did not experience the ongoing growth of new cemeteries and the accumulation of interred bodies.

The great precincts were monuments to permanence, an eternalized time that endured beyond the life cycles and practical rhythms of the communities who visited and maintained the ceremonial compounds. The high walls would have also blocked vistas of the surrounding and ephemeral domestic areas, forming a unique topology that compartmentalized distinctive time frames. In fact, the expansive residential areas surrounding the complexes lack permanent constructions and appear to have been occupied episodically by encampments of visitors, possibly pilgrims. The fleeting temporalities of the everyday were thus juxtaposed with the transcendent, eternal time of the ceremonial enclosures. As possible shrines for oracular huacas, it seems that the compounds of Jatanca materialized an ideology of timelessness and intransience. In other words, the center seems to have created a safe haven of constancy and ontological security where diviners prognosticated and ultimately reduced the uncertainties of the future.

In a similar manner, the compounds of the center appear to have celebrated an ethos of commemoration and conservation (Swenson and Warner 2016). The primordial acts of founding huacas possibly provided a mythic (and architectural) charter for fixed political relationships that were idealized as stable and predictably cyclical (Assmann 2006: 38). In fact, the rigid symmetry and dualistic layout of Jatanca's architecture created a topology of balance, complementarity, and continuity (Swenson and Warner 2016; Warner 2010). The platforms positioned in mirror opposition seem to conform closely to Andean notions of history founded on the dialectic of conflict and complementarity (Salomon and Urioste 1991; see chapter 1). The stages with diametrically opposed platforms encapsulate a liminal tension between union and separation, and they clearly functioned as effective thresholds delineating discrete architectural sectors and moments in time. Jatanca appears then to have materialized a space-time of reversible or alternating dualisms (Isbell 1982: 354; Salomon 2002; see also Gose 1996 and introduction [chapter 1], this volume). The architectural design of the site suggests that the ritual cycling between complementary opposites, possibly materialized in the pairing of lineage groups, chiefs, architectural complexes, celestial bodies, and genealogically related huacas, as well as the past and present, provided a model of succession, political order, and the anticipated cycling of time. The oracular game of *pichca* in the modern-day community of Pacota was realized through the spatialization of such an alternating dualism, and the ritual "allowed for the regeneration of political order in face of climactic and historical imponderables" (Salomon 2002: 2; my translation). Indeed, the intensely dualistic configuration of the Pacota ceremonial *cancha* offers a fascinating point of comparison with Jatanca's rigidly dichotomized ritual complexes (Salomon 2002: fig. 3). The eastern platforms at Jatanca tend to be consistently if slightly larger than their western counterparts (Swenson 2015: 695; Warner 2010: 522–36), and they may have been associated with the rising sun, an upper moiety, an ascendant huaca, and the present/future. By extension, the western platform may have related to the sun's setting, a lower moiety, a subsidiary huaca, the past, and possibly the female life principle. Among the Eastern Timbira of the Amazon, the passage of time was also dramatized in terms of the alternation between the eastern and western halves of the plaza, a rite that played a fundamental role in introducing initiates to the mythic history of the society (Nimuendajú 1946: 89–92).

Jatanca's architecture was clearly gridded to the cardinal directions, and the movement from the northern plazas to the southern inner sanctums of the compounds probably held temporal, astronomical, and religious significance. Exiting the compounds could also have entailed the return from a mythic space-time to the world of everyday temporal routines. It is also worth considering that multiple compounds (or plaza groups within individual precincts) were active only at given days, seasons, or calendar rounds, as characterized the Inca *ceque* system (see introduction [chapter 1], this volume). In the end, the replicated compounds of Jatanca suggest that time as movement and change was achieved through the synchronic alternations of peoples, places, events and celestial bodies, a model that obviously differs from diachronic philosophies of history.

This particular historical ideology founded on an eternal time of juxtaposition and oscillation may have guided the coming and goings of the celebrants and the establishment of the temporary residences of Jatanca. As mentioned, the ephemeral and shifting nature of habitation at the site contrasts with the permanence of the ceremonial architecture. The few known Formative Period hamlets in Jequetepeque are small and dispersed, and they do not appear to have been occupied year round (Dillehay et al. 2009). In fact Warner's (2010: 613) analysis of the distribution of Late Formative settlements in Jequetepeque indicates that Jatanca not only served as the most important ceremonial center during the Late Formative, but that it formed an important node of exchange and distribution for the valley as a whole. The ephemeral domestic areas surrounding the main monumental constructions point to the periodic but large gathering of peoples who originated from smaller satellite settlements in order to partake in festivals, fairs, and religious ceremonies. These gatherings may have occurred during seasons coinciding with downtimes in the growing season. Jatanca thus served as a place of religious and economic convergence for both coastal and mid-valley groups, and different communities may have identified with one of the social or religious corporations materialized by one of the six monumental compounds (Swenson and Warner 2016: 12). During other times of the year, celebrants may have returned to smaller hamlets to clean canals and tend to agricultural fields. In the Late Formative Period, the dead also undertook a final journey to the nearby mortuary site of Puémape, located at the edge of the Pacific Ocean (Warner, personal communications). Carlos Elera (1998) describes Puémape, a site contemporaneous with Jatanca, as having been occupied by the Cupisnique, Salinar, and Gallinazo groups. He further noted that Puémape served mainly as a pan-regional mortuary center.

In the end, it remains difficult to determine how the fixing of mythic history at Jatanca structured the circulation of peoples in Late Formative populations in the valley. However, this peculiar topology of memory and time no doubt shaped everyday temporal regimes and the scheduling of the region's religious and economic calendars.

Time Activated in Place: The Folded and Kinetic Topology of Huaca Colorada

Located just two kilometers northeast of Jatanca, the ceremonial center of Huaca Colorada was established more than 500 years after the abandonment of the Late Formative Center (figure 6.1). It is the largest Late Moche settlement located in the southern bank of the valley, and the elaborate adobe constructions dominating the central and highest prominence of the pyramid point to the presence of high-status individuals. Excavations reveal that the huaca was a setting for elaborate ceremonies, including copper production, feasting, human sacrifice, and cyclical rites of architectural renovation (Swenson 2012b; Swenson and Warner 2012, 2016; Swenson 2015) (see also Spence Morrow [chapter 7], this volume). The center likely served as the religious headquarters of a powerful polity that dominated the southern region of a politically divided valley (Castillo 2010; Dillehay 2001; Swenson 2012b). However, a significant number of ceremonial centers proliferated in Jequetepeque during the Late Moche Period, including San José de Moro, the seat of the famed priestess cult located in the Chaman drainage twenty-five kilometers to the north (Castillo 2001, 2010; Castillo and Donnan 1994). Rural population also increased in Jequetepeque during the early years of the Middle Horizon, and a number of sites with ceremonial architecture were built throughout the region (Castillo 2010; Dillehay 2001; Dillehay, Kolata, and Swenson 2009; Swenson 2006, 2008, 2014). The sheer variety of settlement types indicates a high degree of political decentralization. However, it also points to seasonal transhumance and transient settlement between different sites (Swenson and Warner 2012).

Huaca Colorada is dominated by an elongated platform built on a modified hill of sand. This central construction measures 390 meters × 140 meters and rises twenty meters at its central highest point. This ample elevation is differentiated into three principal sectors that delimit Huaca Colorada's principal ceremonial and manufacturing/residential areas (figure 6.5). The main ceremonial complex of the site (Sector B) was built in the central and highest prominence of the mound. It comprises a series of platforms, corridors, patios, stairs, and ramps and formed the arena for elaborate feasting and sacrificial ritual. Huaca Colorada was constructed to stage specific kinds of ceremonies unique to the region as indicated by the site's exceptional configuration, location, and artefactual associations. Beyond foundation sacrifices and the discovery of a few household burials in the lower production zones, no elite tombs have been unearthed in the monumental sector, and it is clear that the settlement did not function as a mausoleum as did San José de Moro and other prominent Moche centers in the region including Pacatnamú. The recovery of copper ornaments, tools, and production detritus in makeshift domestic constructions indicates that metallurgy constituted an important pursuit of the settlement's inhabitants (Swenson and Warner 2012). The discovery of offerings of prills and finished copper objects in conjunction with foundation sacrifices in the ceremonial zone of the site might even suggest that Huaca Colorada elites were skilled artisans who based their authority on their guarded ritual knowledge of the metallurgical arts. However, most evidence of production derives from the expansive residential areas of Sectors A and C, and nonelite artisans visited the huaca during specific seasons under the patronage of the site's leaders. Copper items may have been produced and exchanged during fairs and feast days. The high quantity of decorated feasting vessels unearthed in the domestic and ceremonial sectors also strongly suggests that artisans, pilgrims, and full-time residents of the site were compensated with corn beer (Swenson 2012a). As will be described in greater detail below, the expansive domestic areas of Sectors A and C were largely characterized by ephemeral but seasonal and recurring occupations.

The layout and architectural history of the Huaca reveal a tradition of affective time and social memory that differed significantly from Jatanca's unique timescape. Indeed, the underlying spatial grammar and political context of ritual performance changed significantly between the two sites (Swenson 2015). Unlike the disorienting maze of walls at Jatanca that divided and occluded, the elevated pyramid afforded broad vistas of the surrounding landscape. In fact, excavation reveals that quotidian taskscapes at the center were more tightly interwoven with the timing of ritual events than at Jatanca. As described earlier, Jatanca was distinguished by the juxtaposition of divergent but parallel temporalities; the unchanging and timeless monumental compounds coexisted with the ephemeral constructions of the surrounding residential area. In contrast to Jatanca's negentropic compounds, which were built to last, Huaca Colorada's decidedly entropic precincts were built to change, and platforms, chambers, and walls were regularly terminated and ritually rededicated (Swenson 2015). Therefore, the cyclical reconstructions of the monument may have been more closely synchronized with the comings and goings of pilgrims. The ritualized and sacrificial nature of reconstruction is exemplified by the decommissioning of eight individual ceremonial platforms, all of which were painstakingly sealed under hard floors or buried under thick deposits of clean sand (see figure 6.6, Spence Morrow, this volume [chapter 7]). Unlike the undying tapia walls at Jatanca, the structures at Huaca Colorada were built of smaller adobe bricks, an architectural medium that facilitated rapid, improvised construction and demolition. In fact,

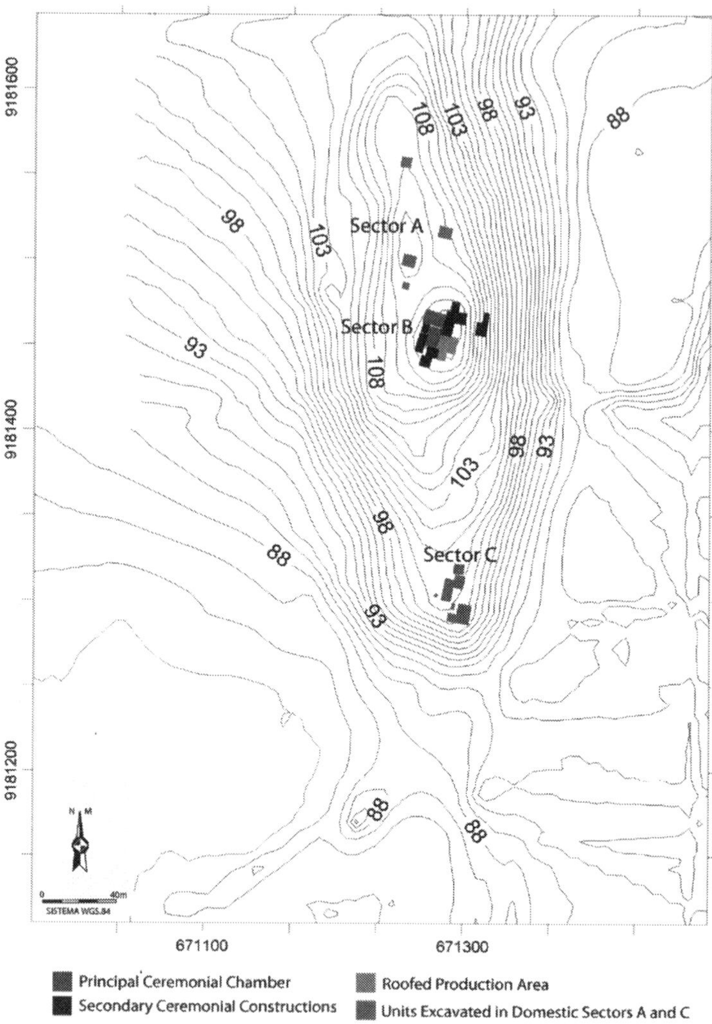

FIGURE 6.5. *Topographic map of Huaca Colorada. The location of the main ceremonial chamber within the monumental core of Sector B is indicated along with the lower domestic and copper production zones of Sectors A and C.*

adobes were commonly robbed from walls and recycled at the site (Swenson 2015: 700). This technology of wall assembly was also predicated on a different temporal regime than the construction of tapia perimeters. The stacking and mortaring of bricks could be executed by multiple workers in a more piecemeal and even improvised manner, and work could be interrupted and resumed with little difficulty. Adobe brick thus allowed for the Huaca to grow or contract in

FIGURE 6.6. *The adobe brick astronomical apparatus and a human sacrificial victim excavated in the monumental sector of Huaca Colorada.*

a more incremental fashion (see below). In contrast, pouring wet adobe into larger cane frames to construct the large tapia segments at Jatanca would have demanded the close coordination and timing of tasks—with little wiggle room for modifying preconceived layouts (Campana 2000). Therefore, the periodic rededication of ceremonial architecture, as perhaps dictated by a ceremonial calendar, defined the dynamic architectural history of the site.

The succession of renovations corresponds directly with the sacrifice of mainly female victims, whose death likely served to ensure rebirth and the continuation of life. Human and animal sacrifices accompanied the renovation of eight distinct ceremonial platforms, and an additional sacrifice was discovered incorporated in construction fill associated with incremental reductions of the main chamber prior to its termination. Time was enlivened and somatically renewed in the sacrificial reconstructions of the site's monumental nucleus. In engineering the consubstantiation of people and vibrant architecture, Huaca Colorada acted as monumental chronotope, emanating a specific temporality centered on gestation and cycles of growth, destruction, and renewal (Swenson 2015: 691). In contradistinction to the solidity and constancy of Jatanca's tapia perimeters, Huaca Colorada's adobe-brick walls were built to live, die, and regenerate, and their material fabric was more analogous to flesh or vegetation.

The proliferation informal graffiti at Huaca Colorada also expresses evanescence and transience, with the complexion of the plaster walls forever changing, like the constantly transforming human body (Swenson 2012a: 18; Swenson 2015: 698). The graffiti are often directly superimposed over older etchings, further suggesting that the Moche did not perceive the structure as a timeless, inviolable monument. Jatanca's heavily plastered walls are largely devoid of any decorative embellishments, evidence that further points to the undying and immortal character of its labyrinthine compounds. Our extensive excavations at Jatanca reveal that sacrificial victims were not incorporated into architectural constructions. The two contemporaneous examples were found buried in the floor of chambers. They were interred in shallow and intrusive pits placed in the room after the precinct and plaster floor were constructed, and architectural rededication did not occur in direct conjunction with a foundation sacrifice. Ritual practices that materially synchronized the life cycle of people and buildings were notably absent at Jatanca, pointing to the distinct temporal ontologies of the two centers (Swenson 2015: 700).

The palimpsest of shifting floors, sand fill, human offerings, and superimposed chambers suggest that Huaca Colorada was celebrated for its kinesthetic force and was perhaps even perceived as a living, metabolizing organism (Swenson 2015: 692). In fact, the ambivalence between inside and outside, frame and content, and natural terrain and built environment resonates with Gilles Deleuze and Felix Guattari's particular notion of a topology or folded space-time (Deleuze and Guattari 1987; Deleuze 1992; see also DeLanda 2005: 84–86; Harris 2005: 45–51; Lynn 2004: 10, 40–42; Witmore 2007: 195–96). Folding architecture is defined by a "fluid logic of connectivity that integrates unrelated elements within a continuous mixture" (Harris 2005: 44). It thus designates a place of pure possibilities celebrated for its creative virtuosities. Following Deleuze and Guattari (1987), "virtual" refers, then, to an abstract temporal condition, an emergent force "that has the possibility of becoming actualized, often in a variety of possible configurations" (Lynn 2004: 10). The Huaca was built on a modified and stationary sand-dune, and sand was used to both inter ritually killed spaces and to provide a foundation for sacrifices and later structural additions or reductions. Therefore, the geometry of the site's architecture was remarkably pliable with the sandy exterior folded in the inside of the building and the monument folded to the outside—just as sacrificial victims and copper offerings were enfolded within the literal pleats of the edifice's monumental body (figure 6.6) (Swenson 2015: 693). The huaca thus appears to have enfolded time and space into a dynamic and vitalizing material matrix (Bennett 2010: 50). The massive middens surrounding the ceremonial complex were filled with food remains including llama bones, peanut, shellfish, and high-quality feasting bowls in the Cajamarca style. In fact, an alimentary time of ingestion, digestion, and growth may provide an apt metaphor to describe the ideologies of social and cosmic change at the site (Swenson and Warner 2016: 23). Both the monument and celebrants were continually transformed by ingesting the outside world of food, corn beer, human offerings, and sacred altars. As another point of contrast, feasting did not play an important role in the ritual economy of Jatanca (Swenson 2011; Warner 2010).

Therefore, the organic and vertically accretional biography of the huaca contrasts from the rigidly linear and horizontally configured compound architecture characteristic of Jatanca. An extraordinary "mark of repetition" also characterized Huaca Colorada's numerous renovations, and the sequential reconstructions followed a predictable pattern with middens superimposed on middens, ceremonial architecture built over altars, and kitchens overlaying earlier kitchens. Once again, this architectural history contrasts starkly with the Jatanca's immutable and timeless religious landscape.

The affective time of the huaca thus departs from the feelings of anticipation and suspense engendered by the horizontal and labyrinthine compounds of Jatanca. Although procession and encounters with huacas probably underwrote certain rituals at Huaca Colorada, the sacrifices and cyclical demolition of ceremonial architecture must have created a potent sense of expectation. The emphasis on killing and destruction no doubt intensified feelings of insecurity and of the impermanence of individual lives, whether architectural, human, or cosmic. Anticipation of such events no doubt affected the outlook and subjective position of participants. Affective time commonly springs from "disruptions in the field of being, like death, grief, and violence," an experience of rupture defined by "suspension of the present, a folding of past in the present in ways that do not fit" (Locke 2010: 147). Such intensely felt moments of time often entail a discontinuity of selves, communities, and meaningful places (even if deemed essential for their reconstitution). In addition, the ingested platforms and buried bodies no doubt invoked strong (even involuntary) memories for celebrants who revisited Huaca Colorada during set seasons.

The sacrificial connotations of reconstruction, likely following principles of Moche political theology and cosmology, are exemplified by the fastidious interment of ceremonial architecture in tons of sand, resulting in the remarkable preservation of the stucco columns of one prominent platform that survived to a more than one meter in height (see figure 7.2, in Spence Morrow this volume [chapter 7]). A similar structure to the east was burnt and buried under the remains of its monumental roof (Swenson et al. 2015) Comparable architectural practices have been documented at other centers including Huacas de Moche (Uceda 2001, 2010), and the sacrificial termination of space appears to have been widespread in the Moche world (Swenson 2011). In this light, Huaca Colorada can be interpreted as a monumental timepiece, a medium for charting and controlling the passage of time through the materialization of temporal cycles. Of course, the geocalendrical and temporal significance of the huaca finds parallels in the cosmographic layout of many other pre-Columbian centers (see Mock 1998; Sugiyama 2004). Therefore, the huaca served as both a "sign in history" and "sign of history" as theorized by Richard Parmentier (1987). Signs in history refer to "value-laden objects implicated in social strategies that focus attention on specific historical processes," while signs of history are those that directly reify and objectify the past—an understanding implied in the etymology of our word monument (cited in Preucel 2010: 86).

The discovery of features that appear to have served astronomical functions further points to the central importance of ritualized time reckoning at Huaca Colorada. An unusual adobe astronomical device (1.42 m long × 1.18 m wide and 18 cm high) made of worked bricks was discovered on a high terrace to the northeast corner of the monumental zone of the huaca (figure 6.6). It is composed of nine bricks of varying size that together form a roughly elongated, pentagonal shape. Three of the smaller bricks are exceptional in that they were carefully cut into pointed, diamond-shaped wedges. All three points of the apparatus align with different peaks of neighboring mountain ranges to the east and north of the Huaca. Moreover, the central point (oriented 45 degrees east of north) aligns with the sunrise on the Winter Solstice (June 21). The peculiar shape of the installation, forming multiple lines of sight with prominent features of the surrounding topography, strongly suggests that it functioned in rituals related to astronomy, time reckoning, and sacred landscapes (Swenson and Warner 2016: 25).

As mentioned, the architectural renovations directly incorporated the sacrifice of human victims, whose death likely served to ensure the revitalization of space and time. Major junctures in the production and festive cycle appear to have coincided with the literal sacrifice and renovation (rebirth) of sacred space at Huaca Colorada. In this regards, the sacrifice of buildings and people rendered physically real the abstract, cyclical, and dialectical movement of time peculiar to the Moche. Huaca Colorada's monumental core was charged with life and time's recursive flows. Similar to the religious objectives of human sacrifice, ritualized architectural construction appears to have been envisioned as a tool of social *engineering*, an attempt to control people, their life cycles, and the unfolding of time itself. Indeed, time and society were given a concrete form in the cyclical reconstructions of the site's monumental nucleus. The dedication and termination rites at Huaca Colorada demonstrate that the Moche made explicit their understanding that "time and space are integral to each other" and that in the "lived world, spatial and temporal dimensions cannot be disentangled," a point that has long fascinated philosophers of time (see Munn 1992: 94–95). Huaca Colorada's ceremonial architecture was thus grounded in an aesthetic of violence that celebrated creation, rebirth, and fertility. In other words, sacrifice activated the movement of time itself (see Uceda 2010: 152), and "time was not ontically prior to the existence of matter and its changes" (Arthur 2010: 45).

Divergent Temporalities in the Moche World

On first inspection, the explicit spatialization of time at Huaca Colorada, as reflected in the ritual interment of earlier edifices, finds parallels at other well-known Moche polities. For instance, in describing Huaca de la Luna, Uceda notes that "its construction was not the outcome of a single construction project but rather the result of a constant renovation of the Old Temple. During each building episode, quite likely as part of a ceremonial cycle, the Old Temple was purposefully and carefully buried, creating a more spacious and elevated platform" (Uceda 2010: 149–50). Excavations at Huaca de la Luna have confirmed five separate and superimposed temple buildings, with subsequent reconstructions increasing the edifice 3–4 meters in height and 4–5 meters on each side (Canziani Amico 2010: 168–70; Uceda 2010: 150). Each major renovation was perhaps dictated by the close (death) and inauguration (rebirth) of a new "Moche" century, possibly analogous to the calendrical significance of monumental reconstruction in Mesoamerica. Ritual renovations also characterized other prominent sites including Cao Viejo (seven distinct building episodes) and Dos Cabezas in Jequetepeque (Franco et al. 1994; 2010: 121; Bourget, personal communications). The termination and rededication of religious architecture no doubt conformed to particular cosmologies and religious traditions, a structure of the Moche *longue-durée* (Canziani Amico 2010: 168).

In light of these commonalities, an intersite comparison of the rate, frequency, and shifting context of reconstruction events should improve interpretations of the changing political fortunes and religious functions of different Moche centers. In other words, closer inspection of the archaeological data reveals subtle if important differences in Moche ideologies of time and history. For instance, Seoane (2013; Seoane and Culquichicón in this volume [chapter 8]) has recently argued that at the onset of the Middle Horizon, which coincided with the final decommissioning of the temple of Huaca de Luna (AD 600), architecture at Huacas de Moche was reconstructed in alignment with the Solar Transit, a celestial point of reference that deviated from the previous sidereal orientation of Huaca de la Luna and earlier buildings configured to the movement of the Southern Cross. Therefore, in the Middle Horizon, the New Temple (which is thought to have replaced Huaca de Luna as the focus of state religious observances) and the alignment of the avenues of the booming urban nucleus were built to harmonize with a solar as opposed to a defunct nocturnal calendar defined by the movement of the stars. Seoane's theory would seem to corroborate Santiago Uceda's (2010) argument that dramatic changes in the architectural layout of Huacas de Moche corresponded with the fundamental political reordering of the urban center during the early Middle Horizon Period (see also Koons and Alex 2014). This political reorganization might very well have entailed major shifts in the construction of time and history. Interestingly, these major reconfigurations in architecture were not accompanied by pronounced changes in the ceramic styles of the city (Chapdelaine 2002; Roddick this volume [chapter 3]). It is also intriguing to note that the Late Moche monuments of Pampa Grande in the Lambayeque Valley were not ritually terminated and regenerated (Shimada 1994). The distinctive chamber-and-fill technique allowed for rapid construction, and pyramids such as Huaca Fortaleza clearly did not serve as animated "timepieces" in the same manner as Huaca de La Luna or even Huaca Colorada. Such discrepancies likely indicate different political strategies to codify and control "public time" (Munn 1992: 102).

Returning to Huaca Colorada, it is intriguing that the monument was not expanded in volume or enlarged in size with each subsequent phase of ritual renovation. This finding contrasts notably from the sequential amplification of Huaca de La Luna and other Moche temples. Stratigraphic analysis reveals that the main ceremonial chamber was incrementally reduced in area with successive stages of renovation. The main precinct with ceremonial dais was much larger in earlier phases but was scaled back in size with the construction of three northern walls built at different time intervals. These walls moved roughly two to three meters to the south with each successive construction event (see figure 7.1; Spence Morrow, this volume [chapter 7]). The northernmost walls are significantly deeper and associated with the earliest-use floors identified throughout the chamber, while the southern and shorter perimeter walls were built with increasingly later flooring episodes. As mentioned above, one of these episodes of chamber reduction was associated with the sacrifice of a female victim. Each reduction corresponded with the plugging of a door on the east wall of the chamber that linked this precinct to a higher terrace housing similar daises. The complex meanings of this construction history elude full interpretation. However, the incremental diminution of the central chamber was most likely also expressive of the distinctive religious meanings and temporalizing functions of the monument. The main chamber was eventually completely decommissioned, and the periodic reduction of space may have signaled a countdown of sorts to mark the end of a political reign or of the influence of a particular huaca. Although the ritual termination of space is evident at both Huacas de Moche and Huaca Colorada, their markedly different execution and architectural context point to contrasting ideologies of time reckoning and history.

The distinctive rituals of time measurement at Huaca Colorada can only be properly understood in relationship to the archaeological signatures of more quotidian temporal rhythms. Did the peculiar rites of mythohistorical commemoration at Huaca Colorada structure more encompassing, nonritualized temporalities? Were these different temporal modalities congruent or possibly out of synch (Gosden 1994: 124)? As intimated above, the shifting and overlapping floor constructions identified throughout the site's expansive domestic zones suggest that the bulk of Huaca Colorada's population only inhabited the site seasonally or periodically. In the expansive domestic zone of Sectors A and C, the intercalation of thin lenses of aeolian sand, indexing short-term abandonment, with both compacted sand surfaces and plaster floors demonstrates that a good number of the Huaca's inhabitants only lived there during certain periods of the year. Periodic occupation is further indicated by the rapid conversion of living spaces into middens and by the irregular destruction of ephemeral living spaces (characterized by no more than one or two floors) by later building events that deviated from the layout of earlier floor-plans (Swenson and Warner 2012). Returning inhabitants obviously were not compelled to reestablish their residential structures in exactly the same locale, but expediently built comparable floor units during subsequent reoccupations of the settlement.

As mentioned previously, the evidence suggests that communities descended on Huaca Colorada during set seasons in order to fulfill tribute obligations and partake in associated religious festivals. During other temporal intervals, perhaps associated with the growing season, these groups may have resided in smaller hamlets found in association with agricultural fields surrounding Huaca Colorada, or in the hillside settlements distinguished by domestic terraces and stone ceremonial architecture (Dillehay, Kolata, and Swenson 2009; Swenson 2006, 2007, 2012b).

It is significant that the Jequetepeque Valley as a whole was characterized by notable variability in Moche residential constructions during the Late Moche Period. Whether *aldeas temporales* built in the vicinity of agricultural fields, terrace structures at fortified ceremonial sites in the countryside, or the adobe-footed quincha shelters in the nonelite domestic zone of Huaca Colorada, it appears that separate residential constructions do not correspond to distinct ethnic groups or socioeconomic classes. Instead, there seems to have been an expectation for communities of possibly changing constitution and size to maintain multiple residences in accordance with the *rhythms* of economic, religious, and possibly military cycles (Swenson 2012b). The Jequetepeque Valley was distinguished by the regular internal movement of people, and certain value was placed on transient habitation and impermanent domesticity, at least among a large segment of the population during the Late Moche Period (Swenson 2012a; Swenson and Warner 2012, 2016). The priestess center of San José de Moro is also largely devoid of domestic contexts, and people congregated here in large numbers only at specific moments in time to participate in funerary and feasting rites (Castillo 2010; Castillo et al. 2009).

Shifting and overlapping floors intercalated with short episodes of abandonment further point to periodic occupations at a number of settlements in the valley; cyclical residence at agricultural hamlets, fortified ceremonial sites, and religious centers such as San José de Moro, Pacatnamú, or Huaca Colorada was perhaps dictated by a ceremonial calendar or related economic and military schedules (Dillehay, Kolata, and Swenson 2009; Swenson 2012b). The pulsating social landscape of Jequetepeque, defined by the regularized comings and goings of people, thus seems to have defined both the Late Formative and Middle Horizon Periods in the valley. Interestingly, the numerous graffiti of litter bearers engraved on the walls of Huaca Colorada provide additional evidence of the value placed on transhumance in the greater Jequetepeque region: they appear to depict the transportation of the honored dead, perhaps to funerary sites including San José de Moro. This ultimate journey finds parallel in the final migration of the dead to the site of Puémape during the Late Formative Period.

It is especially significant that the temporalities of the Jequetepeque landscape appear to have changed dramatically between the Middle and Late Moche Periods. The Middle Moche Period (AD 450–600) is characterized by fewer, larger, and more permanent settlements, as the research of Chris Donnan and now Steve Bourget would indicate at the massive urban center of Dos Cabezas (Donnan 2007). Research conducted by Christopher Donnan (1997: 12) and his team further demonstrates that the central core of Pacatnamú was more intensely occupied during the Middle Moche Period (in the vicinity of Huaca 1 and Huaca 31). During this particular era of the Early Intermediate Period, this site is marked by more sustained settlement, cemeteries, and dwelling structures. In contrast, the Late Moche occupation at Pacatnamú is distinguished by ephemeral habitation and the lack of permanent constructions (McClelland 1997). The valley-wide survey of Dillehay and Kolata further confirms that settlement was concentrated and centralized in the Middle Moche Period, a scenario that differs notably from the

Late Moche Period characterized by a dispersion and proliferation of intermediate centers, small residential sites, and agricultural encampments (Dillehay 2001; Dillehay, Kolata, and Swenson 2009; see also Castillo and Donnan 1994). In other words, it would seem that communities were politically more integrated in the Middle Moche Period as well as more sedentary than either in the Late Formative Period or subsequent Late Moche populations. Of course, future research is required at other settlements to confirm these emerging patterns and to identify the possibly distinct and changing temporal signatures of specific sites.

CONCLUSION

An examination of architectural and settlement data as distinct topologies ("timescapes") permitted an interpretation of remarkably pronounced differences in the creation and experience of time in the Jequetepeque Valley. The study also revealed that rituals of history, myth, and social memory varied not only within the Jequetepeque Valley but between contemporaneous Moche polities of the greater North Coast. Therefore, nuanced reconstructions of sociopolitical change must combine standard chronological analyses with investigations of the archaeological signatures of everyday temporal rhythms and more conspicuous rituals of history and memory.

The above comparison strongly suggests that the Moche endeavored to channel or harness the power of chaos, flux, and death, whereas the celebrants at Jatanca attempted to deny or displace the entropic processes of time's passage. Time for the Moche was more about rupture and metabolic transformation, whereas at Jatanca time seems to have been understood as duration and the balanced alteration of complementary opposites. The permanency of the Jatanca monuments points to an eternal, perhaps ancestral time that defied procreative temporalities and mortal life-cycles. On the other hand, Huaca Colorada was not designed to express a timeless monumentalism but seems to have acted as an engine for time's movement and perpetual re-creation (Swenson 2015: 704). Despite these pronounced differences in ideologies of history and memory between the two sites of the case study, it is intriguing that Late Moche temporalities more closely resembled those of the Late Formative Period than the intervening Middle Moche phase. The evidence may well support theories that the onset of the Middle Horizon Period was characterized by major sociopolitical transformations. More important, it reveals that much is lost when historical explanation relies simply on charting chronological shifts in ceramic sequences, religious iconography, and architectural styles.

ACKNOWLEDGMENTS

I wish to express my sincere gratitude to the Wenner-Gren Foundation for Anthropological Research, the National Geographic Society, the Social Sciences and Humanities Research Council of Canada, the Canadian Foundation for Innovation, the University of Lethbridge, and the University of Toronto for their generous financial support of my recent research in the Jequetepeque Valley. I also extend my warmest thanks to Jorge Chiguala Azabache, Francisco Seoane, and John Warner, my codirectors of many years.

REFERENCES

Abercrombie, Thomas A. 1998. *Pathways of Memory and Power: Ethnography and History among an Andean People*. Madison: University of Wisconsin Press.

Alcock, Susan E. 2002. *Archaeologies of the Greek Past: Landscapes, Monuments, and Memories*. Cambridge: Cambridge University Press.

Allen, Catherine J. 1988. *The Hold Life Has: Coca and Cultural Identity in an Andean Community*. Washington, DC: Smithsonian Institution Press.

Anonymous. [1594] 2008. *De las costumbres antiguas de los naturales del Pirú*. Edited by Chiara Albertin. Madrid: Iberoamericana; Frankfurt am Main: Vervuert.

Arthur, Richard T. W. 2010. "Materialist Theories of Time." In *Chrono-topologies: Hybrid Spatialities and Multiple Temporalities, Critical Studies*, vol. 32. ed. Leslie Kavanaugh, 43–61. Amsterdam: Rodopi B.V. https://doi.org/10.1163/9789042031425_005.

Assmann, Jan. 2006. *Religion and Cultural Memory*. Stanford, CA: Stanford University Press.

Barrett, John C. 2004. "Temporality and the Study of Prehistory." In *Time and Temporality in the Ancient World*, ed. Ralph M. Rosen, 11–27. Philadelphia: University of Pennsylvania Museum of Archaeology and Anthropology.

Bennett, Jane. 2010. *Vibrant Matter: A Political Ecology of things*. Durham, NC: Duke University Press.

Bergson, Henri. 2001. *Time and Free Will: An Essay on the Immediate Data of Consciousness*. Trans. F. L. Pogson. Mineloa: Dover Publications.

Bloch, Ernst. 1991. *Heritage of Our Times*. Translated by Neville and Stephen Plaice. Berkeley: University of California Press.

Boyer, M. Christine. 2010. "Societies of Control and Chrono-Topologies." In *Chrono-topologies: Hybrid Spatialities and Multiple Temporalities, Critical Studies*, vol. 32. ed. Leslie Kavanaugh, 203–22. Amsterdam: Rodopi B.V. https://doi.org/10.1163/978 9042031425_014.

Bradley, Richard. 2002. *The Past in Prehistoric Societies*. New York: Routledge.

Campana, Cristobal. 2000. *Tecnologías constructivas de tierra en la Costa Norte Prehispanica*. Trujillo: Instituto Nacional de Cultura.

Canziani Amico, José. 2010. "City Planning and Architecture in the Analysis of Moche Social Formation." In *New Perspectives on Moche Political Organization*, ed. Jeffrey Quilter and Luis Jaime Castillo, 159–80. Washington, DC: Dumbarton Oaks.

Castillo, Luis Jaime. 2001. "The Last of the Mochicas." In *Moche Art and Archaeology in Ancient Peru*, ed. Joanne Pillsbury, 307–32. New Haven, CT: National of Gallery of Art and Yale University Press.

Castillo, Luis Jaime. 2010. "Moche Politics in the Jequetepeque Valley: A Case for Political Opportunism." In *New Perspectives in Moche Political Organization*, ed. Jeffrey Quilter and Luis Jaime Castillo, 1–24. Washington, DC: Dumbarton Oaks.

Castillo, Luis Jaime, and Christopher B. Donnan. 1994. "Los nochica del norte y los mochica del sur." In *Vicús*, ed. Krzysztof Makowski and Chrisopher B. Donnan, 143–81. Lima: Colección Arte y Tesoros del Perú, Banco de Crédito.

Castillo, Luis Jaime, Julio Rucabado Y., Martín del Carpio P., Katiusha Bernuy Q., Karim Ruiz R., Carlos Rengifo Ch., Gabriel Prieto B., and Carole Fraresso. 2009. "Ideología y poder en la consolidación, colapso y reconstitución del estado mochica del Jequetepeque: El Proyecto Arqueológico San José de Moro (1991–2006)." *Ñawpa Pacha* 29:2–73.

Castillo, Luis Jaime Butters, and Santiago Uceda Castillo. 2008. "The Mochicas." In *Handbook of South American Archaeology*, ed. Helaine Silverman and William H. Isbell, 707–29. New York: Springer.

Chapdelaine, Claude. 2002. "Out in the Streets of Moche: Urbanism and Sociopolitical Organization at a Moche IV Urban Center." In *Andean Archaeology I: Variations in Sociopolitical Organization*, ed. William H. Isbell and Helaine Silverman, 53–88. New York: Kluwer Academic/Plenum Publishers. https://doi.org/10.1007/978-1-4615-0639-3_3.

Chapdelaine, Claude. 2011. "Recent Advances in Moche Archaeology." *Journal of Archaeological Research* 19 (2): 191–231. https://doi.org/10.1007/s10814-010-9046-6.

Connerton, Paul. 1989. *How Societies Remember*. Cambridge: Cambridge University Press. https://doi.org/10.1017/CBO9780511628061.

Cutright, Robyn. 2009. "Between the Kitchen and the State: Domestic Practice and Chimú Expansion in the Jequetepeque Valley, Peru." PhD dissertation, University of Pittsburgh.

DeLanda, Manuel. 2005. "Space: Extensive and Intensive, Actual and Virtual." In *Deleuze and Space*, ed. Ian Buchanan and Gregg Lambert, 80–88. Toronto: University of Toronto Press. https://doi.org/10.3366/edinburgh/9780748618743.003.0005.

Deleuze, Gilles. 1992. *The Fold: Leibniz and the Baroque*. Trans. Tom Conley. Minneapolis: University of Minnesota Press.

Deleuze, Gilles, and Felix Guattari. 1987. *A Thousand Plateaus: Capitalism and Schizophrenia*. Minneapolis: University of Minnesota Press.

Dillehay, Tom D. 2001. "Town and Country in Late Moche Times: A View from Two Northern Valleys." In *Moche Art and Archaeology in Ancient Peru*, ed. Joanne Pillsbury, 259–84. New Haven, CT: National of Gallery of Art and Yale University Press.

Dillehay, Tom, Alan Kolata, and Edward Swenson. 2009. *Paisajes culturales en el Valle del Jequetepeque: Los yacimientos arqueológicos*. Trujillo: Ediciones Sian.

Donnan, Christopher B. 1997. "Introduction." In *The Pacatnamú Papers*, vol. 2. ed. Christopher B. Donnan and Guillermo Cock, 9–16. Los Angeles: Fowler Museum of Cultural History, University of California Press.

Donnan, Christopher B. 2007. *Moche Tombs at Dos Cabezas*. Los Angeles: Costin Institute of Archaeology, University of California Press.

Donnan, Christopher, and Guillermo Cock, eds. 1997. *The Pacatnamú Papers*. vol. 2. Los Angeles: Fowler Museum of Cultural History, University of California Press.

Eeckhout, Peter. 1999. *Pachacamac durant l'Intermédiaire Récent: Étude d'un site monumental Préhispanique de la Côte Central du Pérou. BAR International Series 747*. Oxford: British Archaeological Reports.

Elera, Carlos. 1998. "The Puemape Site and the Cupisnique Culture: A Case Study on the Origins and Development of Complex Society in the Central Andes, Peru." PhD dissertation, University of Calgary.

Franco, Régulo, César Gálvez, and Segundo Vásquez. 1994. "Arquitectura y decoración mochica en la Huaca Cao Viejo, Complejo el Brujo: Resultados preliminares." In *Moche: Propuestas y perspectivas*, ed. Santiago Uceda and Elías Mujica, 147–80. Lima: Travaux de l'Institut Français d'Études Andines 79. https://doi.org/10.4000/books.ifea.2394.

Franco, Régulo, César Gálvez, and Segundo Vásquez. 2010. "Moche Power and Ideology at the El Brujo Complex and in the Chicama Valley." In *New Perspectives on Moche Political Organization*, ed. Jeffrey Quilter and Luis Jaime Castillo, 110–31. Washington DC: Dumbarton Oaks.

González-Ruibal, Alfredo. 2014.*An Archaeology of Resistance: Materiality and Time in an African Borderland*. New York: Rowman and Littlefield.

Gosden, Christopher. 1994. *Social Being and Time*. Oxford: Blackwell.

Gose, Peter. 1996. "The Past is a Lower Moiety: Diarchy, History, and Divine Kingship in the Inka Empire." *History and Anthropology* 9 (4): 383–414. https://doi.org/10.1080/02757206.1996.9960887.

Gurvitch, Georges. 1964. *The Spectrum of Social Time*. Trans. M. Korenbaum. Dordrecht: Reidel.

Harootunian, Harry. 2005. "Some Thoughts on Comparability and the Space-Time Problem." *Boundary 2* 32 (2): 23–52. https://doi.org/10.1215/01903659-32-2-23.

Harris, Paul A. 2005. "To See with the Mind and Think through the Eye: Deleuze, Folding Architecture, and Simon Rodia's Watts Towers." In *Deleuze and Space*, ed. Ian Buchanan and Gregg Lambert, 36–60. Toronto: University of Toronto Press. https://doi.org/10.3366/edinburgh/9780748618743.003.0003.

Heidegger, Martin. 1985. *History of the Concept of Time*. Bloomington: Indiana University Press.

Hodder, Ian, and Peter Pels. 2010. "History Houses: A New Interpretation of Architectural Elaboration at Çatalhöyük." In *Probing Religion at Çatalhöyük in its Regional Context*, ed. Ian Hodder, 163–86 Cambridge: Cambridge University Press.

Hyland, Sabine. 2011. *Gods of the Andes: An Early Jesuit Account of Inca Religion and Andean Christianity*. University Park: Pennsylvania State University.

Ingold, Tim. 2000. *The Perception of the Environment: Essays in Livelihood, Dwelling, and Skill*. New York: Routledge. https://doi.org/10.4324/9780203466025.

Isbell, Billie Jean. 1982. "Culture Confronts Nature in the Tropics." In *Ethnoastronomy and Archaeoastronomy in the American Tropics*, ed. Anthony F. Aveni and Gary Urton, 353–63. Annals of the New York Academy of Sciences 385. New York: New York Academy of Sciences.

Kavanaugh, Leslie. 2010a. "Introduction." In *Chrono-Topologies: Hybrid Spatialities and Multiple Temporalities, Critical Studies*, vol. 32. ed. Leslie Kavanaugh, 3–8. Amsterdam: Rodopi B.V. https://doi.org/10.1163/9789042031425_003.

Kavanaugh, Leslie. 2010b. "The Time of History / The History of Time." In *Chrono-topologies: Hybrid Spatialities and Multiple Temporalities, Critical Studies*, vol. 32. ed. Leslie Kavanaugh, 91–124. Amsterdam: Rodopi B.V. https://doi.org/10.1163/978 9042031425_008.

Koons, Michele L., and Bridget A. Alex. 2014. "Revised Moche Chronology Based on Bayesian Models of Reliable Radiocarbon Dates." *Radiocarbon* 56 (3): 1039–55. https://doi.org/10.2458/56.16919.

Koselleck, Reinhart. 2004. *Futures Past: On the Semantics of Historical Time.* Translated and introduced by Keith Tribe. New York: Columbia University Press.

Lefebvre, Henri. 1991. *The Production of Space*. Trans. Donald Nicholson-Smith. Oxford: Blackwell.

Locke, Patricia. 2010. "Intermittences: Merleau-Ponty and Proust on Time and Grief." In *Chrono-topologies: Hybrid Spatialities and Multiple Temporalities, Critical Studies*, vol. 32. ed. Leslie Kavanaugh, 146–58. Amsterdam: Rodopi B.V. https://doi.org/10.1163 /9789042031425_010.

Lucas, Gavin. 2005. *The Archaeology of Time*. London: Routledge.

Lynn, Greg. 2004. *Animate Form*. Princeton, NJ: Princeton Architectural Press.

Mackey, Carol.2009."Chimú Statecraft in the Provinces."In *Foundations of Andean Civilization: Papers in Honor of M. E. Moseley*, ed. Joyce Marcus, Charles Stanish, and P. R. Williams, 325–49. Los Angeles: Cotsen Institute of Archaeology, University of California.

McClelland, Donna. 1997. "Moche Fineline Ceramics at Pacatnamú." In *The Pacatnamu Papers*, vol. 2. ed. Christopher B. Donnan and Guillermo Cock, 265–82. Los Angeles: Fowler Museum of Cultural History, University of California Press.

McFadyen, Lesley. 2006. "Building Technologies, Quick and Slow Architectures and Early Neolithic Long Barrow Sites in Southern Britain." *Archaeological Review from Cambridge* 21 (1): 115–34.

Merleau-Ponty, Maurice. 1962. *Phenomenology of Perception*. Trans. Colin Smith. London: Routledge.

Mills, Barbara J., and William H. Walker, eds. 2008. *Memory Work: Archaeologies of Material Practice*. Albuquerque: SAR Press.

Mock, Shirley Boteler, ed. 1998. *The Sowing and the Dawning: Termination, Dedication, and Transformation in the Archaeological and Ethnographic Record of Mesoamerica*. Albuquerque: University of New Mexico Press.

Munn, Nancy. 1992. "The Cultural Anthropology of Time: A Critical Essay." *Annual Review of Anthropology* 21 (1): 93–123. https://doi.org/10.1146/annurev.an.21.100192.000521.

Nimuendajú, Curt. 1946. *The Eastern Timbira*. Trans. and ed. Robert H. Lowie. Publications in American Archaeology and Ethnology, Volume 41. Berkeley: University of California.

Olivier, Laurent. 2004. "The Past of the Present: Archaeological Memory and Time." *Archaeological Dialogues* 10 (2): 204–13. https://doi.org/10.1017/S1380203804001254.

Olivier, Laurent.2011. *The Dark Abyss of Time: Archaeology and Social Memory*. Lanham: Altamira.

Ossio, Juan M. 2002. "Contemporary Indigenous Religious Life in Peru." In *Native Religions and Cultures of Central and South America*, ed. Lawrence E. Sullivan, 200–20. New York: Continuum International Publishing.

Parmentier, Richard J. 1987. *The Sacred Remains: Myth, History, and Polity in Belau*. Chicago: University of Chicago Press.

Patterson, Tom C. 1985. "Pachacamac: An Andean Oracle under Inca Rule." In *Recent Studies in Andean Prehistory and Ethnohistory*, ed. D. Peter Kvietok and Dan H. Sandweiss, 159–76. Ithaca: Cornell University Press.

Preucel, Robert W. 2010. *Archaeological Semiotics*. New York: Wiley-Blackwell.

Quilter, Jeffrey, and Luis Jaime Castillo, eds. 2010. *New Perspectives on Moche Political Organization*. Washington, DC: Dumbarton Oaks.

Quilter, Jeffrey, and Michelle L. Koons. 2012. "A Critique of Claims for South America's First State." *Latin American Antiquity* 23 (2): 127–43. https://doi.org/10.7183/1045-6635.23.2.127.

Randall, Robert. 1982. "Qoyllur Rit'i, an Inca Fiesta of the Pleiades: Reflections on Time and Space in the Andean World." *Bulletin de L'Institut Français d'Etudes Andines* 11 (1–2): 37–81.

Roddick, Andrew. 2013. "Temporalities of the Formative Period Taraco Peninsula, Bolivia." *Journal of Social Archaeology* 13 (3): 287–309. https://doi.org/10.1177/1469605313485396.

Salomon, Frank. 2002. "'*Huayra Huayra Pichcamanta!*' Augurio, risa y regeneración en la política tradicional (Pacota, Huarochirí).'" *Boletín del Instituto Francés de Estudios Andinos* 31 (1): 1–22. https://doi.org/10.4000/bifea.6900.

Salomon, Frank, and George L. Urioste, eds. 1991. *The Huarochirí Manuscript: A Testament of Ancient and Colonial Andean Religion*. Austin: University of Texas Press.

Seoane, Francisco. 2013. "Del atardecer a la mañana: Replanteando el orden del mundo en la sociedad Moche durante el Horizonte Medio." Universidad Nacional de Trujillo. *Revista del Museo de Arqueología, Antropología e Historia* 12:41–52.

Serres, Michel, and Bruno Latour. 1995. *Conversations on Science, Culture, and Time*. Trans. Rozanne Lapidus. Ann Arbor: University of Michigan Press. https://doi.org/10.3998/mpub.9736.

Shimada, Izumi. 1994. *Pampa Grande and the Mochica Culture*. Austin: University of Texas Press.

Sugiyama, Saburo. 2004. "Governance and Polity at Classic Teotihuacan." In *Mesoamerican Archaeology*, ed. Julia A. Hendon and Rosemary A. Joyce, 97–123. Malden, MA: Blackwell.

Swenson, Edward. 2006. "Competitive Feasting, Religious Pluralism, and Decentralized Power in the Late Moche Period." In *Andean Archaeology III: North and South*, ed. William H. Isbell and Helaine Silverman, 112–42. New York: Springer.

Swenson, Edward. 2007. "Adaptive Strategies or Ideological Innovations? Interpreting Sociopolitical Developments in the Jequetepeque Valley of Peru during the Late Moche Period." *Journal of Anthropological Archaeology* 26 (2): 1–30.

Swenson, Edward. 2008. "San Ildefonso and the Popularization of Moche Ideology in the Jequetepeque Valley." In *Arqueología mochica: Nuevos enfoques*, ed. Luis Jaime Castillo, Hélène Bernier, Greg Lockard, and Julio Rucabado, 411–32. Lima: Fondo Editorial Pontificia Universidad Católica del Perú.

Swenson, Edward. 2011. "Stagecraft and the Politics of Spectacle in Ancient Peru." *Cambridge Archaeological Journal* 21 (2): 283–313. https://doi.org/10.1017/S095977431100028X.

Swenson, Edward. 2012a. "Moche Ceremonial Architecture as Thirdspace: The Politics of Place-Making in the Ancient Andes." *Journal of Social Archaeology* 12 (1): 3–28. https://doi.org/10.1177/1469605311426548.

Swenson, Edward. 2012b. "Warfare, Gender, and Sacrifice in Jequetepeque Peru." *Latin American Antiquity* 23 (2): 167–93. https://doi.org/10.7183/1045-6635.23.2.167.

Swenson, Edward. 2014. "Los Fundamentos Cosmológicos de las Interacciones Moche-Sierra durante el Horizonte Medio en Jequetepeque." In *Los rostros de Wari: Perspectivas interregionales sobre el Horizonte Medio*, ed. Luis Jaime Castillo and Justin Jennings, 79–104. Lima: Boletín de Arqueología PUCP 16.

Swenson, Edward. 2015. "The Materialities of Place-Making in the Ancient Andes: A Critical Appraisal of the Ontological Turn in Archaeological Interpretation." *Journal of Archaeological Method and Theory* 22 (3): 677–712. https://doi.org/10.1007/s10816-014-9202-2.

Swenson, Edward. 2017. "Timing Is Everything: Religion and the Regulation of Temporalities in the Ancient Andes." In *Religion and Politics in the Ancient Americas*, ed. Sarah B. Barber and Arthur A. Joyce, 210–33. New York: Routledge.

Swenson, Edward, Jorge Chiguala, and John Warner. 2008. *Proyecto de investigación arqueológica jatanca 2007, Valle de Jequetepeque*. Lima: Technical Report submitted to the Instituto Nacional de Cultura, Perú.

Swenson, Edward, Jorge Chiguala, and John Warner. 2009. "Proyecto de investigación arqueológica Jatanca 2008, Valle de Jequetepeque." Technical Report submitted to the Instituto Nacional de Cultura, Perú, Lima.

Swenson, Edward R., Francisco Seoane, and John P. Warner. 2015. "Informe Final Proyecto Arqueológico Jatanca-Huaca Colorada, Valle Jequetepeque Temporada 2014." University of Toronto, Pacasmayo. Unpublished report submitted to the Instituto Nacional de la Cultura, Perú, Lima.

Swenson, Edward, and John Warner. 2012. "Crucibles of Power: Forging Copper and Forging Subjects at the Moche Ceremonial Center of Huaca Colorada, Peru." *Journal of Anthropological Archaeology* 31 (3): 314–33. https://doi.org/10.1016/j.jaa.2012.01.010.

Swenson, Edward, and John Warner. 2016. "Landscapes of Mimesis and Convergence in the Southern Jequetepeque Valley, Peru." *Cambridge Archaeological Journal* 26 (1): 23–51. https://doi.org/10.1017/S0959774315000165.

Uceda, Santiago. 2001. "Investigations at Huaca de la Luna, Moche Valley: An Example of Moche Religious Architecture." In *Moche Art and Archaeology in Ancient Peru*, ed. Joanne Pillsbury, 47–68. New Haven, CT: National Gallery of Art and Yale University Press.

Uceda, Santiago. 2010. "Theocracy and Secularism: Relationship between the Temple and Urban Nucleus and Political Change at the Huacas de Moche." In *New Perspectives on Moche Political Organization*, ed. Jeffrey Quilter and Luis Jaime Castillo, 132–58. Washington, DC: Dumbarton Oaks.

Van Dyke, Ruth M., and Susan E. Alcock, eds. 2003. *Archaeologies of Memory*. Malden, MA: Blackwell. https://doi.org/10.1002/9780470774304.

Warner, John P. 2010. "Interpreting the Architectonics of Power and Memory at the Late Formative Center of Jatanca, Jequetepeque Valley, Peru." PhD dissertation, University of Kentucky, Lexington.

Witmore, Christopher. 2007. "Landscapes, Time, Topology: An Archaeological Account of the Southern Argolid, Greece." In *Envisioning Landscape: Situations and Standpoints in Archaeology and Heritage*, ed. Dan Hicks, Laura McAtackney, and Graham Fairclough, 194–225. Walnut Creek, CA: Leftcoast Press.

Zori, Colleen, and Ilana Johnson, eds.2011.*From State to Empire in the Prehistoric Jequetepeque Valley, Peru.*British Archaeological Reports International Series 2310. Oxford: BAR.

SEVEN

Scaling the Huaca

Synecdochal Temporalities and the Mimetic Materialization of Late Moche Timescapes

GILES SPENCE MORROW

"The cleverer I am at miniaturizing the world, the better I possess it. But in doing this, it must be understood that the values become condensed and enriched in miniature. Platonic dialectics of large and small do not suffice for us to become cognizant of the dynamic virtues of miniature thinking. One must go beyond logic in order to experience what is large in what is small."

—Gaston Bachelard (1958: 150)

INTRODUCTION

In a short commentary on the status of scale models in contemporary architectural and artistic practice titled "Models are Real," Danish artist Olafur Eliasson argues that our ability to understand, inhabit, and evaluate space is dependent on recognition that "space does not simply exist *in* time; it is *of* time" (Eliasson 2007: 19). As the actions of users continually re-create its structure, Eliasson suggests, this temporal aspect of the built environment is often forgotten or repressed in Western society. This repression is the due to a persistent understanding of space

DOI: 10.5876/9781607326427.c007

as static and atemporal (Eliasson 2007: 19). Acknowledging the manner in which human populations model and remodel their modes of engagement with the lived-in-world, Eliasson advocates a theory that the modeling process, entailing the creation of concrete models themselves, plays an essential role in the co-production of space and society. The relationship between model and "reality" constitutes a recurring mimetic dialogue, each referencing the other in a cycle of infinite regression. Just as two parallel facing mirrors produce an infinite series of receding images, models and their referents are reflections of each other, nested interdependent realities that together produce our experience of the world.

Modeling the world is an inherently temporal process involving a combination of accumulated knowledge, skill, and our unique capacity to adapt and project action into the future. The construction of the built environment relies especially on our ability to carefully consider the interplay between memories of past projects and plans for current and future tasks. These projects come about through our relationship to the repertoire of tools available as objects to think with (Lévi-Strauss 1966). The process in which thoughts and meanings are attached to things is not arbitrary; rather, artifacts are active components of thought, ideas embodied in their fabric and made explicit in their use (Henare et al. 2007; Jones 2013; Malafouris 2013). Some kinds of things might be considered "cognitive objects" that do more than represent and reference other things but serve as a forum by which ideas are concretized. These things, whether manufactured or merely selected from the environment, are essential in "aiding, enhancing or improving cognition" (Hutchins 1999: 126). These objects allow individuals not just to think *about* their world but think *with* their world (Birth 2012: 6). As will be argued in this chapter, cognitive objects often play a critical role in temporalizing practices, especially in terms of how humans understand and regulate the passage of time. For instance, Kevin Birth (2012) argues that cognitive artifacts such as clocks and calendars serve to mediate our consciousness of cycles and patterns, concealing logics and processes that the users of artifacts do not understand or even know, but innately accept (Birth 2012).

In this chapter I analyze a specific kind of cognitive artifact—Moche architectural miniatures, or *maquetas*, of religious buildings—focusing in particular on their relationship to specific places in the Jequetepeque Valley, Peru, that date to the early Middle Horizon Period (AD 650–800). These architectural models were manufactured of unbaked clay and were placed in tombs at the great Moche cult center of San José de Moro in the north bank of the Jequetepeque Valley. I argue that their mimetic faculty, in which the architectural copy is manipulated to exert influence on original, larger buildings, worked in a temporal capacity as much as a spatial one (Taussig 1993). The miniatures functioned not simply to commemorate or memorialize a space but to store and reactivate a "time" that was understood as inseparable from the space in question.

Thinking through the miniature is contingent on a process of abstraction in which the fidelity of the model always relies on the deliberate exclusion of detail. As efficient forms of "off-line" reasoning (the habitual thought processes that internalize abstract ideas), miniatures and models present "surrogate situations" that more effectively allow us to come to understand the world (Clark 2010: 26). According to Andy Clark, the suppression of detail inherent to the creation of miniatures is a process of "selective concretization" that comes about through a fundamental "temporal relaxation" required when confronting the miniature. By "relaxing" the temporal restraints of action and reaction, contemplation of the miniature is reliant on episodic memory systems and explicit stored knowledge (Clark 2010: 26). The symbolic compression of miniatures improves understandings of abstract relations (Knappett 2010, 2012; Clark 2010). In this chapter, I will show that notions of time were inextricably embedded in the maquetas (architectural miniatures) recovered from the cult center of San José de Moro in the Jequetepeque Valley and the spaces they referenced. Ultimately, Moche temporalities were structured by architectural interventions in space on two distinct scales.

Ultimately, this chapter examines the temporal dimensions underwriting relationships linking humans, maquetas, and meaningful places in past Andean life worlds. I argue that for the Moche of the North Coast of Peru, acts of symbolic compression and miniaturization served to reanimate specific times, known ceremonial locales, and the attendant social identities created and reaffirmed in these places. The ritual efficacy of the architectural simulacra rests in their mimetic power to condense and transfer the potency of ritually charged spaces and their human and other-than-human attendants across time (Freedberg 1991). I interpret clay models, or maquetas, of ceremonial space, placed in tombs at the religious center of San José de Moro, as playing an important role in rites of death and renewal as well as serving as chronometric markers of temporal breaks. However, a comparison of these architectural models in relation to a remarkable sequence of ritualized reconstructions of architecture found at the Late Moche site of Huaca Colorada (AD 650–800) in the Jequetepeque Valley reveals that they served much more than as passive mementos of deceased leaders and revered places. Instead, they condensed a latent temporal or generative energy that at once stored but could potentially reactivate or remake ritually charged landscapes as active *timescapes*. In other words, scalar representations and transformations of place as exemplified in the maquetas and building renovations did not simply chart events in conformance with Moche ideologies of time, calendrics, and history. They also served as direct interventions and reinventions of time, place, and identity particular to the Moche of the early Middle Horizon Period. More specifically, the renovation of ceremonial architecture is interpreted as demarcating periods of temporal rupture that allowed for simultaneous acts of place making and identity reconstitution. Miniaturization in both monumental and modeled architectural spaces effectively materialized time's passage, serving as synecdochal "timepieces" that concretized Moche temporalities in material form through a nested sequence of spatial compressions of ceremonial locales on two distinct yet intimately referential scales.

Therefore, the role of portable miniature architectural models of ceremonial edifices represent more than depictions of lived space but should be interpreted as "objects-in-time" that served as symbolic referents to the passage of time (Birth 2012; Hutchins 1999). A comparison of the form of these models to the particular setting of ceremonial activity at Huaca Colorada will connect both scales as devices in a mimetic process that literally created and moved time in space, linking events and locales across the Jequetepeque Valley and further afield across the Moche sphere of influence. I argue that the ritualized acts of architectural renovation that accorded with Late Moche temporal ideologies were directly manifested in these models, and should not be approached as representations of mythic scenes, but were in fact "selective concretizations" of the specific lived experience of particular events in known locales (Clark 2010; Castillo et al. 2011; Swenson 2011a and 2011b; Uceda 2010; Wiersema 2015a). These models cannot be seen merely as souvenirs of particular phases of a structure's biography, but as instruments for the transference of particular temporal forces that are materialized in full-scale *huacas* (adobe brick edifices) themselves. Following a discussion of the utility and importance of miniaturization in Andean ethnographic and archaeological case studies, the setting of Huaca Colorada will be presented prior to a comparison to the maquetas discovered at San José de Moro.

In Small Things Remembered: Mind and Miniaturization

The form of communication presented by any miniature is iconic in nature: such objects are scalar models of the things they reference (Peirce 1984). However, there is a deeper purpose to this iconicity beyond simple representation. In his famous discussion of miniatures, Claude Lévi-Strauss asks a fundamental question: "What is the virtue of reduction either of scale or in number of properties?" (Lévi-Strauss 1966: 23). He concludes that miniaturization facilitates instant apprehension of the whole and that this produces a "very profound aesthetic emotion" by forcing us to reverse the normal process of comprehension (Lévi-Strauss 1966: 15). Although manipulation of scale allows us to consider miniaturized objects as less "formidable," qualitatively simplified through scalar diminution, the relationship between the maker and consumer of scalar artifacts is far more complex than is immediately apparent. Such a scalar transposition allows us to extend and diversify "our power over a homologue of the thing, and by means of it the latter can be grasped, assessed, and apprehended at a glance" (Lévi-Strauss 1966: 16). It is the immediacy of recognition that enlivens miniatures, drawing the viewer into these objects. Through the miniature, a sort of conceptual victory can be had, "which gratifies the intelligence and gives rise to a sense of pleasure," an intrinsic value of scale models that "compensates for renunciation of sensible dimensions by the acquisition of intelligible dimensions" (Lévi-Strauss 1966: 17). As I will argue later in this chapter, the models effectively shift the experience of time by creating a place that is a condensation and recombination of memory and imagination.

The scaled-down artifact offers instant totality and a subsequent unraveling of detail. The effort to make a model having been undertaken, its reception as an object requires almost no effort at all, allowing comprehension to be experienced immediately. If the space, object, or individual (in the case of figurines) depicted in miniature is familiar to a viewer as a component of her or his lived experience or memory, then appreciation is primary and directly evocative. Even if the viewer has no such previous experience of the context or form depicted in miniature, the totalizing effect of viewing such objects allows for a much more nuanced and complete understanding of the whole. To extend the aphorism, if a "picture is worth a thousand words," then a model must be worth a thousand pictures (Smith 2010). The communicative potential of the miniature is unrivalled in terms of the immediacy of its transmission, something clearly recognized since the earliest phases of the creation of portable art during the Neolithic "revolution" to our current fascination with three-dimensionally printed objects. Such a relationship can be established in the contemplation of any miniature, perhaps explaining the considerable body of literature focused on the role of anthropomorphic figurines in cultures worldwide (Bailey 2005; Joyce 2007; Halperin 2014; Hubert 2014, 2016; Lesure 2011; Meskell et al. 2008). Although architectural models are far less prevalent than that of figurines in the archaeological record, the study of miniature depictions of the built environment can provide a different kind of understanding of past worlds, especially how spaces were experienced in time. By focusing specifically on miniature depictions of interior architectural spaces, I will discuss the potential impacts that such models held as temporal devices that invoked symbolically charged places.

Within the context of literary criticism, Gaston Bachelard, suggests that the small conjures up the infinite more easily than the large. Indeed, the small, in certain instances, is the only way to create the sublime or to express abstract ideas. An attempt to understand the nature of time's passage is no doubt among the most profound and enduring conceptual abstractions that human populations face. Only through the miniature can we really think about the truly enormous aspects of both material realities and conceptual infinities such as time, an experience that Bachelard describes as "intimate immensity." Critiquing the "majestic" notions of "being-in-the-world," Bachelard argues that an individual feels "more at home in miniature worlds which, for me, are dominated worlds" (Bachelard

1958: 161). The miniature densely compresses information, collapsing elements of space and time that form the fundamental aspects of memory. Psychological studies suggest that miniaturization affects time perception, concentrating experience as space is compressed, suggesting that spatial scale may be a principal mediator in the experience of time (De Long 1981, 1983). If such a temporal compression is fundamental to the consideration of scalar objects, viewing the miniature invites the subject to know something without yet really knowing it, to circumvent the limitations of experience and transport the viewer into a time beyond time. Such considerations illuminate the psychological efficacy of miniatures but still fall short of explaining the role of maquetas in Andean societies.

Andean Approaches to Scale

Bachelard's point that miniatures engage and enliven memory of lived experience of place and time is particularly clear in Andean ethnographic studies. In her work on the role of miniatures in the contemporary Andean community of Sonqo, Catherine Allen explicitly states that miniatures do more than simply delight their owners; "they change the lived-in world" (Allen 1997). Through her investigation of the "pebble game" conducted at the destination of the ritual pilgrimage of the annual festival of Qoyllur Rit'i, Allen shows that the act of constructing elaborate miniature houses and corrals serves as a performative conduit through which humans communicate with the divine. Allen argues that the underlying motivation for such a form of communication in contemporary Andean ritual practice is based on the idea that all beings (animate or otherwise) are interconnected through *ayni*, the fundamental reciprocal "give-and-take" that controls and circulates vitality in the Andean world (Allen 2002; Bray 2009; McEwan 2015). As a cosmology that does not separate between mind and matter, this form of "consubstantiality" relies upon the coexistence of this vitality within all substances, thereby attributing agentive animacy to material objects. This reciprocal consubstantiality also relies on a sense of envelopment or synecdoche, with parts standing for the whole, and the whole standing for the part. Such synecdochal frameworks conceptualize the world in terms of "mutually enveloping homologous structures that act upon each other: ayllus within ayllus, places within places, every potato field has its own vertical ecology, every huaca a mountain and every mountain a huaca." (Allen 1997: 76). Within such a worldview Allen suggests that there is a tendency for relationships between microcosms and macrocosms to collapse and expand back and forth endlessly, with each a reference to and creator of the other. In that both scales of objective reality imply each other concretely, Allen suggests that "a powerful miniature informs the cosmos with its own form, and in turn the cosmos is not only referenced but instantiated in the objects at the same moment" (Allen 1997: 81; 2011, 2014a, 2014b; Swenson 2015: 701–2; Swenson and Warner 2015: 26).

I argue that this conceptual framework of enfolded material identities can be extended to time in the ancient Andes, a "synecdochal temporality" that differs from both linear and cyclical temporalities so often discussed in the literature. Accordingly, a "temporal synecdoche" is based on a dynamic yet nested flow between points of memory, perception, and expectancy that mirror Edmund Husserl's tripartite structure of temporal cognition. For Husserl, the perception of any moment always contains elements of "retention" and "protention" of what has just passed, and what is about to occur (Husserl 1992). Within this structure, the "now" or the present moment cannot be pulled out of the sequence of the immediate past and the immediate future, but is formed as a shifting frame of experience informed by knowledge and prospect.

A musical analogy can illuminate both Husserl's theory of time and Andean synecdochal historicities. It is impossible to hear and appreciate a melody if our immediate apprehension of the note before our ears was not accompanied by our "memory" of the note just before and in expectation of the note to follow (Hoffman 2009). If the melody is familiar, the structure known, each individual note simultaneously prompts memory and action, each nested within the other, referencing each other in a cycle of constant revelation. However, as Tim Ingold states in his work on temporality, "in music as in social life, there is not one rhythmic cycle, but a complex interweaving of many concurrent cycles" (Ingold 1993: 160). It is in the efforts to find resonances between these cycles that "timework" is conducted, the meaningful actions that prevent and produce particular temporal experiences (Flaherty 2011). By marking moments of change through repeated, iterative action fundamental to most ritual, we create and define the tempo of social life through the manipulation of temporal experience (Leach 1961; Rappaport 1992). Ritual actions often serve as "social technologies of time," methods by which we seek to comprehend and attempt to control the seemingly relentless passage of time (Østergaard 2013; Willerslev et al. 2013; see introduction and chapter 1 in this volume).

An example of this sort of "timework" is clearly expressed in the manipulation of miniature quotidian elements such as animals and domestic tools in the "pebble game" of the Qoyllur Rit'i. Scale models of homesteads are populated with carefully selected stones representing desired livestock in order to control and influence possible futures (Meddens 1995; Sillar 1994, 2009; Stenstrud 2010). Seen as tiny storehouses of prosperity and well-being, these miniature figures are carefully tended, animals "fed," and houses maintained to create miniature "texts" that express materials desires in full view of the deities at the end of a difficult pilgrimage (Allen 1997: 81–88; McEwan 2015: 268). Although occurring at different spatial scales, the materialized control of the passage of time is also evident in acts of ritual termination and dedication of architecture at earlier Moche sites discussed in the next section. Iterative renovations memorialized the phasing, scheduling, and sequencing of ritual events and activated an awareness of temporal rhythms.

In the end this synecdochal temporality was expressed in the spatialization of time's passage through the combined acts of architectural reconstruction and miniaturization that appear to link the contemporaneous Moche sites of Huaca Colorada and San José de Moro. Following a description of the specific construction history of Huaca Colorada in the next section, I will outline the tradition of including architectural models of iconic ritual locales alongside elite individuals interred at San José de Moro as acts not only of memorialization, but as efforts to control and direct the flow of time within this particular community.

Architecture, Time, and Power in Moche Religious Ideology

The Moche refers to a shared political and religious ideology that persisted throughout the desert North Coast of Peru during the Andean Early Intermediate and Middle Horizon Periods (AD 100–800) (Bawden 1996; Castillo et al. 2008; Quilter 2002; Quilter and Castillo 2010; Shimada 1994; Uceda and Mujica 1994, 2003). Often considered to represent one of the earliest state polities in the Americas, Moche society was defined by a highly stratified social structure (Bawden 1996; Benson 2012; Shimada 1994; Quilter 2002). Moche ideology and power appear to have been founded on elite ritual spectacles, usually focused on interactions between individuals of elevated status who are engaged in ritual, warfare, and mythic events (Hocquenghem 1987; Jackson 2008). There are clear indications that the known suite of Moche iconographic scenes were reenacted by elite ritual practitioners who were subsequently buried in their regalia and accompanied by the same material symbols of their status depicted in the iconography (Alva and Donnan 1993; Castillo and Donnan 1994a, 1994b; Golte 2009; Quilter 1997).

Defined by cycles of warfare, prisoner capture, and human sacrifice, the activities depicted likely served to legitimate religious authority; however, the relationship between Moche political theology and cosmogonic myths is poorly understood and the subject of considerable debate (Alva and Donnan 1993; Bourget 2006; Donnan 1978; Golte 2009; Hill 1998; Swenson 2003). The ritual control of human life, death, and regeneration were realized through acts of destructive creation, a belief that applied as viscerally to the built environment as it did to human subjects. (Bourget 2006; Prieto Burmester 2008; Swenson 2003; Swenson and Warner 2012). Death was a liminal and necessary phase in the process of creation and becoming. Therefore, the sacrificial transformation of the human body or architectural space regenerated life, cosmos, time, authority, and ultimately place itself (see Swenson, this volume [chapter 6]). This sacrificial ontology appears to have persisted across the Moche sphere, with dedication and termination rites at a number of centers indicating that the Moche conceived of architectural spaces as vital, living entities in their own right (Swenson 2015; Uceda 2010).

The Moche are famed for their expressive and conceptually nuanced sculptural and painted ceramic traditions, and of particular importance to the present study are depictions of stepped ceremonial platforms in both two and three dimensions (Castillo, Cusicanqui, and Mauricio 2011; Wiersema 2010, 2011, 2015a, 2015b). The ubiquity of these representations strongly suggest the central importance of these structures in Moche religious ideology, a hypothesis corroborated by the archaeological investigation of numerous monumental adobe brick edifices, known as huacas, that dot the landscape of the North Coast of Peru. These platforms are often situated in chambers, as also depicted in the ceramic corpus (Wiersema 2010, 2011, 2012, 2015a, 2015b). Just such a highly remodeled and carefully curated room complex containing a stepped platform was recently excavated at Huaca Colorada. This complex has proved to be one of the closest physical analogs to the ritual platforms depicted in Moche iconography (Swenson et al. 2010, 2011, 2012).

Adobe brick huaca mounds have been interpreted as mimetic simulations or miniaturizations of the mountains that form the foothills of the Andes along the Pacific Coast, relating closely to the pan-Andean veneration of mountain peaks as both deified ancestors and loci of supernatural and generative power (Bastien 1978; Bawden 1996; Dean 2007; Kolata 1996; Quilter 2001; Swenson and Warner 2015; Uceda 2001a, 2010). For instance, Huaca de la Luna was located at the base of the coastal massif known as Cerro Blanco and was the primary temple for the urban settlement in the toponymic Moche Valley during the early phases of occupation (Uceda 2010). The ongoing excavations at the monumental complex of plazas, ramps, and platforms at this site have provided critical information for our interpretations of Moche religious and political ideology (Uceda 2001a, 2001b; see Swenson [chapter 6] and Seoane and Culquichicón [chapter 8] this volume).

Architectural reconstructions at Huaca de la Luna repeat and mimic earlier building phases and reinvigorated these structures as living timepieces, "chronotopic" spaces that underlined the conservatively maintained ideals expressed through the monument (Bakhtin 1981: 84; Swenson 2015: 689; see Roddick this volume [chapter 3]). The physical act of construction and completion of a particular phase created the tempo fundamental to acts of renovation.

Sitting above tiers of brightly painted friezes, the summit of Huaca de la Luna supports a series of elevated interior chambers and platforms that served as central stages for state sacrificial rituals. The ceremonial events orchestrated here may have been scheduled to natural annual cycles, and the five documented reconstructions of the huaca—involving the complete envelopment of earlier monuments by larger and later edifices—appear to have materialized and set time in motion (Bourget 2001; Hocquenghem 2008; Swenson 2012a). In fact, the likely calendrically timed expansion of temple mounds, involving the curated termination and rededication of preexisting monuments, has also been documented at Huaca Cao Viejo in the Chicama Valley and Dos Cabezas in Jequetepeque (see Swenson [chapter 6], this volume). The enveloped nesting of the platform mounds points to a notion of temporality as synecdochal, in which older, terminated temples (previous wholes) become an active and energizing "part" of the later, "present" whole (Canziani 2009: 206; Uceda 2010: 150). Excavations of the ceremonial precinct at the summit of Huaca Colorada have identified elements of this core Moche ideology, but as Swenson (this volume [chapter 6]) has discussed, the reductive as opposed to amplified architectural terminations point to a local tradition of architectural time-reckoning. Although the sequence of renovation events at Huaca Colorada may parallel elements of this clearly enduring architectural tradition, the incremental compression of the internal platform chamber within the core of the ritual precinct presents a unique opportunity to consider variations in how the Moche made and regulated time in space (see also Swenson, this volume [chapter 6])

Marking Time at Huaca Colorada

The Late Moche (Moche V) Period (ca. AD 600–850) emerged after a period of social and environmental disruptions that ushered in major transformations in Moche society (Castillo 2001, 2003; Quilter and Koons 2012). Evidence from ice cores extracted from the Quelccaya glacier in south-central Peru indicates that a severe drought associated with El Niño—Southern Oscillation (ENSO) events occurred between AD 593 and 594, the effects of which extended through the end of the sixth century AD (Shimada et al. 1991). Therefore the Late Moche Period was an era marked by demographic and ideological restructuration that saw a diversification and popularization of Moche religious culture across the North Coast region including the Jequetepeque Valley (Dillehay and Kolata 2004; Dillehay, Kolata, and Swenson 2009; Swenson 2004, 2008).

Huaca Colorada was the most prominent Late Moche ritual locale in the southern Jequetepeque Valley (see Swenson this volume [chapter 6]). The center is located approximately 100 kilometers north of the Huacas de Moche, situated at the base of the Cerro Cañoncillo mountain range within an arid region known as the Pampa de Mojucape (see Swenson, this volume [chapter 6] figure 6.1). Surrounded by a largely agricultural settlement covering approximately twenty-four hectares, Huaca Colorada is dominated by an elongated adobe brick platform structure built atop a modified sand dune, and it measures approximately 390 meters from north to south by 140 meters from east to west and rises nearly 20 meters at its highest point above the relict agricultural fields that surround it. Differentiated into three distinct sectors, the principal ceremonial precinct under investigation, Sector B, is located at the peak of the structure above and between two manufacturing and residential areas on the lower tiers situated to the north and south, Sectors A and C respectively (see figure 6.5,

Swenson this volume [chapter 6]). Possibly serving as ceremonial and political headquarters of a powerful polity, the principal religious constructions of the monumental core of Huaca Colorada consist of eight daises or altars, all of which were intentionally interred under floors or construction fill (see Swenson, this volume [chapter 6]).

This central ceremonial district is now understood to have comprised two complementary but opposed halves, each anchored by a stepped ritual platform at the southern end of each respective precinct (which will be referred to as the western chamber and the eastern terrace). The ritually-terminated and well-preserved western chamber measures ten meters (NS) by seven meters (EW) and contains a two-stepped platform or dais that served as a stage for acts of ritual performance (Swenson et al. 2010, 2011). Excavations have revealed that this western platform chamber was repeatedly remodeled, with each renovation concurrently raising the level of the previous floor while simultaneously diminishing the overall length of the room, compressing the space both laterally and vertically, incrementally reducing the original area from 154 square meters to 70 square meters prior to termination and abandonment (figure 7.1). Measuring two meters long by four meters wide, the two-stepped platform located in the southern end of this western chamber maintained its position and dimensions across all phases of construction, indicating that this component clearly served as the point of focus for the activities that took place in this interior space throughout all occupation phases (figure 7.2). As mentioned, the ultimate phase of use of the internal platform chamber had compressed the space to cover an area of less than half the original dimensions, and it was extraordinarily well preserved due to an intentional decommissioning episode that saw the entire chamber filled with upward of 180 cubic meters of clean sand fill. This singular termination event encased and preserved two plaster-coated wooden pillars found rising from the platform, highly curated architectural components that once supported a simple gable roof commonly depicted in Moche iconography (figure 7.2).

In total, we have discovered at least seven distinct phases of renovation of the western chamber each incrementally reducing the chamber in size, compressing the layout of the original dimensions while carefully maintaining and reiterating fundamental components of its spatial organization (figure 7.1). Entrances into the chamber through the eastern, western and southern walls of the room were carefully re-created in each phase, maintaining access patterns into and out of the space through time. These reductions clearly defined the use of this space as focused on the gable roofed structure found at the southern end of the western chamber. Certain phases of these architectural reductions were punctuated by the incorporation of human sacrificial victims beneath each successive floor and within the construction fill behind various reductions of the northern wall of the chamber (Swenson et al. 2010, 2011, 2012, 2013, 2015) (figure 7.3).

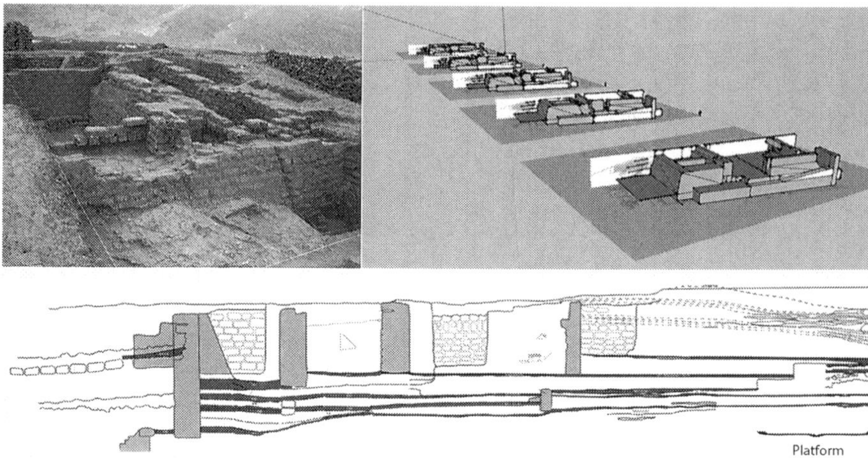

FIGURE 7.1. *Sequential reductions of northern wall of western platform chamber at Huaca Colorada, modeled three-dimensionally and compiled as a combined profile of all construction phases (bottom).*

Excavation of construction fill that composed the penultimate reduction of the chamber uncovered one particularly noteworthy sacrificial burial, an individual who was sprawled across the area between the newly constructed northern wall and the wall that preceded it, tossed into the rubble while the space was being closed (Swenson et al. 2012). Directly beneath this individual were the fragile remains of a large wooden post approximately thirty centimeters in diameter and nearly two meters long, laid in the adobe rubble in an almost identical orientation and position as the sacrificial victim (figure 7.4). Considering their proximity, relative size, and orientation, this post and the human offering were interred together in a single event underscoring the symbolic link between acts of human and architectural sacrifice.

Atop this construction fill was a circular alignment of adobe bricks surrounding the remnants of a post under which a spondylus shell had been placed, a common dedicatory offering in the construction of Late Moche architecture (Millaire 2016; Shimada 1994). This particular architectural element was similar to a series of other adobe brick shaft-like post emplacements that dominate the Eastern half of the ceremonial nucleus. Constructed of levels of mortared adobe bricks arranged in roughly conical forms, these circular shafts tapered downward from approximately one meter in diameter to fifty centimeters at their bases, cutting through numerous earlier floors, linking earlier phases of construction otherwise obscured. Often over two meters deep, these well-like adobe structures were clearly associated with floors on the higher levels of the huaca east of the western ceremonial chamber. Continued excavation in

FIGURE 7.2. *Iconographic depiction of ceremonial platform from Moche ceramics in comparison to penultimate phase of use of the western platform chamber at Huaca Colorada with clay-covered wooden columns in situ.*

Sector B uncovered alignments of these features that proved their function to be the supporting bases for large wooden posts, continually built up over time as new floors were added to the huaca with each cycle of ceremonial reconstruction (Swenson et al. 2012).

Once the interior ceremonial chamber had been filled and abandoned in the latest periods of occupation, it was this higher eastern area that continued to be an important focus for ceremonial activity. The uppermost eastern terrace of the huaca was covered by a roofed veranda that sheltered a stepped platform and ramp complex, the counterpart to the dais in the chamber to the west (figure 7.5). Renovation in this higher eastern half of Sector B was based on vertical as opposed to horizontal repetition; post emplacements were built on top of older adobe post bins allowing the huaca to change and grow while maintaining its elements original positions relative to the publically visible ceremonial terrace. Thus space was renewed with vertical shifts rather than horizontal reductions, signaling a difference in architectural construction between the eastern and western half of the ceremonial sector of the huaca. This desire to maintain the location of posts over time could easily have been constructed in other ways, but

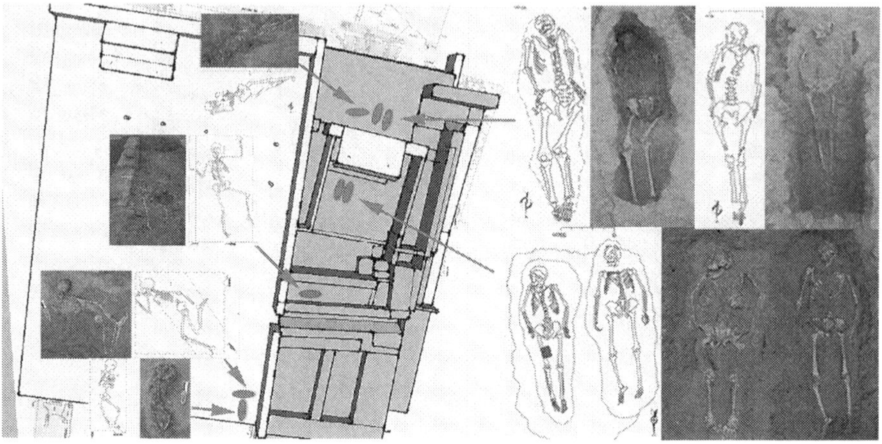

FIGURE 7.3. *Three-dimensional model of the ceremonial complex of Huaca Colorada showing the locations and contexts of the sacrificial victims found in situ.*

it seems that the continuity of post emplacements cited the previous construction sequences, creating physical conduits through which an association with the past is maintained, and affirming a continual connection to the earliest iteration of the structure and its ancestral inhabitants. The transference of cultural knowledge though the process of removing, preparing, and resetting these posts connected the different generations of construction teams at Huaca Colorada (Pauketat and Alt 2005: 217). These construction efforts marked and created temporal rhythms at Huaca Colorada, and the envelopment of past floors and posts under new constructions speaks to this synecdochal nesting of time in space.

As mentioned above, excavations of this eastern terrace has presented a complex sequence of remodeling episodes that both paralleled but differed from renovations in the lower western chamber below. Our recent excavations in 2014 and 2016 exposed a slightly larger (2.2 m × 4.4 m) but otherwise identical two-step platform here in exactly the same north-facing direction as the equivalent platform in the western chamber (Swenson et al. 2015). Unlike the latter, the dais of the eastern terrace was accessed by a central two-meter-wide ramp immediately north of the platform, with this ritual stage complex located at the very eastern edge of Sector B, and would have been visible from the open plaza below that stretches eastward from the base of Huaca Colorada toward the later Middle Horizon Period ruins of the site of Tecapa (Swenson et al. 2015). Investigation of this public platform revealed that it was enlarged laterally at least once before being intentionally destroyed in an intense burning episode. This dramatic termination event effectively "fired" the floor surfaces of both the platform and surrounding area as well as the clay-coated gabled roof that once stood over

FIGURE 7.4. *Sacrificial victim in the rubble of a northern wall reduction (left), with location of decommissioned and interred post directly beneath the burial (right).*

the eastern ramped platform, which had collapsed to seal the surface of the entire southeastern terrace of the Huaca before being carefully encapsulated by subsequent floors (figure 7.6). Considering the size of the platform, the clay-covered roof would have been of considerable weight, requiring the support of large wooden columns, the burnt bases of which remained deeply embedded in the floor of the platform. The concentration of burnt rubble and flooring found across the entire eastern platform area indicates that it was intentionally immolated, likely requiring considerable volumes of combustible material to be amassed in preparation for this act of architectural termination.

Stratigraphic comparison of the complete construction sequence of Sector B has shown that the eastern platform was built contemporaneously to the earliest phases of the western chamber, indicating that both public and private performances would have been held on these two ritual stages, linked by the phased sequence of entrances through the central dividing wall separating the eastern and western sectors (Swenson et al. 2015). As the private western chamber was reduced in length and volume, each of these doorways were carefully bricked up and plastered over, actively maintaining the accessibility of the two spaces over time. While the higher and eastern public platform would have been visible

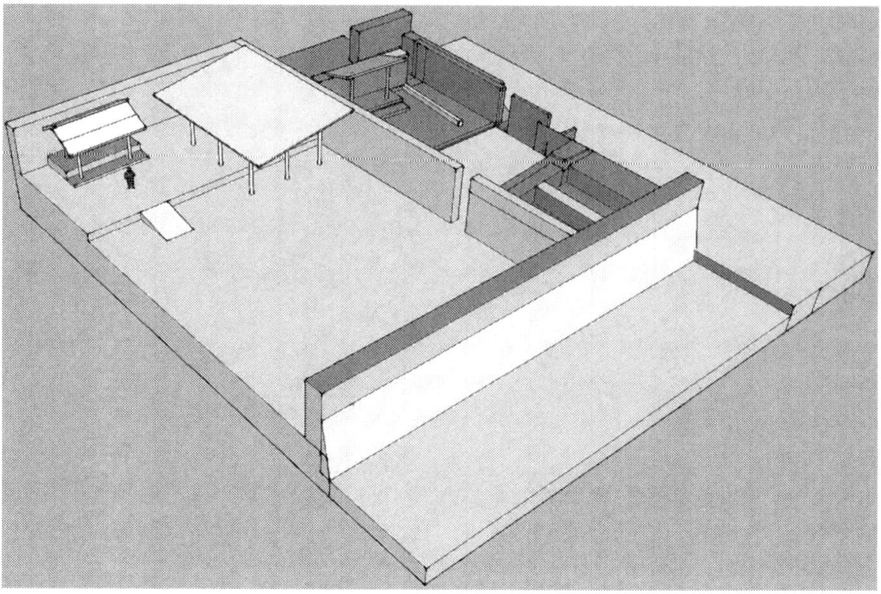

FIGURE 7.5. *Isometric three-dimensional model of relationship between western chamber platform and eastern terrace platform.*

to audiences who could have gathered in the plaza at the base of the huaca, the western, private chamber was an intimate space wherein access and movement were highly controlled. This incremental compression of ceremonial space raises the question of the degree to which this particular architectural choice was possibly influenced by the nested sequence of interconnected chambers at the nearby site of Jatanca as discussed by Swenson in chapter 6. The dual ceremonial chambers found atop Huaca Colorada may have referenced and possibly commemorated the dyadic organization of ritual platforms and nested precincts at Jatanca, which may have been revered as a sacred, ancestral space.

Moreover, the dualistic configuration of the ceremonial nucleus of Huaca Colorada speaks precisely to the synecdochal construction of space-time in that the whole was conceived through the operation of two separate but dependent and complementary parts (see introduction to this volume [chapter 1]). Some of the oppositions between the eastern and western halves of the huaca, anchored by comparable stepped platforms, are certainly intriguing and raise many questions. The differences include the following: the east terrace is public, higher, renovated through vertical reconstructions, and was terminated by fire; the west chamber is private, lower, characterized by phased, horizontal reductions, and was terminated by interment in rubble and sand. The incremental and repeated architectural terminations on both the east and west precincts provide further

FIGURE 7.6. *Eastern terrace platform of Huaca Colorada in situ bearing evidence of extreme burning prior to decommissioning.*

testament to this synecdochal creation and movement of time—in which wholes were continually divided into parts, similar to Allen's *ayllus* or potato fields referenced above (Allen 1997).

The complex mutigenerational construction of the ceremonial architecture at Huaca Colorada clearly functioned to "anchor people in space and to link them in time" (Gillespie 2007). Excavation and analysis of the sequence of repeated, intentional construction phases have provided an increasingly more nuanced architectural biography punctuated by specific cycles or ritual construction and pilgrimage to the site (see Swenson, this volume [chapter 6]). Communal acts of meaningful renovation that took place at Huaca Colorada served to "embody a collective memory about the past, a reference to origins that often forms a salient bond" that united members of this particular community (Gillespie 2007: 3). With each of the sequential spatial reductions of the western chamber at Huaca Colorada, coinciding with acts of human and animal sacrifice, subjects were literally incorporated into the fabric of such sacred spaces, consubstantially enlivening the built environment through sacrifice (Swenson 2015, this volume [chapter 6]). In this way, the human and animal sacrifices at Huaca Colorada can be seen as synecdochal referents of the community as a whole at a given point in time. As will be discussed in the next section, this simultaneous, incremental ("partial") sacrifice of the architectural edifice, accompanied by a representative human (in)*dividual* at Huaca Colorada, is repeated in reverse within the tombs of San José de Moro through the incorporation of architectural models within mortuary spaces dedicated to interred elite individuals. The interment of the models in elaborate chamber tombs at San José de Moro continued to enliven and replenish distant sacred locales such as Huaca Colorada.

The process of reduction and miniaturization of the ceremonial platform chamber at Huaca Colorada finds parallel in the maquetas for the latter served to synecdochally fix or "store" specific architectural complexes and their associated moments in time at the great Priestess center of San José de Moro. Interconnected acts of architectural and elite burial were spatially separated yet fundamentally linked efforts to achieve the same "timework," possibly during periods of social and communal rupture or possibly to commemorate the cessation or beginnings of celestial cycles (Willerslev et al. 2013). In other words, both the interment of models in the burials and architectural terminations and rededications served to mark the passage of time and the renewal of social order (Østergaard 2013). Although separated by a distance of twenty-five kilometers within the Jequetepeque Valley, the interconnection between the ceremonial core of Huaca Colorada and San José de Moro is evident due to notable similarities of the artefactual assemblages. However, this is most apparent in parallels between the formal design features and access patterns of some of the architectural models interred in the tombs in relation to specific phases of occupation of the eastern and western halves of Huaca Colorada.

(Re)presentations of Moche Architecture and the Maquetas of San José de Moro

Over the past twenty-five years, the extraordinary maquetas of the great funerary center of San José de Moro have been extensively analyzed, but undertheorized (Castillo et al. 2011; Gavazzi 2012; McClelland 2010; Pillsbury 2015; Wiersema 2010, 2011, 2012, 2015a and 2015b). Donald McClelland (2010) has recently outlined the complex interrelation of maquetas to the particular artifact assemblages of each tomb context investigated during the 1991 and 1992 excavation seasons. It is from within this particular data set that my current comparison to the building phases and the associated ceramic assemblages of Huaca Colorada will be drawn. My archival research conducted under the supervision of Prof. Luis Jaime Castillo Butters at the San José de Moro project collections at the Pontifica Universidad Católica de Perú in Lima allowed for further consideration of the artifact assemblages associated with the specific tomb contexts that contained maquetas. Maquetas were manufactured of either unfired or very low-fired clay and were found in extremely poor states of preservation. Careful consolidation and recording in situ have allowed a small number of examples to be fully conserved and reconstructed, with detailed notes, drawings, and photographs serving as the foundation for the plans of those maquetas that were damaged beyond repair.

Until quite recently, depictions of architecture in in the Moche artistic corpus were simply interpreted as a descriptive shorthand to communicate the specific setting of ritual activity within painted and modeled scenes. This line of thinking also extends to the Moche maquetas of architecture excavated at San José de Moro (Castillo 2000a, 2000b, 2001, 2003; Castillo, Nelson, and Nelson 1997; Castillo et al. 2008; McClelland 2010). A collection of forty-four such models have been documented and conserved, with many of the elaborate tombs containing numerous architectural models in a single context (Castillo et al. 2011; Castillo et al. 1997). The extant maquetas are all less than fifty square centimeters and are constructed and finished in much the same way buildings of the period would have been, rectangular in plan and slipped and painted with red, black, and white pigments (Castillo et al. 1997: 127). Three of the so-called Priestess burials for which San José de Moro is famed contained maquetas that were seemingly made locally, perhaps even constructed within the tomb itself during mortuary rituals (Castillo, Nelson, and Nelson 1997; Castillo, Cusicanqui, and Mauricio 2011; Wiersema 2010, 2011, 2012, 2015a, 2015b).

Due to the close similarity of one particular excavated maqueta to a small platform structure at the site of Portachuelo de Charape in the hinterland of the Jequetepeque Valley, Edward Swenson convincingly argued that the maquetas from the tombs at San José de Moro were representations of real rather than imagined spaces (Castillo, Nelson, and Nelson 1997; Johnson 2008, 2012; McClelland 2010; Swenson 2004; 2008: 415–21). Portachuelo is an example of one of many relatively small ramped structures called *tablados* found throughout the Jequetepeque Valley thought to represent scaled-down minor regional versions of the massive platform mounds found at Late Moche sites of more central importance such as Huaca Colorada (Swenson 2004, 2011a; Wiersema 2010). Given this correspondence, it seems that maquetas may have served as symbolic tomb substitutes for the full-scale *tablados*, possibly as markers of community identity. If these models served as referential substitutes for the ritual architecture associated with the deceased, this speaks to the social connection that individuals likely held vis-à-vis these meaningful locales. If this were the case, having a representation of the specific layout of the gable-roofed complex of their "home" huaca included in their tombs may have served as a powerful emblem of their political and geographic affiliation (Swenson and Warner 2015: 46).

As an important component of the funerary assemblages found in many of the tombs of San José de Moro, the maquetas have been found in the same context as fine-line painted stirrup vessels bearing iconographic depictions of identical though schematized ceremonial locales that are enlivened with characters in action. In stark contrast, the maquetas are devoid of evidence of any such personages; merely empty spaces, or ad hoc containers for miniature ceramic vessels known as *crisoles* and, in two notable cases, collections of miniature copper artifacts (McClelland 2010). Despite the fact that these models would have served as ideal stages for the placement of figurines comparable to a remarkable wooden Chimú Period maqueta found at Huaca de la Luna, none of the Moche maquetas bear any evidence of their inclusion (Pillsbury 2015; Uceda 1999). Accordingly, I suggest that these models were not merely depictions of ceremonial locales, but should be seen as individualized and animated representatives of the locales in question. As instantiations of particular powerful places in specific moments, the identity of the full-scale ceremonial locale they reference is entirely compressed, transferred, and contained in these miniatures as generative seeds filled with potential energy to connect and collapse space and time (see also Swenson 2012a:186).

The architectural similarity of two particular maquetas from separate tomb contexts (M-U30 and M-U41) to the basic layout of the western chamber at Huaca Colorada are particularly relevant to this analysis. Specific investigation of the artifact assemblages associated with both of these tomb contexts show stylistic and chronological continuities with the ceramics found in the feasting middens of Huaca Colorada, pointing to additional connections between the sites. Both of the tombs contained multiple models, with a total of seven found in tomb M-U30, and a total of five in tomb M-U41. It is worth noting that there is a direct one-to-one relationship between the number of maquetas and the number of individuals buried in both tombs, suggesting that each interred individual was associated with a particular model and thus a specific geographic locale.

The layouts of Model 7 of Tomb M-U30 and Model 15 of Tomb M-U41 bear remarkable similarity to the enclosed space of the western chamber at Huaca Colorada (figure 7.7). Indeed, it deserves consideration that these models are directly referencing the site. Both of these particular maquetas are divided into two sections within an enclosing perimeter wall, composed of exterior forecourts with a raised platform partially covered with a single gable roof, and a series of interior chambers connected to the forecourt by a doorway on the right hand side of each platform in exactly the same manner as at Huaca Colorada. Considering that the entrances from both the east and west sides of the northern limits of the internal platform chamber at Huaca Colorada were reestablished anew at each phase of construction, it is notable that the doorway immediately to the east of the platform through the southern wall of the chamber itself remained unchanged throughout the entire sequence of renovations that compressed the chamber toward the south. Excavations of the complex of divided rooms located behind this southern access point immediately behind the platform at Huaca Colorada closely mimics the layout and organization of the interior chambers found in the two models. Models 7 and 15 both have a singular central access through the perimeter wall through a stepped threshold located immediately opposite the platform. Again, this access pattern is remarkably

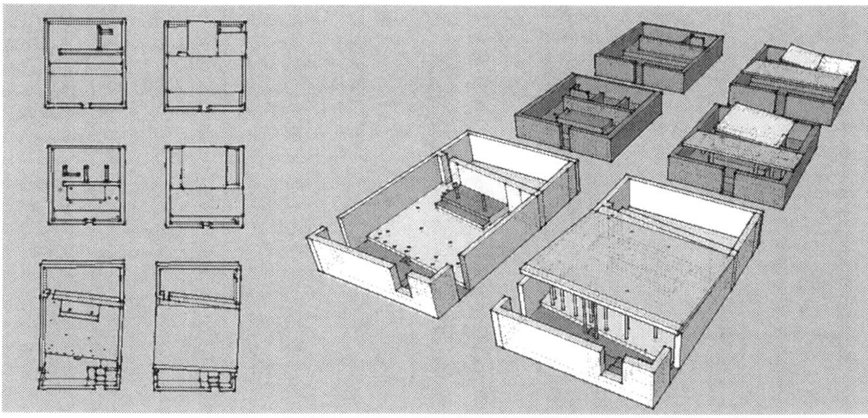

FIGURE 7.7. *Plan views (left panel) and isometric views (right panel) of three-dimensional models of western platform chamber of Huaca Colorada (bottom) and two maquetas from San José de Moro, Model 7 of Tomb M-U30 (top) and Model 15 of M-U41 (center). In both panels, the models on the left show the chambers without any covering roofing material or supporting posts, and on the right, hypothetical roofs are added based on those found in situ with Model 7 of M-U30 as reported in McClelland 2010.*

similar to the spatial organization of the internal chamber at Huaca Colorada. Excavation of the floor covering the penultimate phase of structural compression of the interior chamber revealed a heavily worn adobe brick step located immediately in front the platform. As this pattern of controlled center-line access into these chambers is repeated at both scales, it suggests that movement and vision were intentionally directed toward the ceremonial platform and the activities that took place upon them.

Based upon analysis of the associated ceramics, the specific chamber tomb contexts in which Models 7 and 15 were encountered were dated to the same Late Moche Period contemporaneous to that of Huaca Colorada (Castillo et al. 1997; McClelland 2010). The similarities of a number of the artifacts found in both tomb M-U30 and tomb M-U41 bear striking resemblance to many of the ceramics found in direct association to significant deposits of ceramics located in close proximity to the western chamber at Huaca Colorada. Fine-ware ceramics and enormous amounts of organic remains deposited in significant feasting events were recovered from large middens that surrounded the entire western and northern perimeters of the western chamber (Lynch 2013; Swenson et al. 2011, 2012). Representing the highest concentration of modeled and Cajamarca feasting ceramics of any area excavated at Huaca Colorada, the density of material in the middens contrasts with the clean and relatively sterile contexts found within the chamber itself. The ceramics recovered from these middens are stylistically identical with those found intact within the two specific tombs from San José de Moro discussed above. The cooccurrence of Cajamarca kaolin plates in both of the tombs and in midden contexts at Huaca Colorada is especially noteworthy (Castillo et al. 2008; Lynch 2013). The only other site in the Jequetepeque Valley where Cajamarca plates have been found in secure context is the contemporaneous site of Cerro Chepén, a fortified hilltop site thought to be directly associated with San José de Moro (Castillo, Cusicanqui, and Mauricio 2011). Remarkably, many of the stone structures found atop Cerro Chepén bear striking resemblance to other maquetas found at San José de Moro, further suggesting linkages between these three sites (Castillo et al. 2008; Castillo et al. 2011; Swenson 2012b). The presence of Cajamarca fine-ware and the architectural correspondences between the maquetas and these two contemporaneous sites point to a shared ceremonial tradition.

This inclusion of such "architectural portraits" in elite Moche burials at San José de Moro suggests meaningful connections between the individual interred and the original spaces they replicate in miniature. By comparing the specific layout of two particular maquetas to that of the western chamber at Huaca Colorada, I argue that the scalar relationship between these objects complement the architectural renovations of Huaca Colorada by reaffirming the nested synecdochal temporal ideologies at play during the Late Moche Period. Understood as cognitive artifacts or "synecdochal timepieces," maquetas not only served as representations of particular events and places but materialized and froze moments in time that connected individuals with specific places and ceremonies. Nested within an array of meaningful and interrelated symbols within the tombs at San José de Moro, the maquetas indexed distant locales, connecting within a synecdochal network the tomb and the interred individual with a specific meaningful place. As miniature "wholes," forming "parts" of a larger mortuary assemblage, the maquetas thus reified fleeting but essential phases (and peoples) associated with sacred places of memory and identity distributed across the valley. Indeed, the maquetas served as powerful historiographic media—making and remaking the history of likely diverse groups of people in the region. If the ceremonial locales were perceived as living entities with their own life cycles, the "timework" involved in maintaining, changing, and memorializing these spaces through the acts of renovation could also explain why the maquetas were buried in tombs at San José de Moro. By choosing to model specific architectural phases in miniature, specific moments of the life histories of ceremonial structures were transferred and "selectively concretized" (Clark 2010). Buried in tombs (just as daises and chambers were carefully interred at Huaca Colorada), these frozen moments of space-time would continue to shape and animate their architectural macrocosms. The presence of representative emblems of distant locales within the tombs of San José de Moro thus speaks to a concerted effort to at once assert social difference but still align local histories with the religious and temporal ideologies of the Priestess center. The tombs of San José de Moro should thus be considered as partially archival in nature, preserving and crystallizing the specific temporal rhythms of communities or authorities associated with the structures represented in miniature.

CONCLUSION

Both the incremental ritual renovations at Huaca Colorada and the creation and inclusion of maquetas in the tombs of San José de Moro served to make and order time. Indeed, both the models and architectural interventions provide a means to regulate the temporalization of practices in material form. In fact, the relationship between the two scales of architectural expressions served simultaneously as both "Signs *in* History" as well as "Signs *of* History" as discussed by Richard Parmentier (1987). As "Signs *in* History" the maquetas are value-laden material objects implicated in social strategies that focus attention on specific historical processes. As "social technologies of time" they secured the places they present as the loci of historical intentionality and agency (Parmentier 1987; Preucel 2006; Willerslev, Christensen, and Meinert 2013). As "Signs *of* History" both the renovations and the models they objectified acted to materialize discourses of history, especially how historical knowledge was materially recorded and transmitted. As built objects, monumental structures such as huacas served to mark the longue dureé, as their very existence mediated and legitimized the political order linked to the deep cosmological history framing mythic time, ordering the present and planning for the future. Robert Preucel notes that in traditional, nonliterate societies "Signs *of* History" are often simultaneously "Signs *in* History," wherein material objects acting as historicizing signs are often directly implicated in the manipulation, contestation, and concealment of meaning (Preucel 2006: 86). At Huaca Colorada, the act of preparing building materials and of construction itself were embodied, incorporating Signs of History, while the finished spaces and the maquetas that reference them can be understood as material Signs in History (Connerton 1989). As physical, habitable, and subjectified "cognitive artifacts" embedded in the landscape, full-scale ceremonial loci such as Huaca Colorada were the conceptual stages on which Moche senses of time were created, preserved, and continually reaffirmed (Moore 2010).

To conclude, the ceremonial platform structures and their miniature depictions were synecdochal referents to particular moments in the architectural biographies of important ritual locales in the Moche world. The act of constructing and renovating Huaca Colorada may have been memorialized in miniature in the creation of maquetas to synecdochally concretize and store the particular spatial identity of a singular phase of construction. These powerful objects stood not only as marker of affinity with a particular community or region but as symbols of particular events in time. Of course, religiously charged places such as Huaca Colorada must have marked particularly charged liminal events of encapsulated acts of destruction and renewal. These events at once marked, materialized, and effectively controlled the passage of time.

ACKNOWLEDGMENTS

I would like to express my gratitude to the Social Sciences and Humanities Research Council of Canada and the University of Toronto for their generous financial support of my research in the Jequetepeque Valley. I also extend many thanks to Dr. Luis Jaime Castillo Butters for providing access to his vast archive of research data on San José de Moro in Lima and of course to Edward Swenson and Andy Roddick for their invaluable editorial comments and suggestions over the course of preparing this volume.

REFERENCES

Allen, Catherine J. 1997. "When Pebbles Move Mountains: Iconicity and Symbolism in Quechua Ritual." In *Creating Context in Andean Cultures*, ed. Rosaleen Howard-Malverde, 73–84. Oxford: Oxford University Press.

Allen, Catherine J. 2002. *The Hold Life Has: Coca and Cultural Identity in an Andean Community*. Washington, DC: Smithsonian Institution Press.

Allen, Catherine J. 2011. *Foxboy: Intimacy and Aesthetics in Andean Stories*. Austin: University of Texas Press.

Allen, Catherine J. 2014a. "Ushnus and Interiority." In *Inca Sacred Space: Landscape, Site and Symbol in the Andes*, ed. F. M. Meddens, K. Willis, C. McEwan, and N. Branch, 71–77. London: Archetype Publications.

Allen, Catherine J. 2014b. "The Whole World Is Watching: New Perspectives on Andean Animism." In *Archaeology of Wak'as: Explorations of the Sacred in the Pre-Columbian Andes*, ed. Tamara L. Bray, 23–44. Boulder: University of Colorado Press.

Alva, Walter, and Christopher B. Donnan. 1993. *Royal Tombs of Sipán*. Los Angeles: University of California, Fowler Museum of Cultural History.

Bachelard, Gaston. [1994] 1958. *The Poetics of Space*, Vol. 330. Trans. M. Jolas. Boston: Beacon Press.

Bailey, Douglass W. 2005. *Prehistoric Figurines: Representation and Corporeality in the Neolithic*. London: Routledge. https://doi.org/10.4324/9780203392454.

Bakhtin, Mikhail. 1981. *The Dialogic Imagination: Four Essays*. Trans. C. Emerson and M. Holquist (ed.). Austin: University of Texas Press.

Bastien, Joseph W. 1978. *Mountain of the Condor: Metaphor and Ritual in an Andean Ayllu. American Ethnological Society Monograph No. 64*. St. Paul: West Publishing.

Bawden, Garth. 1996. *The Moche*. Oxford: Blackwell.

Benson, Elizabeth P. 2012. *The Worlds of the Moche on the North Coast of Peru*. Austin: University of Texas Press.

Birth, Kevin K. 2012. *Objects of Time: How Things Shape Temporality*. Basingstoke: Palgrave Macmillan. https://doi.org/10.1057/9781137017895.

Bray, Tamara L. 2009. "An Archaeological Perspective on the Andean Concept of Camaquen: Thinking through Late Pre-Columbian Ofrendas and Huacas." *Cambridge Archaeological Journal* 19 (3): 357–66. https://doi.org/10.1017/S0959774309000547.

Bourget, Steven. 2001. "Rituals of Sacrifice: Its Practice at Huaca de la Luna and its Representation." In *Moche Iconography in Moche Art and Archaeology*, ed. J. Pillsbury, 89–109. Washington, DC: National Gallery of Art.

Bourget, Steven. 2006. *Sex, Death and Sacrifice in Moche Religion and Visual Culture*. Austin: University of Texas Press.

Canziani, José C. 2009. *Ciudad y territorio en los Andes: Contribuciones a la historia del urbanismo prehispánico*. Lima: Fondo Editorial de la Pontificia Universidad Católica del Perú.

Castillo, Luis Jaime. 2000a. *La ceremonial del sacrificio: Batallas y muerte en el arte mochica*. Lima: Museo Arqueológico Larco Rafael Herrera.

Castillo, Luis Jaime. 2000b. "La Presencia Wari en San José de Moro." In *Huari y Tiwanaku: Modelos vs. evidencias*, ed. Peter Kaulicke and William H. Isbell. *Boletín de Arqueología PUCP* 4:143–79. Lima: Pontificia Universidad Católica del Perú.

Castillo, Luis Jaime. 2001. "The Last of the Mochicas." In *Moche Art and Archaeology in Ancient Peru*, ed. J. Pillsbury, 307–32. Studies in the History of Art 63. Center for Advanced Studies in the Visual Arts, Symposium Papers XL. Washington, DC: National Gallery of Art.

Castillo, Luis Jaime. 2003. "Los æultimos mochicas en Jequetepeque." In *Moche hacia el final del milenio*, vol. 2, ed. Santiago Uceda and Elías Mujica, 65–123. Lima: Fondo Editorial; Trujillo: Pontificia Universidad Católica del Perú and Universidad Nacional de Trujillo.

Castillo, Luis, Jaime S. Cusicanqui, and A. C. Mauricio. 2011. "Las maquetas de San José de Moro: Aproximaciones a su contexto y significado." In *Modelando el mundo: Imágenes de la arquitectura precolombina*, ed. José Canziani, Paulo Dam, and Luis Jaime Castillo, 112–43. Lima: MALI.

Castillo, Luis Jaime, and C. B. Donnan. 1994a. "La Ocupación Moche de San José de Moro, Jequetepeque." In *Moche: Propuestas y Perspectivas*, ed. Santiago Uceda and Elias Mujica. Lima: Travaux de l'Institut Français d'études Andines. https://doi.org/10.4000/books.ifea.2391.

Castillo, Luis Jaime, and C. B. Donnan. 1994b. "Los Mochica del norte y los Mochica del sur." In *Vicús*, ed. K. Makowski and C. B. Donnan, 143–81. Lima: Colección Arte y Tesoros del Perú, Banco de Crédito.

Castillo, Luis Jaime, Andrew Nelson, and Chris Nelson. 1997. "Maquetas Mochicas de San José de Moro." *Arkinka* 22:120–28.

Castillo, Luis Jaime, Y. Julio Rucabado, P. Martín del Carpio, Q. Katiusha Bernuy, R. Karim Ruiz, C. Carlos Rengifo, B. Gabriel Prieto, and C. Fraresso. 2008. "Ideología y poder en la consolidación, colapso y reconstitución del Estado Mochica del Jequetepeque: El Proyecto Arqueológico San José de Moro (1991–2006)." *Ñawpa Pacha* 29 (1): 1–86. https://doi.org/10.1179/naw.2008.29.1.001.

Clark, Andy. 2010. "Material Surrogacy and the Supernatural: Reflections on the Role of Artefacts in 'Off-Line Cognition." In *The Cognitive Life of Things: Recasting the Boundaries of the Mind*, ed. Lambros Malafouris and Colin Renfrew, 23–28. Cambridge: Cambridge University Press.

Connerton, Paul. 1989. *How Societies Remember*. Cambridge: Cambridge University Press. https://doi.org/10.1017/CBO9780511628061.

Dean, Carolyn. 2007. "The Inka Married the Earth: Integrated Outcrops and the Making of Place." *Art Bulletin* 89 (3): 502–18. https://doi.org/10.1080/00043079.2007.10786358.

De Long, Alton J. 1981. "Phenomenological Space-Time: Toward and Experiential Relativity." *Science* 213 (4508): 681–83. https://doi.org/10.1126/science.7256273.

De Long, Alton J. 1983. "Spatial Scale, Temporal Experience and information Processing: An Empirical Examination of Experiential Reality." *Man-Environment Systems* 13:77–86.

Dillehay, T. D., and A. L. Kolata. 2004. "Long-Term Human Response to Uncertain Environmental Conditions in the Andes." *Proceedings of the National Academy of Sciences of the United States of America* 101 (12): 4325–30. https://doi.org/10.1073/pnas.0400538101.

Dillehay, Tom D., Alan L. Kolata, and Edward Swenson. 2009. *Paisajes culturales en el Valle del Jequetepeque: Los yacimientos arqueológicos*. Trujillo: Ediciones SIAN de Luis Valle Alvarez.

Donnan, Christopher B. 1978. *Moche Art of Peru*. Los Angeles: Museum of Culture History, University of California.

Eliasson, Olafur. 2007. "Models Are Real." In *Models* 306090:19.

Flaherty, Michael G. 2011. *The Textures of Time: Agency and Temporal Experience*. Philadelphia: Temple University Press.

Freedberg, David. 1991. *The Power of Images: Studies in the History and Theory of Response*. Chicago: University of Chicago Press.

Gavazzi, Adine. 2012. *Microcosmos: Visión andina de los espacios pre hispánicos*. Lima: Apus Graph Ediciones.

Gillespie, Susan D. 2007. "When Is a House?" In *The Durable House: House Society Models in Archaeology*, ed. Robin A. Beck Jr., 25–50. Center for Archaeological Investigations Occasional Paper No. 35. Carbondale: Southern Illinois University.

Golte, Jürgen. 2009. *Moche cosmología y sociedad: Una interpretación iconográfica*. Lima: IEP Instituto de Estudios Peruanos.

Halperin, Christina T. 2014. *Maya Figurines: Intersections between State and Household*. Austin: University of Texas Press.

Henare, Amiria, Martin Holbraad, and Sari Wastell, eds. 2007. *Thinking through Things*. London: Routledge.

Hill, Erica. 1998. "Death as a Rite of Passage: The Iconography of the Moche Burial Theme." *Antiquity* 72 (277): 528–38. https://doi.org/10.1017/S0003598X00086944.

Hocquenghem, Anne Marie. 1987. *Iconografía mochica*. Lima: Pontificia Universidad Católica del Perú.

Hocquenghem, Anne Marie. 2008. "Sacrifices and Ceremonial Calendars in Societies of the Central Andes: A Reconsideration." In *The Art and Archaeology of the Moche: An Ancient Andean Society of the Peruvian North Coast*, ed. Steve Bourget and Kimberly L. Jones, 23–42. Austin: University of Texas Press.

Hoffman, Eva. 2009. *Time*. New York: Picador.

Hubert, Erell. 2014. "Moche Colonial Identity in the Santa Valley, Peru." PhD dissertation. University of Cambridge, Department of Archaeology and Anthropology.

Hubert, Erell. 2016. "Figuring Identity in Everyday Life." *Journal of Anthropological Archaeology* 44:1–13. https://doi.org/10.1016/j.jaa.2016.07.015.

Husserl, Edmund. 1992. *On the Phenomenology of the Consciousness of Internal Time, 1893–1917*. Trans. John Barnett Brough. Dordrecht: Kluwer Academic.

Hutchins, Edwin. 1999. "Cognitive Artifacts." *The MIT Encyclopedia of the Cognitive Sciences* 126:127.

Ingold, Tim. 1993. "The Temporality of the Landscape." *World Archaeology* 25 (2): 152–74. https://doi.org/10.1080/00438243.1993.9980235.

Jackson, Margaret A. 2008. *Moche Art and Visual Culture in Ancient Peru*. Albuquerque: University of New Mexico Press.

Johnson, Ilana. 2008. "Portachuelo de Charcape: Daily Life and Political Power in the Hinterland during the Late Moche Period." In *Arqueología mochica: Nuevos enfoques*, 261–74. Lima: Fondo Editorial, Pontificia Universidad Católica del Perú.

Johnson, Ilana. 2012. "The Development of Semi-Autonomous Communities in the Late Moche Period (AD 600–900)." In *From State to Empire in the Prehistoric Jequetepeque Valley, Peru*, ed. C. M Zori and I. Johnson, 51–64. BAR International Series, 2310. Oxford: British Archaeological Reports.

Jones. Andrew. 2013. "In Small Things Remembered: Scale, Materiality and Miniatures in the British Early Bronze Age." In *Counterpoint: Essays in Archaeology and Heritage Studies in Honour of Professor Kristian Kristiansen*, ed. Sophie Bergerbrant and Serena Sabatini. BAR International Series 2508. Oxford: Archaeopress.

Joyce, Rosemary A. 2007. "Figurines, Meaning, and Meaning-Making in Early Mesoamerica." In *Image and Imagination: A Global Prehistory of Figurative*

Representation, ed. Colin Renfrew and Iain Morley, 107–116. Cambridge: McDonald Institute of Archaeological Research.

Knappett, Carl. 2010. "Communities of Things and Objects: a Spatial Perspective." In *The Cognitive Life of Things: Recasting the Boundaries of the Mind*, ed. L. Malafouris and C. Renfrew, 81–89. Cambridge: McDonald Institute for Archaeological Research.

Knappett, Carl. 2012. "Meaning in Miniature: Semiotic Networks in Material Culture." In *Excavating the Mind: Cross-Sections through Culture, Cognition and Materiality*, ed. M. Jessen, N. Johanssen, and H. J. Jensen, 87–109. Aarhus: Aarhus University Press.

Kolata, Alan L. 1996. "Mimesis and Monumentalism in Native Andean Cities." *Res: Anthropology and Aesthetics* 29–30 (29/30): 223–36. https://doi.org/10.1086/RES vn1ms20166951.

Leach, Edmund. 1961. *Rethinking Anthropology*. London: Althone.

Lesure, Richard G. 2011. *Interpreting Ancient Figurines: Context, Comparison, and Prehistoric Art*. Cambridge: Cambridge University Press. https://doi.org/10.1017/CBO97805 11973376.

Lévi-Strauss, Claude. 1966. *The Savage Mind*. Chicago: University of Chicago Press.

Lynch, Sally. 2013. "A Ceramic Based Analysis of Feasting and Power at the Moche Site of Huaca Colorada." Master's thesis. University of Toronto.

Malafouris, L. 2013. *How Things Shape the Mind: A Theory of Material Engagement*. Cambridge: MIT Press.

McClelland, Donald. 2010. "'Architectural Models in Late Moche Tombs.' *Ñawpa Pacha*." *Journal of Andean Archaeology* 30 (2): 209–30.

McEwan, Colin. 2015. "Ordering the Sacred and Recreating Cuzco." In *Archaeology of Wak'as: Explorations of the Sacred in the Pre-Columbian Andes*, ed. Tamara L. Bray, 265–91. Boulder: University of Colorado Press.

Meddens, Frank. 1995. "Mountains, Miniatures, Ancestors and Fertility: The Meaning of a Late Horizon Offering in a Middle Horizon Structure in Peru." *Bulletin of the Institute of Archaeology London*.

Meskell, Lynn, Carolyn Nakamura, Rachel King, and Shahina Farid. 2008. "Figured Lifeworlds and Depositional Practices at Çatalhöyük." *Cambridge Archaeological Journal* 18 (2): 139–61. https://doi.org/10.1017/S095977430800022X.

Millaire, Jean-François. 2016. "Posts and Pots: Propitiatory Ritual at Huaca Santa Clara in the Virú Valley, Peru." In *Reconstructing Ritual Killing on the North Coast of Peru: Diachronic and Multidisciplinary Explorations of Sacrifice, Violence, and Belief, 200 B.C–A.D.1532*, ed. Haagen D. Klaus and J. Marla Toyne, 342–358. Austin: University of Texas Press.

Moore, Jerry D. 2010. "Making a Huaca: Memory and Praxis in Prehispanic Far Northern Peru." *Journal of Social Archaeology* 10 (3): 398–422. https://doi.org/10.1177/1469605310381550.

Østergaard, Jesper. 2013. "Pilgrimage Landscape as Material Anchor of Time." In *Taming Time, Timing Death: Social Technologies and Ritual*, ed. Dorthe Reflsund Christensen and Rane Willerslev, 157–174. Burlington, VT: Ashgate.

Parmentier, Richard. J.1987. *The Sacred Remains: Myth, History, and Polity in Belau*. Chicago: University of Chicago Press.

Pauketat, Timothy R., and Susan M. Alt. 2005. "Agency in a Postmold? Physicality and the Archaeology of Culture-Making." *Journal of Archaeological Method and Theory* 12 (3): 213–37. https://doi.org/10.1007/s10816-005-6929-9.

Peirce, Charles Sanders. 1984. *Peirce Edition Project Writings of Charles S. Peirce: A Chronological Edition*. Vol. 2., 1867–71. Bloomington: Indiana University Press.

Pillsbury, Joanne. 2015. "Building for the Beyond: Architectural Models from the Ancient Americas." In *Design for Eternity: Architectural Models from the Ancient Americas*, ed. J. Pillsbury, P. J. Sarro, J. Doyle, and J. Wiersema, 3–30. Metropolitan Museum of Art.

Prieto Burmester, Gabriel. 2008. "Rituales de enterramiento arquitectónico en el núcleo urbano Moche: Una aproximación desde una residencia de elite en el valle de Moche." In *Arqueología mochica: Nuevos enfoques*, ed. Luis Jaime Castillo Butters, Hélène Bernier, Gregory Lockard, Julio Rucabado Yong. Lima: Fondo Editorial Pontifica Universidad Católica del Perú.

Preucel, Robert W. 2006. *Archaeological Semiotics*. Oxford: Wiley-Blackwell. https://doi.org/10.1002/9780470754962.

Quilter, Jeffrey. 1997. "The Narrative Approach to Moche Iconography." *Latin American Antiquity* 8 (2): 113–33. https://doi.org/10.2307/971689.

Quilter, Jeffrey. 2001. "Moche Mimesis: Continuity and Change in Public Art in Early Peru." In *Moche Art and Archaeology in Ancient Peru*, ed. Joanne Pillsbury, 21–45. New Haven, CT: Yale University Press.

Quilter, Jeffrey. 2002. "Moche Politics, Religion, and Warfare." *Journal of World Prehistory* 16 (2): 145–95. https://doi.org/10.1023/A:1019933420233.

Quilter, Jeffrey, and Luis Jaime Castillo. 2010. *New Perspectives on Moche Political Organization*. Washington, DC: Dumbarton Oaks Research Library and Collection.

Quilter, Jeffrey, and Michelle Koons. 2012. "The Fall of the Moche: A Critique of Claims for South America's First State." *Latin American Antiquity* 23 (2): 127–43. https://doi.org/10.7183/1045-6635.23.2.127.

Rappaport, Roy A. 1992. "Ritual, Time, Eternity." *Zygon* 27 (1): 5–30. https://doi.org/10.1111/j.1467-9744.1992.tb00996.x.

Shimada, Izumi. 1994. *Pampa Grande and the Mochica Culture*. Austin: University of Texas Press.

Shimada, Izumi, Crystal B. Schaaf, Lonnie G. Thompson, and Ellen Mosley-Thompson. 1991. "Cultural Impacts of Severe Droughts in the Prehistoric Andes: Application of a 1500-Year Ice Core Precipitation Record." *World Archaeology* 22 (3): 247–70. https://doi.org/10.1080/00438243.1991.9980145.

Sillar, Bill. 1994. "Playing with God? Children, Play, and the Ritual Use of Miniatures in the Andes." *Archaeological Review from Cambridge* 13 (2): 47–63.

Sillar, Bill. 2009. "The Social Agency of Things? Animism and Materiality in the Andes." *Cambridge Archaeological Journal* 19 (3): 367–77. https://doi.org/10.1017/S0959774309000559.

Smith, Albert C. 2010. *Architectural Model as Machine: A New View of Models from Antiquity to the Present Day*. Oxford: Architectural Press.

Stenstrud, Astrid B. 2010. "Los peregrinos urbanos en Qoyllurit'I y el juego mimético de miniaturas." *Anthropoligica* 28:39–65.

Swenson, Edward R. 2003. "Cities of Violence: Sacrifice, Power, and Urbanization in the Andes." *Journal of Social Archaeology* 3 (2): 256–96. https://doi.org/10.1177/1469605303003002006.

Swenson, Edward R. 2004. "Ritual and Power in the Urban Hinterland: Religious Pluralism and Political Decentralization in Late Moche Jequetepeque, Peru." PhD dissertation. University of Chicago.

Swenson, Edward R. 2008. "San Ildefonso and the 'Popularization' of Moche Ideology in the Jequetepeque Valley." In *Arqueología mochica: Nuevos enfoques*, ed. Jaime Castillo Butters Luis, Hélène Bernier, Gregory Lockard, and Julio Rucabado Yong, 411–31. Lima: Fondo Editorial Pontifica Universidad Católica del Perú.

Swenson, Edward R. 2011a. "Stagecraft and the Politics of Spectacle in Ancient Peru." *Cambridge Archaeological Journal* 21 (2): 283–313. https://doi.org/10.1017/S095977431100028X.

Swenson, Edward R. 2011b. "Architectural Renovation as Ritual Process in Late Intermediate Period Jequetepeque." In *From State to Empire in the Prehistoric Jequetepeque Valley, Peru*, ed. C. M Zori and I. Johnson, 129–48. BAR International Series 2310. Oxford: British Archaeological Reports.

Swenson, Edward R. 2012a. "Moche Ceremonial Architecture as Thirdspace: The Politics of Place-Making in the Ancient Andes." *Journal of Social Archaeology* 12 (1): 3–28. https://doi.org/10.1177/1469605311426548.

Swenson, Edward R. 2012b. "Los fundamentos cosmológicos de las interacciones Moche-Sierra durante el Horizonte Medio en Jequetepeque." *Boletín de Arqueología PUCP* (16): 77–102.

Swenson, Edward R., and John P. Warner. 2012. "Crucibles of Power: Forging Copper and Forging Subjects at the Moche Ceremonial Center of Huaca Colorada, Peru." *Journal of Anthropological Archaeology* 31 (3): 314–33. https://doi.org/10.1016/j.jaa.2012.01.010.

Swenson, Edward R. 2015. "The Materialities of Place Making in the Ancient Andes: A Critical Appraisal of the Ontological Turn in Archaeological Interpretation." *Journal of Archaeological Method and Theory* 22 (3): 677–712. https://doi.org/10.1007/s10816-014-9202-2.

Swenson, Edward R., Jorge Y. Chiguala, and John P. Warner. 2010. *Informe Final Proyecto Arqueológico Jatanca-Huaca Colorada, Valle Jequetepeque Temporada 2009*. Pacasmayo: University of Toronto; Lima: Unpublished report submitted to the Instituto Nacional de la Cultura, Perú.

Swenson, Edward R., Jorge Y. Chiguala, and John P. Warner. 2011. "Informe final de la temporada de investigación 2010." University of Toronto, Pacasmayo, Peru. Unpublished report submitted to the Instituto Nacional de la Cultura, Perú, Lima.

Swenson, Edward R., Jorge Y. Chiguala, and John P. Warner. 2012. "Informe Final Proyecto Arqueológico Jatanca-Huaca Colorada, Valle Jequetepeque Temporada 2011." University of Toronto, Pacasmayo. Unpublished report submitted to the Instituto Nacional de la Cultura, Perú, Lima.

Swenson, Edward R., Jorge Y. Chiguala, and John P. Warner. 2013. "Informe Final Proyecto Arqueológico Jatanca-Huaca Colorada, Valle Jequetepeque Temporada 2012." University of Toronto, Pacasmayo. Unpublished report submitted to the Instituto Nacional de la Cultura, Perú, Lima.

Swenson, Edward R., Francisco Seoane, and John P. Warner. 2015. "Informe Final Proyecto Arqueológico Jatanca-Huaca Colorada, Valle Jequetepeque Temporada 2014." University of Toronto, Pacasmayo. Unpublished report submitted to the Instituto Nacional de la Cultura, Perú, Lima.

Swenson, Edward R., and J. P. Warner. 2015. "Landscapes of Mimesis and Convergence in the Southern Jequetepeque Valley, Peru." *Cambridge Archaeological Journal*: 1–29.

Taussig, Michael. 1993. *Mimesis and Alterity: A Particular History of the Senses*. London: Routledge.

Uceda, Santiago. 1999. "Esculturas en miniatura y una maqueta en madera: El Culto a los Muertos y a los ancestros el la época Chimú." *Beiträge zur Allgemeinen und Vergleichenden Archäologie* 19:259–302.

Uceda, Santiago. 2001a. "Investigations at Huaca de la Luna, Moche Valley: An Example of Moche Religious Architecture." In *Moche Art and Archaeology in Ancient Peru*, vol. 63. ed. Joanne Pillsbury, 47–67. Studies in the History of Art. Washington, DC: National Gallery of Art.

Uceda, Santiago. 2001b. "El nivel alto de la Plataforma I de Huaca de la Luna: Un espacio multifuncional." *Arkinka* 67:90–95.

Uceda, Santiago. 2010. "Theocracy and Secularism: Relationships between the Temple and Urban Nucleus and Political Change at the Huacas de Moche." In *New Perspectives on Moche Political Organization*, ed. Jeffrey Quilter and Luis Jaime Castillo, 32–58. Washington, DC: Dumbarton Oaks.

Uceda, Santiago, and Elias Mujica. 1994. *Moche: Propuestas y perspectiva*. Lima: Travaux de l'Institut Français d'études Andines. https://doi.org/10.4000/books.ifea.2367.

Uceda, Santiago, and Elias Mujica. 2003. *Lima: Fondo Editorial, Pontificia Universidad Católica del Perú*. Vols. 1 and 2. Moche Hacia el Final Del Milenio. Trujillo: Universidad Nacional de Trujillo.

Wiersema, Juliet B. 2010. "The Architectural Vessels of the Moche of Peru (C.E. 200–850): Architecture for the Afterlife." PhD dissertation. University of Maryland Department of Art History and Archaeology, College Park, MD.

Wiersema, Juliet B. 2011. "La Relación simbólica entre las representaciones arquitectónicas prehispánicas." In *Modelando el mundo: Imágenes de la arquitectura precolombina*, ed. Cecilia Pardo, 164–91. Lima: Asociación Museo de Arte de Lima.

Wiersema, Juliet B. 2012. "Vasijas arquetectónicas Moche: Pequeñas estructuras, grandes consecuencias." In *Microcosmos: Visión andina de los espacios pre hispánicos*, ed. Adine Gavazzi, 96–111. Lima: Apus Graph Ediciones.

Wiersema, Juliet B. 2015a. *Architectural Vessels of the Moche: Ceramic Diagrams of Sacred Space in Ancient Peru*. Austin: University of Texas Press.

Wiersema, Juliet B. 2015b. "The Art of Ancient Andean Architectural Representations." In *Design for Eternity: Architectural Models from the Ancient Americas*, ed. J. Pillsbury, P. J. Sarro, J. Doyle, and J. Wiersema, 55–79. Metropolitan Museum of Art.

Willerslev, Rane, Dorthe R. Christensen, and Lotte Meinert. 2013. "Introduction: Taming Time, Timing Death." In *Taming Time, Timing Death*, ed. Rane Willerslev and Dorthe Refslund, 1–21. London: Ashgate.

EIGHT

Hitching the Present to the Stars

The Architecture of Time and Space in the Ancient Andes

FRANCISCO SEOANE AND MARÍA JOSÉ CULQUICHICÓN-VENEGAS

INTRODUCTION

When archaeologists study other societies in a comparative framework, they tend to downplay problems of alterity and differences in cultural practice and worldview. Of course, this presents a perennial problem in anthropology: how to understand what is exactly different, especially among cultures that have been silent for centuries. As scientists, we try to explain culture and historical change following the criteria of cause and effect, but we have been blind to a very important aspect of this relationship: how people of different cultures experienced the relation of cause and effect within time and space (Spengler 1989). This relation has been simplified by Alfred Gell (1992) in his breakdown of time into two categories, and how they are marked by internal and external representations configuring the way in which we think about time (see introduction [chapter 1], this volume). How societies experience time tells us a good deal about how they relate to space, and this relation is vital to understanding differences in cultural practices and constructions of history. Interpreting how societies differently conceptualized and felt time offers novel analogies by which to interpret

DOI: 10.5876/9781607326427.c008

archaeological data sets and will prevent the imposition of our own conceptions on the archaeological record that would thus distort past realities (Gosden 1994; see introduction [chapter 1], this volume).

Constructions and experiences of time have varied considerably in world history. Certain cultures live totally in the present—and though they are aware of the past and future—they prefer to live in that present, transforming the past into myth (a kind of nontime) and rejecting the future by avoiding anything that would invoke it, including certain traditions of astronomy, time computing, and history (Assmann 2006; Spengler 1989; see Wilkinson and D'Altroy this volume [chapter 4]). Even in Europe prior to the eighteenth century, the past recorded by historians was commonly perceived not as distant but as intimate and immediate (Lowenthal 1985).

In general, Andean worldviews concerning myth or history have been studied mainly through anthropological and historical analysis of colonial chronicles, folkloric tales, legends from coast and highland (Pease 1973), and iconographic analysis (Golte 2009), but there have been scarcely any direct contributions by archaeologists. Most of the historical sources point to a cyclical view of time in the Andes, where humanity is created and destroyed by wrathful deities and a new humanity is created—as described in the myths of Pachacamac and other Andean deities (Toro 1990; see also introduction to this volume [chapter 1]). This cycling of time is distinguished by a succession of eras, each defined by its own kind of humanity (Guamán Poma 1956; Ossio 2014) and each coming to an end as the result of a natural disaster. Through careful analysis of myths and legends, several authors propose that Andean ideologies of epochal change were not characterized by alternating cycles of creation-destruction, but were distinguished instead by a succession of reconfigurations, consisting of the reshaping or the realigning of the world (see Pease 1973).

Most of these tales and myths, recorded by Spaniards or by heavily Europeanized noble Indians and Mestizos, including Guamán Poma or Garcilaso, are beset by varying degrees of adaptations, distortions, and Eurocentric "opinions" (Pease 1994: 90). Nevertheless, notions deriving from Andean thought were obviously recorded as well (Zuidema 1990). This is one of the major obstacles we encounter, for many of the sources we are using to unravel Andean worldviews of time and historical change are tainted by European conceptions, personal opinions, and misunderstandings (Zuidema 1990). These viewpoints supposed a simplified cause-effect relation often as determined by ecological, economic, or political factors, presupposing that under similar circumstances, similar organizations developed in the Old and New World. As Patricia Netherly (1990, 462) stated, there is a great difference between the purpose of Spanish and Andean tales, for the latter were more interested in structural relations than in linear causalities.

In other words, Andean myths and tales cannot be read as chronology or causal "history" in the Western sense (Zuidema 1964, 1990). Instead, they are intended to be read as a set of complex relations between people, nature, supernatural forces and objects, not as a "list of kings," as say in the tradition of the dynastic succession of pharaohs in ancient Egypt. For instance, when we read the tale of Naymlap, the founder of the Lambayeque Kingdom of the North Coast of Peru, or the Ayar Brothers, who founded the Inca Empire, we cannot understand these figures as founding leaders or dynastic lineages, but as expressions of structural rather than causal relations (Netherly 1990: 462). Many Andean people had no interest in the flow of time in this sense (see below). Instead the Naymlap and Ayar tales express how space is organized, divided and managed *in the present* (see also Salomon 2002 and the introduction [chapter 1], this volume). Therefore, anthropologists have wrongly assigned a chronological value to the tales in question where none is actually present (Kosok 1965; Rowe 1948). Another difficulty comes from how Andean people understood space itself. As Franklin Pease (1994: 92) writes, space in the Andes was understood socially and even ethnically, not geographically. This is also clear in the use of demonyms (inhabitants of a place) in the Andes as described by Pulgar-Vidal (1987): more than indicating a place of origin or birth, they express instead the vital place of and for the living. Furthermore, borders and boundaries in the Andean world were not "lines" dividing spaces as we understand them nowadays, instead, they were understood as concrete, living spaces of different scales and qualities: a limit could be a mountain, a plain, a corral, and so forth (Pease 1994: 93). This created many difficulties for the Spaniards in the administration of people and resources in the Andes, and conceptions of "property" and "territory" varied significantly between the two civilizations. Later investigations—based on misinterpretations of Andean conceptions of narrative, identity, and landscape—facilitated the application of foreign temporal and spatial models to the Andes.

The Spaniards also tried to explain Andean political hierarchy in terms of a monarchic structure, transforming ruler lists into stories of dynastic succession, as mentioned above with the case of Naymlap or the Ayar Brothers (Gose 1996). Indian rulers under colonial rule quickly adopted European notions of history to improve their positions under the Spaniard administration (Pease 1994; Zuidema 1964, 1990). A common mistake was to extend the validity of these late colonial cases back into previous periods.

Finally, standard methodological or theoretical approaches have distorted our understanding of Andean temporal schemes. One clear example pertains to art historical interpretations of iconography, especially as pertains to the study of the Moche culture. The problem derives partially from a methodological procedure and partly from a theoretical one. Traditionally, iconographic studies have proceeded through the "unfolding" and two-dimensional flattening of images depicted on three-dimensional surfaces such as ceramics (see Davis 1990; Panofsky 1991; and critique by Golte 2009: 26). Thus Moche scenes of warfare, ritualism, or mythic events were originally painted on bottles with volume, and this volumetric medium was no doubt critical to understanding the symbolism of a particular pictorial representation. However, this meaning is completely lost in common two-dimensional iconographic representations (see critique by Weismantel 2004). Certainly, the unfolded image facilitates our reading and interpretation, for it permits the creation of linear narration in the historical sense as critiqued above. However, this method inevitably subverts and distorts the intended meaning of the iconographic representation in question. As Jürgen Golte (1994) explains, the volumetric contextualization of various scenes was designed to convey structural relations rather than recount a historical tale. Nevertheless, the same author is unwittingly guilty of imposing a linear historical trajectory on Moche iconography, for he also orders a sequence of images into a single and integrated narration (Golte 1994). In the end, we agree with Zuidema and others that if we wish to understand Andean cultural schemas, including constructions of time, we have to think in structures, not in sequences (Golte 2009; Netherly 1990; Pease 1994; Zuidema 1964, 1990).

The other limitation emerges from a theoretical bias. As Kent Flannery (2006: 11) states, postprocessual archaeology offered much, but generated "mediocre field data enhanced by the addition of postmodern phrases." These deficiencies have led many anthropological archaeologists to return to a generalizing archaeology that tends to downplay issues of cultural worldview or alternate ontologies and historicities (Flannery 2006: 12). As Flannery mentions, different postprocessual approaches still failed to answer the questions originally considered unimportant in processual archaeology. As a consequence, a good number of archaeologists have "fallen back" to a type of processualism that either ignores idealism completely or simply incorporates postmodern phraseology to compensate for poor archaeological analysis. This development is especially true in current Peruvian archaeology. Unfortunately, these approaches, in their attempt to avoid conflict and criticism, are incapable of finding solutions to real anthropological problems and are guilty of casting aside the most significant elements of human culture: art, worldview, and religion. Indeed, an analysis of the latter domains is critical to a proper understanding of indigenous conceptions of time and historicity.

What we present here is to understand Andean conceptions of time and space through the study of art and architecture, focusing on the changing astronomical function of Preceramic, Formative and Moche ceremonial constructions. More specifically, we make use of Andean linguistic and cultural mediations of time to offer an interpretation that later Andean cultures conceived of time as an intimate and continuous present that needed constant regeneration through ritual and astronomical work. This distinctive and familiar flow of time was based on an ethos of reciprocity and sociospatial/cosmic complementarity. It thus differs from Mircea Eliade's notion of the "eternal return," the ritual reenactment of the mythic acts of primordial creation—for the idea of a distant, "primordial" past was not in play in the later pre-Columbian historical context of the Andes (Eliade 2011). By the Early Intermediate Period and Middle Horizon, Andean people were concerned with constantly "rebooting" and maintaining a reciprocal and inherently spatialized present—as exemplified in the master symbol of the *tinkuy* (for discussions of space, time, and tinkuy, see Bouysse-Cassagne and Harris 1987; Dean 2007). Thus time was not understood as a linear progression between past, present, and future, but it was construed as a present (moral order) that had to be sustained to ward off a condition of "non-present" (chaos, a feared transition to a new, immoral present). However, an important objective of our chapter is to show this particular notion of time only seems to have coalesced during the Middle Horizon Period, as will be exemplified by our cases study from Huacas de Moche at the end of the chapter. In contrast, our research of the astronomical alignments of Preceramic and Initial Period architecture reveals that earlier Andean people seemed to have developed ideologies of historical rupture and great temporal distances. Ultimately, our chapter intends to show that history in the Andean context was not simply a product of chance, climatic changes, or economic and technological transformations—the engines of history so valued in modernist historiography. Instead, we demonstrate that Andean communities actively sought to regulate time and history in distinct architectural and astronomical projects that no doubt played an important role in managing environmental and social transformations.

TIME IN ANDEAN WORLDVIEWS

Andean worldview as regards to time is generally understood as cyclic and dualistic, consisting of the oscillating movement between the present and a nonpresent (Earls and Silverblatt 1978; Netherly 1990; Pease 1973, 1994). In this sense, time in the Andes was felt as immanent and intimate, with the past and future merged into a phenomenal continuum. Mythic or historical time was thus never deemed separated or "far" from "human" or lived time (see Wilkinson and D'Altroy, this volume [chapter 4]). In other words, it was rarely conceived as a "foreign country" (see Lowenthal 1985 and the introduction [chapter 1], this volume). This conception of the mortal realm and the otherworld in constant communion and needed renewal marks Andean thought. It is clearly expressed in the concept of tinkuy, or the idea of "meeting" between opposed but complementary entities (Earls and Silverblatt 1978; see also Swenson, this volume [chapter 6]), shaping and configuring physical and social interaction and the greater world order. The use of temporal-space concepts in Quechua is also revealing. Words such as *nawpa* (ancient, before, ahead) and *qipa* (following, next, behind) reveal how Andean peoples understood time and space. As Earls and Silverblatt explain, the past can lie ahead, while the future may reside "behind" the present (see Wilkinson and D'Altroy this volume [chapter 4]). Other concepts, such as *hanan* and *uku*, are poorly translated in Western languages, where they have no equivalents. Generally they are translated as "up" and "down," but in Quechua they can also mean "beyond" (the visible) and "within" (Earls and Silverblatt 1978). In Quechua and other Andean languages, there is not even a word for "death" as we understand it. Instead, this is understood as a transition or a voyage to the other world that does not entail the complete separation from the here and now.

Quechua is also a peculiar in that the present tense is privileged, even when speakers are referring to the past or the future (see Wilkinson and D'Altroy, this volume [chapter 4]; personal communication, Eduardo Pajuelo, 2000). As reflected linguistically, it is as if Andean thought rejects or negates the flow of time. Quechua and other languages such as Aymara avoid referring to the non-present, at least to describe everyday life, and this can be also appreciated in the treatment of the dead. Cemeteries as places separated from everyday space are virtually unheard of in traditional Andean societies. Instead, sepulchers in the Andes were kept open to maintain a relation between the living and the dead in *chulpas*, or in platforms that remained part of the human present (Isbell 1997).

In the end, the dead played an active part in society, and their continued interaction ensured the order necessary for the existence of the world (Dillehay 1995; Hocquenghem 1989). Indeed, it is widely known that Andean people were constantly in communication with their dead, reopening tombs, moving bodies, adding or subtracting objects, or altering in different ways funerary areas, as also has been reported for the Moche of the North Coast. The Spaniards famously described how the mummies of past rulers were kept and cared for at the Koricancha (the Sun temple in Cusco). This is also known for Chimú times, as the finding of a wooden model at Huaca de la Luna depicts. This remarkable model shows a mortuary ceremony centered on a mummy that was taken out from its resting place to participate in a feast with the living (Uceda 2008). As is well known, then, the dead in the Andean world did not abandon this world, but they remained an essential part "within" it.

In considering Andean linguistic constructions of time, Spanish chronicles also mention how time was counted among the Inca. As they describe, months were counted by moons, and they had no names for the days of the week, or a way to measure time in smaller intervals (Garcilaso de la Vega 1991: 119–20; Guamán Poma 1956: 165, 182). Beyond the measuring of astronomical phenomena for important agricultural events and ceremonies, in the Andes there was generally a lack of interest in tracking time, because time was essentially present for Andean periods. As it flowed, this present could potentially transform itself into a "nonpresent," and the purpose to measure time as discrete units would have served no purpose. In a sense, it seems that Andeans abhorred not the passing of time, but its development into deep history, as exemplified in the notion of "tying the sun," a way to hold the sun in place (Silva 2006). Andean people thus appear to have understood the flow of time not in terms of disruptions, discontinuities, or breaks but in terms of the constant negotiations of intimate, reciprocal exchanges as reflected in the treatment of the dead and in linguistic structures as discussed above (Netherly 1990; Pease 1994; see also introduction [chapter 1], to this volume). The Andean preoccupation with continually re-creating an intimate and familiar time constitutes a master symbol, as expressed in particular in the notion of tinkuy, a subject to which we now turn.

Tinkuy and Andean Notions of Time

It has been already mentioned that several researchers consider time in the Andes as cyclic, an opposition between two spatial entities, including between this world and others. But as we have seen, some of these conceptions may have been tainted by Spaniard chroniclers or theoretical and methodological limitations. As described above, Andean myth has the purpose of informing how the world is divided and organized in the present rather than remembering when or how long ago an event occurred. Implemented in practice—and thus underwriting social, political, religious, and economical pursuits—this structure was the key to maintaining world order. Therefore, Andean peoples thought of the present as something that flows, moves within itself, and dissolves *if* you get too far from it.

An excellent example of this spatiotemporal scheme is revealed by Tom Zuidema (1990, see figs. 7–8) and Netherly (1990) when they refer to the political organization of the Chimú. As Zuidema explains, a myth such as that of Naymlap describes the founding not of a dynasty, but of a world order with its subdivisions of two complementary lines of corulers, the *principal* and *segunda persona* common to the chronicles. Power was shared between two unequal parts, and hierarchy was organized in such a way that no one, even the supreme leader, exercised absolute power. This concept was so alien to the Spanish that they were unable to register it adequately, as Andean hierarchy was not clearly understood. As Netherly (1984, 1990) explains, Andean political organization can be considered a complex textile of relationships in which one's position was conditioned by the way it was connected to others, and these connections needed to be reestablished continuously to keep that fabric's integrity, thus exemplifying a presentist and reciprocal understanding of time (see Mannheim 2015 and introduction [chapter 1], to this volume).

Thus Andean political order needed to be constantly renegotiated as a means to maintain an intimate and ordered time in a constantly flowing present. This is especially evident in Zuidema's interpretation of Andean myth. A "founder" and his successors were remembered only to a depth of four to six generations, primarily as a means to organize space in a series of complementary sections—as exemplified in the notion of Tawantinsuyu and the four quarters of the of the Inca Empire. The maintenance of the reciprocal and hierarchical dependencies of the sections would constantly reboot the present in order to prevent a feared temporal rupture—a break dividing a past and future that would signal a new world (dis)order. Once again, the mythologies point to the merging of the past and future and a delineation between a nonpresent and a present in the Andes (Earls and Silverblatt 1978; see Wilkinson and D'Altroy this volume [chapter 4]). It is probably no coincidence that Naymlap's sons were twelve, the same number as the Incas listed as rulers in Cusco (Zuidema 1990).

This duality evident in myths and sociospatial organization in the Andes should not be interpreted as a social contradiction or an opposition in the spirit of Marxist thinking. Instead, we must understand this dualistic structure as comprising complementary unequal halves that were necessary to the maintenance of the social fabric and temporal world order. This duality has been uncovered in the iconographical analysis of Golte (2008, 2009) for the Early Horizon and Early Intermediate Periods.

To summarize, Andean worldview is based upon the fear of the present dissolving into a nonpresent. Moreover, the rightful creation of time entailed the "separation" of the world in unequal halves, not as strictly opposing entities, but to engender through their union a point of encounter, a crossroad where "time" could be maintained though the continued ritual refounding of the present. This process of spatial and temporal regeneration is captured by the Quechua word tinkuy, a practice and symbol meant to ensure the continuation of the present order of things through the reciprocal cycling of spatially and materially dependent entities. This fear of the nonpresent and the need to continuously reweave the present constitute the liminal tinkuy-time, where human life develops within an unstable and narrow band between the nontime and the reciprocally flowing present. At once a spatial and temporal construct, tinkuy leaves no place for long time counts, dynastic histories, or temporal absolutism in any form. Everything exists only and if they manage to weave *ayllus*, institutions, huacas, and so forth in a flowing and flexible *space* that properly structured the present moral order (Makowski 2008; Oliver-Smith 1994). In fact, the astronomical function of ceremonial architecture points to a long-term concern in the Andes to stave off a "nonpresent" threat to cosmic order, a subject to which we now turn. However, this ideological imperative seems to have only fully coalesced during the Early Intermediate and Middle Horizon Periods.

CHANGING CONCEPTIONS OF TIME AS REFLECTED IN ANDEAN ARCHITECTURE AND ASTRONOMY

In light of our argument that Andean conceptions of time were structural and intensely "presentist," an analysis of architecture proves essential to interpreting how time was variably fixed and how space and sociocosmic relations were regulated by the stars. Andean ceremonial centers with astronomically aligned constructions date as early as Preceramic Caral and the Geoglyph of Chao (Milla 2008). Since the Preceramic, many ceremonial sites were built in alignment with determined axes defined by important celestial bodies, but not always with the same purpose (Benfer 2012; Urton and Aveni 1983). Sometimes they were built to directly observe a celestial body, such as the Southern Cross, whose apparent positions or azimuth seem "fixed" over long periods of time, perhaps to commemorate and reenact mythical times of cosmic origin. In other cases, architecture was used to track the apparent displacement of celestial bodies, especially high-magnitude stars through long periods of time. Such astronomical calculations likely computed the passing of time from a point of origin. The former is mainly used when historic depth is unimportant or explicitly denied, as among the Inca, while in the latter scenario, time is tracked to prepare for the closing of a long cycle that is not apparent in a single human life.

Indeed, our archaeoastronomical research reveals important shifts in time reckoning and conceptions of history in the ancient Andes (see table 8.1). In the Preceramic, many structures were built following important celestial alignments such as the Southern Cross, the Pleiades, the "black constellations" (dark spaces between stars in the Milky Way), a few high-visibility stars, and the Sun (Milla 2008). In fact, the Preceramic is marked by considerable diversity in terms of the astronomical orientations of ceremonial buildings (see also Benfer 2012).

At Caral, for example, the Main Pyramid is aligned to the position of the Milky Way during the June Solstice, with secondary elements built following the Southern Cross, the position of the sunset during the solstice, or other solar positions. The latter appears to have traced and established the *diagonals* of the sacred spaces in question. The same can be said of the temple of the Amphitheater, with slight variations (Shady et al. 2012). But other pyramids at the site (those surrounding the main buildings) seem to have been designed only to chart solar positions or those of the Pleiades, and they are not aligned with or oriented to the Southern Cross or other stars. Our research points to comparable diversity in astronomical alignments at other important Preceramic sites in the valley of Supe including Miraya, Era de Pando, and Pueblo Nuevo. Many of these constructions are aligned with the Milky Way at the time of June Solstice, but other buildings were clearly built to commemorate the Southern Cross, the Pleiades, the rise or setting of the sun, or even the position of stars including A or B Centaurus.

TABLE 8.1. Main Astronomical organizers, Peruvian North Coast, by period

Period	*Main Site Organizer*	*Structure Orientation*	*Season of Interest*	*Sunken Plaza*	*Sites*
Preceramic (3000–1800 BC)	Milky Way (June)	Crux[a] Sun	Dry	Yes	Supe: *Caral, Miraya, Pando*
Initial Period and Early Horizon (1800–200 BC)	Milky Way (June) Solstice (December) Crux (May–August)	Crux Crux (diagonal)[b] Equinox (diagonal)	Dry and wet	Yes	Casma: *Moxeque, Sechin Alto, Cerro Sechín* Lima: *Garagay, Cardal*
Early Intermediate (200 BC–AD 650)	Milky Way (June)	Crux (diagonal) Sunset (June Solstice) Mountains	Dry	No	Moche: *Huaca de la Luna*
Middle Horizon (AD 650–1000)	Solar Transit	Crux (diagonal)	Wet	No	Moche: *Platform III*
Late Intermediate (AD 650–1400)	Milky Way (June) Equinox	Crux and other Stars, Sun, Mountains	Wet	No	Pacatnamú, Manchan, Farfan, Chan Chan

a. The Crux = the Southern Cross

b. The diagonal refers to the orientation of corners or intersecting points of important rooms, open areas, and perimeters in direct alignment with a celestial body (thus bisecting buildings along a diagonal angle). The celestial body in question thus organizes the positioning and orientation of a building and is not observed directly.

Another important astronomical element used to design these early monuments was the apparent angle of the long arm of the Southern Cross (Milla 2008) at 31°30', but this standard is not yet widely used, as demonstrated by the sites mentioned above in the north-central coast of Peru. Instead, orienting buildings to the long arm of the Southern Cross starts to gain popularity with time, and during the Initial and Early Horizon Periods, this alignment becomes *the main* celestial reference for organizing architecture for both ceremonial and mundane purposes. The proliferation and diffusion of the sunken plaza in these latter periods may very well be related to the growing popularity of this alignment and celestial body. Carlos Milla (2008) interprets the sunken plazas and the angle of the staircase they contain as functioning to plot an "intersolsticial stellar angle." As proposed by Milla, narrower angles indicate earlier buildings, while wider ones were constructed in a later era. The sunken plazas appear to have been designed to measure how much time passed from a set point, a "year zero" as perhaps established when the monument was actually built. Construction thus appears to have charted a time of considerable depth, possibly comparable to what is known from ancient Mesoamerica (the "Long Count"), or the old World with its "Platonic Year" (di Santillana and Von

Dechend 1977). This beginning "year zero" may have commemorated the day of world creation or the initiation of a new era, and it is possible that it coincided with the year 4468 BC, a very significant date in ancient astronomy and myth. The year 4468 BC corresponds to the special alignment of the Milky Way and the ecliptic. In many myths, this alignment has been recorded as a remarkable time when the gods descended to live on Earth, through the Milky Way, variably viewed as a bridge, road, tree, or river (Eliade 2011; Freidel et al. 1999; di Santillana and Von Dechend 1977).

Therefore, Initial Period societies appear to have been interested in measuring huge magnitudes of time, as opposed to maintaining and preserving a fleeting present. However, something new—what here we could call "tinkuy-time" or the "reciprocally recreated present" as described above—starts to take shape during the Early Horizon. This is suggested by the abandonment of earlier architectural layouts designed for the measurement of long periods of time, the general replacement of embroidery and stone-like incised pottery with tapestry, and painted and modeled vessels that start to express new notions of time. Here we see the introduction of a new way to design architecture, not abandoning the pattern mentioned, just abandoning those aspects related to the measurement of time, especially the sunken plazas. Moreover, when stellar alignments went out of phase, sites were often rebuilt and realigned to "recenter" them again. Sometimes this could be done on-site (Sakai 1998), or occasionally it required the abandonment and refoundation of settlements elsewhere. This also reflects a concern with maintaining a constant present.

Beginning in the Early Horizon, temple accessibility and visibility also change in important ways (Seoane 2013b). The idea is not yet clearly defined, but we start to see this new design emerging at Chavín de Huántar, Cerro Blanco in Nepeña (Vega-Centeno y Lafosse 2000), Las Aldas in Casma, Huaca de los Reyes in the valley of Moche and many others. This finally leads to the huge mud-and-adobe platforms of the Early Intermediate Period with complex interplay of access, visibility, and volume.

The Case of Huacas de Moche

As has been described by a number of archaeologists, the Moche reinterpreted cultural elements developed in earlier times to produce a totally new cultural discourse (Uceda y Tufinio 2014). For instance, they continued to make similar astronomical observations as in earlier times, but they focused on different celestial bodies (Seoane 2013b).

For example, At Huacas de Moche prior to the Middle Horizon, priests continued to observe the night sky using techniques developed during the Preceramic and Early Horizon, especially chartering the movement of the Southern Cross and Pleiades, but they did not make use of sunken plazas and as a possible means

FIGURE 8.1. *Huacas de Moche's main astronomic alignments in architecture: (1) Milky Way, beginning of the dry season; (2) Southern Cross's setting in August at the beginning of the planting season; (3) the sunset during the June Solstice; (4) Cerro Centinela; (5) Cerro Blanco; (6) The New Temple—Platform III is aligned with the Solar Transit at beginning of the rainy season.*

to record time in a long-count system that may have been developed in the earlier periods (see Milla 2008 and above). The Moche seem to be among the first to consciously employ astronomy not to measure time but to "hold or arrest it." Huaca de la Luna in the Moche Valley, along with the larger site of Huacas de Moche in which it is located, appear to have simulated the *layout* of the sky in a mythical "present" at its possible starting point on the June Solstice (Hocquenghem 1989). This layout involves the alignment of the main platforms of the site, and the residential constructions of the urban zone, as well as the position and dimensions of entries, ramps, secondary structures, main streets, secondary alleys, and even funerary monuments. Everything and everyone is in this way became "anchored" to a celestially inscribed present. As will be described below, the architectural constructions of the city were gridded to the Sun, Southern Cross, mountain peaks, and so forth in an effort to center the meeting point of the Huacas de Moche at the point between the earth and sky (figure 8.1).

Around AD 600, however, we see the closure of the Old Temple (Huaca de la Luna) and the building of a new one (Platform III) on Cerro Blanco, which deviated from the indicated pattern of celestial alignments (Seoane 2013b). This new temple, associated with the last occupations of the site, also lacks the previous murals and iconography of Huaca de la Luna, and is interpreted as an indication

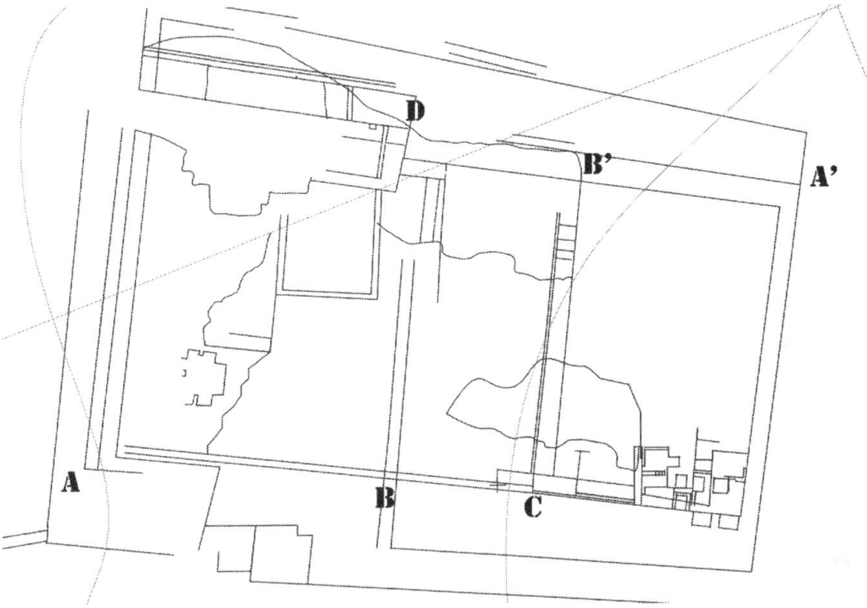

FIGURE 8.2. *Platform III, the New Temple's astronomic alignments: A-A'—Solar Transit diagonal; B-B'—Southern Cross minor arm diagonal, as observed around the June Solstice; A-D and A-C'—Sunset around December Solstice.*

of a power shift from a theocratic model to a more secular one (Uceda 2009). This also corresponds to a shift in the type of the astronomical observations used to design the layout of architecture as Platform III shows (figure 8.2), replacing the traditional night-oriented observations of the Early Horizon and Early Intermediate Periods for the day-oriented ones that regain popularity during the Middle Horizon, but which can be traced as early as Preceramic times. However, we do not view these changes as signaling a radical break in Moche conceptions of time. Rather, we understand this structural reorganization of religion and cult as conforming to an ideological move to recenter the Moche world in a solidified and continually rejuvenated "present." In this sense, figures such as the Late Moche Priestess should be understood as responsible for the resetting of the world in the ordered present of "tinkuy-time" by linking stellar, solar, and lunar positions to that "present."

As Milla (2008) mentions, from the earliest times, the organizing axis for the production of space in the Andes was the Southern Cross. He argues that the ratio between the long and short arm of the constellations determined the system of measurement of architectural construction, the aesthetic proportion for artworks, and the calendar sequencing of the ritual calendar. This symbol and

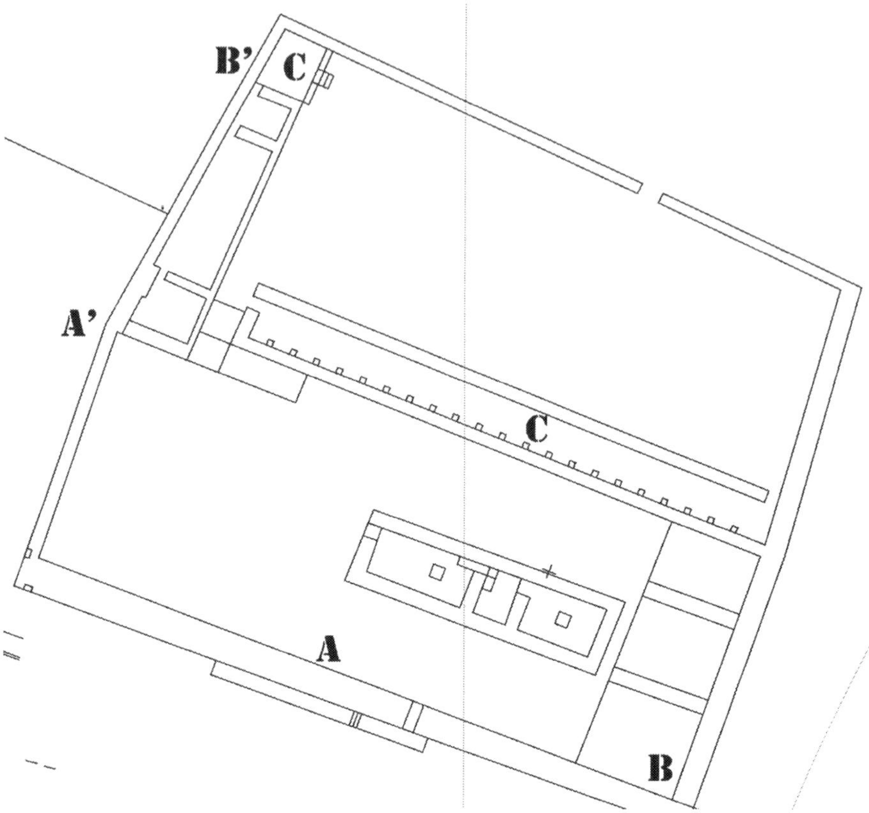

FIGURE 8.3. *Plataforma Uhle: The angle at A' divides the structure in two main sections. The southern half has its west wall aligned to the position of the Southern Cross at the beginning of the planting season (August), while the northern section is aligned to the Milky Way around the June Solstice. Inner architectonic elements are placed following the alignments of the major and minor arms of the Southern Cross. A-A' and B-B' follow the major arms and define the diagonals for the platform. Elements at C indicate alignments to the sunset during the June Solstice.*

spatial axis was in use even in Inca times, and nowadays it survives as the Fiesta de las Cruces in much of Andes.

For our analysis, we considered the alignment of the main celestial bodies in use in the ancient Andes such as the sunset during the Solstices and Equinoxes, the position of the Milky Way on the Solstices, and the inclination of the long arm of the Southern Cross in June and May, basing our calculations on the years between AD 100 and 600. Our study was designed to test the degree to which later and ethnohistorically documented alignments with celestial bodies were recognized and architecturally cited at Huaca de la Luna during Moche times.

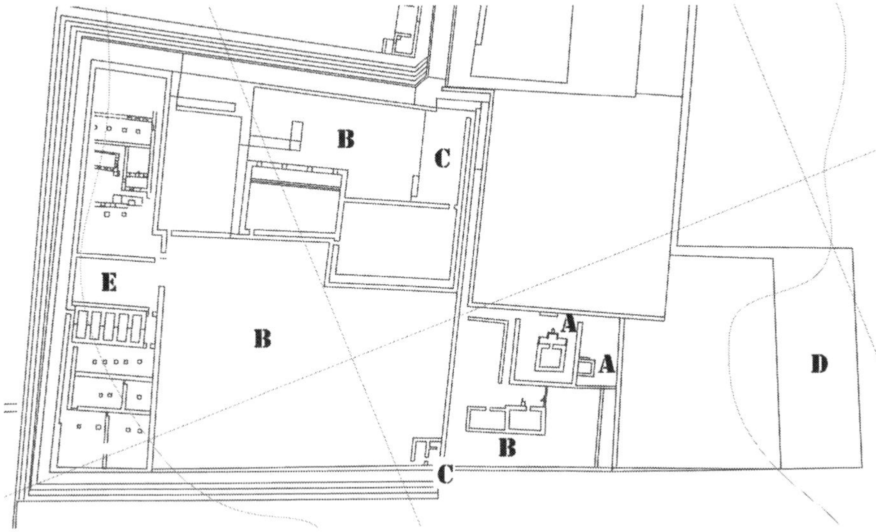

FIGURE 8.4. *Huaca de la Luna, Platform I. The temple was built using the Southern Cross alignments in June as a guideline, mainly to trace the diagonals of different areas or chambers. All alignments relate to the positions of different celestial bodies around the June Solstice: (A) Milky Way, (B) Southern Cross major arm, (C) Southern Cross minor arm, (D) Southern Cross set in August, (E) Sunset.*

The first architectural element attracting our attention was the evident angle or corner in the west wall of the Plataforma Uhle (figure 8.3, A'). Stellar orientations for the site were identified with Redshift 7 software, and the prominent west wall was analyzed in terms of possible alignments with relevant celestial bodies. From this corner we can draw a line with a bearing of 31°30' (figure 8.3, B-B') which crosses the Plataforma Uhle in a diagonal. The bearing of this diagonal defines a set of architectural elements important not only here, but also at Huaca de la Luna (figure 8.4). This diagonal line corresponds to the angle made in the sky by the long arm of the Southern Cross at the time of its appearance during the June Solstice, when the Andean New Year was celebrated, marked by the reappearance of the Pleiades (Milla 2008; Bauer and Dearborn 1998).

At the same time, the southwestern section of the platform houses an area whose diagonal (figure 8.3, A-A') measures 20.40 meters in length (Milla 2008), and thus appears to follow the proportions and measurements of the Southern Cross as it was represented in the geoglyph of Chao (Milla 2008). Moreover, in this area, a corridor (figure 8.3, C) was found aligned with the sunset at the time of the Winter Solstice, and it may also have functioned as a calendric indicator.

At Huaca de la Luna this materialization of the proportions and measurements of the Southern Cross is repeated. Following the wall alignment extending from the prominent corner room decorated with marine motifs (figure 8.4, C), one would approach the main patio of the temple where its diagonal follows the same angle indicated above. The same alignments have been identified for the structures at the sacrificial patios (Plazas 3A and 3B), revealing the great importance it had for Moche religion and ideology. Indeed the Southern Cross was employed as a blueprint to design the inner spaces of the temple. Finally, the east-west structures within the temple also align with the sunset during the June Solstice.

We believe this consistent gridding of architecture to the Southern Cross served to link and orient the temple, with the timing of ceremonial events and rituals processions to a perduring "time-0." The establishment of this continual "present" may have first occurred sometime in the first decades of the second century AD, if this pattern represents something similar to the intersolsticial angle used by Milla (2008) to date Preceramic and Early Horizon architecture. The main difference with these earlier architectural complexes, however, was that the Moche arrangement was not used to keep track of time, but served to constantly fix, sustain and reenact this immanent "time-0." An examination of the Urban Nucleus located between Huaca de la Luna and Huaca del Sol reveals that several of these alignments are repeated and configure the bearings of main streets, alleys, architectural compounds, and other lesser structures that in the last years have been identified at the site (Zavaleta Paredes et al. 2015).The Main street of the center, separating Huaca de la Luna from the urban area itself, Avenida 1, follows the position of the Milky Way (*Hatunmayu*) at the time of its zenith (about 26–28°) around the time of the June Solstice. The southern portion of this street turns slightly and follows the same angle identified on the western wall of the Plataforma Uhle. This angle deviates roughly 10° from the northern section of the street, following instead important alignments in the sky relevant to Andean agricultural practices. This alignment corresponds to the position in the sky when the Southern Cross sets at nightfall late in August, the time that marks the beginning of the planting season and other related activities (see introduction [chapter 1], this volume). This difference in bearing could have symbolically divided two neighborhoods at the site, perhaps based on social differences. This is further supported by the differing nature, characteristics, and architectural elements of the compounds at the center (Culquichicón et al. 2011; Meneses et al. 2014; Seoane et al. 2010). Interestingly, the alleys separating city block from city block, also follow the position of the sunset on the June Solstice (figure 8.1).

Nevertheless, something very different appears to be happening at the New Temple of Platform III (figure 8.2). The long arm of the Southern Cross seems to have lost its preeminence in the design of the layout of the new structure. The structure itself lacks most of the iconographic themes of the Old Temple (Uceda and Tufinio 2014), and it has shifted from the alignments identified at the earlier structure. What we see now is that the alignments of the Solar Transit take precedence in determining the diagonals of the New Temple. However, to a lesser degree, the bearings of the lesser arm and greater arm of the Southern Cross also configure space within this structure. The Solar Transit takes place at this latitude in October and February, and it marks the beginning and ending of the rainy season in the adjacent Andean mountains just to the east.

The new orientation of the New Temple points to profound shifts in certain conceptions of time, life, and society in Moche culture. This shift includes the gradual replacement of night-oriented sky observations in favor of Sun-oriented ones. The Southern Cross and other night bodies were not simply rejected or forgotten—they simply appear to lose their preeminence—while solar observation assume the greatest importance. This change in emphasis appears to take place sometime during the seventh century AD, when the Old Temple was closed, and the city of Huacas de Moche was reorganized at the site (Uceda and Tufinio 2014: 352).

At the same time, the elevation of the Solar Transit indicates that the Moche were now privileging the calendrical commemoration of the wet season when the rains return in the highlands in October. In contrast, prior to the seventh century, the Southern Cross and the celebration of the June Solstice were emphasized, a period coinciding with the beginning of the dry season. This significant transformation in astronomical observations was no doubt related to modifications in mythology and the reconstitution of power relationships. These developments may have even coincided with the polarization of Moche priesthood into competing factions: the traditional night-observers and a possibly new cult that privileged the Sun.

Moreover, we see the appearance or empowerment of religious figures that in the past were merely secondary figures, most notably the priestess and her cult, likely based at San José de Moro. A later version of the famed Moche Presentation Theme may be described by Guamán Poma (1956), who drew a "little devil" presenting a cup full of water to a thirsty Sun during the festival of the Inti Raymi. Astronomically, this is associated with the heliacal apparition of Orion, where a devil (Orion) extends its arms to the Sun as presenting him something to drink, while Eridanus seems to represent the tail of the little devil. This line of interpretation may be supported by the iconography in a Moche bottle presented by Christopher Donnan (1999: 88, fig. 4.27). If this image has any bearing on earlier Moche practices, as Anne Marie Hocquenghem (1989) argues, it helps to understand the inversion in the order proposed by the authors precisely after the Transit. It may even explain the shift in organization between Moche and Chimú times (Netherly 1990), when the northern margin eclipses the southern in political and religious importance during the Late Intermediate Period. The new spatial organization required a mythical-astronomical basis since during the June Solstice, the Sun runs across the sky along the southern horizon (supporting the Moche order), while after the adoption of the solar transit as the main astronomical organizer, the Sun crosses the northern horizon (supporting the later Chimú order). In this context, the June celebrations around Orion, the boat, and the refreshing of the Sun in June needed to be shifted to a date that supported the new order favoring the growing importance of the northern banks of North Coast rivers (Netherly 1990).

We interpret this shift toward daytime astronomy as the final break from the system developed during the Preceramic and Early Horizon (Seoane 2013a), based on the observation of stellar movements to a less elaborate model developed to reestablish celestial positions at "time-0." These new astronomical practices in the Andes underwrote a wide variation of "systems" and observations through which this could be accomplished, and how they were used to configure space and architecture.

The intensification of social, economic, and political interaction by the end of the Early Horizon could account for the widespread importance of the Southern Cross in Andean time-reckoning. However, beginning at the end of this era and intensifying by the Middle Horizon, knowledge related to the measurement of long periods of time is abandoned as part of religious life. However, the use of the Southern Cross as the organizing element of time and space is maintained in most of the Andes. This era coincides with the crystallization of Andean concepts of space and time discussed in this chapter: the rejection of temporal distance in favor of a continually and reciprocally rejuvenated present that was held in space and time by reenacting or "re-establishing" the sky at the time of its creation.

CONCLUSION

An important position of this chapter was that the way people experience time and make history must be taken into account in interpreting differences in worldview and social organization. In the end, we do not think time in the Andes was strictly cyclic, linear, or bilinear. Despite its considerable variation through history, as demonstrated by this chapter, later Andean worldviews rejected any concept of time as sequential and teleological. By "holding" to the present as described in the previous pages, it exemplifies this rejection. In the Andes, time was not cyclical in the way that characterized ancient Middle Eastern conception of *kismet*, a space-time enclosed in a set period of time with an anticipated and cataclysmic end to come. In contrast, we cannot even find a good word in Andean languages to conceptualize death. Moreover, the predilection of Andeans to speak in the present tense (personal communication, Pajuelo, 2000) exemplifies alternate understandings of time's immanence and movement.

During the Early Intermediate Period, the Moche continued a long tradition of archaeoastronomical observations, iconographic representations, ritual practices, and spatial organization that were widespread in the Peruvian North Coast.

However, this continuity did not necessarily imply a continuity in understandings and measurements of time. Preceramic, Initial Period, and Early Horizon societies developed and implemented a type of long-count system to record time, which eventually seems to have been abandoned. Early and Middle Moche built their belief systems, mythology, and social and political organization on this tradition, but ended abandoning it completely by the Early Intermediate Period and Middle Horizon. This change implied the abandonment of a temporal scheme based on a deep history of distances and finite stages. Although the position, apparent angles, and positions of the Southern Cross formed the foundation of the old and new systems, they were used differently.

In the earlier periods, the purpose of the Southern Cross was to establish a main axis against which astronomical calculations could be referenced and time calculated by using structures such as circular plazas and the width of stairs. The new system that started to be employed in the Early Intermediate Period by Moche astronomers and architects employed the same axis to trace the position of celestial bodies vis-à-vis a "time-0." However, instead of recording deep time, this time-0 was fixed to establish a perduring present and world order. In other words, Moche time was not about recording and commemorating long temporal distances as characterized Mesoamerican astronomy. Instead, it entailed holding and maintaining the initial time in the present.

In the Middle Horizon Period, the Moche clearly executed major changes in architecture, myth, and political organization to reestablish its center around this enduring time-0. These changes entailed the founding of new capitals (as Pampa Grande or Galindo) and the "burial" of old temples and sites (Shimada 1994; Uceda 2009). These ideological and political changes were also related to the refounding of important centers through the building of a new temple gridded to novel astronomical alignments. Platform III at Huaca de la Luna exemplified the refounding of the Moche city and the breaking of the present (time-0) from the Southern Cross and its retethering or rehitching to the solar transit. This process was accompanied by the introduction of rituals and mythic figures that supported a new social and political organization. To reiterate, the major change identified at the Moche site was the rejection of the Southern Cross as the principal organizational axis of ceremonial architecture for an axis defined by the solar transit. These wholesale transformations (in astronomical orientation, architectural construction, settlement patterns, iconography, etc.) strongly suggest that Late Moche chose to move the New Year from June (the beginning of the dry season) to October (the beginning of the wet season) (see table 8.1). At the same time, Moche astronomers and religious specialists likely retained certain traditional ceremonies associated with the June Solstice, but their meanings no doubt changed with the establishment of new festivals coinciding with the September Equinox and the Solar Transit.

In Inca times and among contemporary traditional societies in Peru and Bolivia, the beginning of the new year is inaugurated during the June Solstice, and the most celebrated festivities correspond with the movement of the Southern Cross (Inti Raymi, May's Cross), the Pleiades, and a few stellar bodies associated with the Solstice. Hence, these festivals are not coordinated with the Solar Transit or the initiation of the wet season.

Therefore, it would seem that Late Moche architecture and astronomy departed from the general Andean worldview. This new experiment was short lived, lasting maybe two or three hundred years but not much more. Later societies seem to have reverted to the type of astronomical axis favored in earlier times. However, the hypothesized long-count-like calendrical system—argued to have been recorded by Andean societies as early as the Preceramic and Initial Period—had long disappeared. In the end, our exploration of the astronomical meanings and efficacy of ancient monumental architecture reveals both important continuities and important shifts in Andean constructions of time.

REFERENCES

Assmann, Jan. 2006. *Religion and Cultural Memory*. Stanford, CA: Stanford University Press.

Bauer, Brian S., and David S. Dearborn. 1998. *Astronomía e Imperio en los Andes*. Cusco: Editorial Bartolome de las Casas.

Benfer, Robert A., Jr. 2012. "Monumental Architecture Arising from an Early-Astronomical Religious Complex in Peru, 2200–1750 BC." In *Early New World Monumentality*, ed. Richard L. Burger and Robert M. Rosenswig, 313–63. Gainesville: University of Florida Press. https://doi.org/10.5744/florida/9780813038087.003.0012.

Bouysse-Cassagne, Thérèse, and Olivia Harris. 1987. "Pacha." In Torno al pensamiento aymara: In *tres reflexiones sobre el pensamiento andino*, ed. Thérèse Bouysse-Cassagne and Olivia Harris, 11–59. La Paz: HISBOL.

Culquichicón, María José, Roberto Marcelo, Estela Meléndez, Juan Carlos Novoa, and Ofelia Parimango. 2011. *"Distribución y diseño arquitectónico del CA 44: Función y articulación del área oeste durante la última ocupación moche, en el núcleo urbano de huaca de la luna." Student Technical Report*. Universidad Nacional de Trujillo.

Davis, Whitney. 1990. "Style and History in Art History." In *The Uses of Style in Archaeology*, ed. Margaret W. Conkey, 18–31. Cambridge: Cambridge University Press.

Dean, Carolyn. 2007. "The Inka Married the Earth: Integrated Outcrops and the Making of Place." *Art Bulletin* 89 (3): 502–18. https://doi.org/10.1080/00043079.2007.10786358.

Dillehay, Tom D., ed. 1995. *Tombs for the Living: Andean Mortuary Practices*. Washington, DC: Dumbarton Oaks.

di Santillana, Giorgio, and Hertha von Dechend. 1977. *Hamlet's Mill: An Essay on Myth and the Frame of Time*. Boston: David R. Godine.

Donnan, Christopher. 1999. *Moche Fineline Painting: Its Evolution and Its Artists*. Los Angeles: University of California Press.

Earls, John, and Irene Silverblatt. 1978. "La realidad física y social en la cosmología andina." *Proceedings of the XLI International Congress of Americanists* 4:299–325.

Eliade, Mircea. 2011. *El mito del eterno retorno*. Madrid: Alianza Editorial.

Flannery, Kent. 2006. "On the Resilience of Anthropological Archaeology." *Annual Review of Anthropology* 35 (1): 1–13. https://doi.org/10.1146/annurev.anthro.35.081705.123304.

Freidel, David, Linda Schele, and Joy Parker. 1999. *El cosmos maya: Tres mil Años por la senda de los chamanes*. Mexico City: Fondo de Cultura Económica.

Garcilaso de la Vega, Inca. 1991. *Comentarios Reales de los Incas*. Lima: Fondo de Cultura Económica.

Gell, Alfred. 1992. *The Anthropology of Time: Cultural Constructions of Temporal Maps and Images*. Oxford: Berg.

Golte, Jürgen. 1994. *Íconos y narraciones: La reconstrucción de una secuencia de imágenes moche*. Lima: IEP Ediciones.

Golte, Jürgen. 2008. "La modelación de una cosmología." Lima: Centro de Investigación de la Universidad Ricardo Palma. *Scientia* 10 (10): 17–36.

Golte, Jürgen. 2009. *Moche cosmología y sociedad*. Lima: Instituto de Estudios Peruanos: Centro Bartolomé de las Casas.

Gosden, Christopher. 1994. *Social Being and Time*. Oxford: Blackwell.

Gose, Peter. 1996. "The Past Is a Lower Moiety: Diarchy, History, and Divine Kingship in the Inka Empire." *History and Anthropology* 9 (4): 383–414. https://doi.org/10.1080/02757206.1996.9960887.

Hocquenghem, Anne Marie. 1989. *Iconografía mochica*. Lima: Pontificia Universidad Católica del Perú Fondo Editorial.

Guamán Poma de Ayala, Felipe. 1956. *Nueva Crónica y Buen Gobierno*. Lima: Editorial Cultural.

Isbell, William H. 1997. *Mummies and Mortuary Monuments: A Postprocessual Prehistory of Central Andean Organization*. Austin: University of Texas Press.

Kosok, Paul. 1965. *Life, Land and Water in Ancient Peru*. New York: Long Island University Press.

Lowenthal, David. 1985. *The Past Is a Foreign Country*. Cambridge: Cambridge University Press.

Makowski, Krzystof. 2008. "Andean Urbanism." In *Handbook of South American Archaeology*, ed. Silverman Helaine and William Isbell, 633–57. New York: Springer. https://doi.org/10.1007/978-0-387-74907-5_32.

Mannheim, Bruce. 2015. "All Translation Is Radical Translation." In *Translating Worlds: The Epistemological Space of Translation*, ed. Carlo Severi and William F. Hanks, 199–220. Chicago: Hau Books.

Meneses, Jorge, Antonio Robles, Ronny Vega, Fernando Chávez, Debbie Cruz, Liz Rodríguez, Hoover Rojas, and Fressia Tantte. 2014. "Excavaciones en los conjuntos arquitectónicos 5, 46, 48 y 49: Configuración y ordenamiento espacial de los conjuntos del sur." In *Informe Técnico 2013, Proyecto Arqueológico Huaca de la Luna*, ed. Santiago Uceda and Ricardo Morales, 261–316. Trujillo, Peru: Facultad de Ciencias Sociales, Universidad Nacional de Trujillo.

Milla, Carlos. 2008. *Génesis de la cultura andina*. Lima: Editorial Amaru Wayra.

Netherly, Patricia. 1984. "The Management of Late Andean Irrigation Systems on the North Coast of Peru." *American Antiquity* 49 (2): 227–54. https://doi.org/10.2307/280017.

Netherly, Patricia. 1990. "Out of Many, One: The Organization of Rule in the North Coast Polities." In *The Northern Dynasties: Kingship and Statecraft in Chimor*, ed. Michael Moseley and Alana Crody-Collins, 461–87. Washington, DC: Dumbarton Oaks.

Oliver-Smith, Anthony. 1994. "Perú, 31 de Mayo, 1970: Quinientos años de desastre." *Desastres y sociedad: Especial tragedia, cambio y desarrollo* 2 (2): 4–18.

Ossio, Juan. 2014. *El Tahuantinsuyo bíblico: Ezequiel Ataucusi Gamonal y el mesianismo en los israelitas del Nuevo Pacto Universal*. Lima: Fondo Editorial de la Pontificia Universidad Católica del Perú.

Panofsky, Erwin. 1991. *Perspective as Symbolic Form*. New York: Zone Books.

Pease, Franklin. 1973. *El Dios Creador andino*. Lima: Mosca Azul Editores.

Pease, Franklin. 1994. "Para decir como son: Nociones europeas en la explicación andina del Siglo XVI." In El Inca Garcilaso entre Europa y América, ed. Antonio Garrido Aranda, 89–107. Córdova, Spain: Caja Provincial de Ahorros de Córdova.

Pulgar-Vidal, Javier. 1987. *Geografía del Perú: Las ocho regiones naturales del Perú*. Lima: Ed. PEISA.

Rowe, John. 1948. "The Kingdom of Chimor." *Acta Americana* 6 (1–2): 26–59.

Sakai, Masato. 1998. *Reyes, estrellas y cerros en Chimor: El proceso de cambio de la organización espacial y temporal de Chan Chan*. Lima: Editorial Horizonte.

Salomon, Frank. 2002. "'*Huayra Huayra Pichcamanta!*' Augurio, risa y regeneración en la política tradicional (Pacota, Huarochirí)." *Boletín del Instituto Francés de Estudios Andinos* 31 (1): 1–22. https://doi.org/10.4000/bifea.6900.

Seoane, Francisco. 2013a. "Cerro Sechín y el restablecimiento del orden durante el Horizonte Temprano en el Valle de Casma." *Nosotros* V (6): 193–216. Universidad Nacional de Trujillo.

Seoane, Francisco. 2013b. "Del atardecer a la mañana: Replanteando el orden del mundo en la sociedad moche durante el Horizonte Medio." Universidad Nacional de Trujillo. *Revista del Museo de Arqueología, Antropología e Historia* 12:41–52.

Seoane, Francisco, Víctor Campaña, Feren Castillo, Lucy Chumbe, Patricia Gamboa, and Jermi Mejía. 2010. "El conjunto arquitectónico N° 42 y los bloques arquitectónicos en el núcleo urbano moche." In *Informe técnico 2009, Proyecto Arqueológico Huaca de la Luna*, ed. Santiago Uceda and Ricardo Morales, 305–95. Trujillo: Facultad de Ciencias Sociales. Universidad Nacional de Trujillo.

Shady, Ruth, Daniel Cáceda, Aldemar Crispín, Marco Machacuay, Pedro Novoa, and Edna Quispe. 2012. *Caral la Civilización más antigua de las Américas: 15 años develando su historia*. Lima: Rapimagen S.A.

Shimada, Izumi. 1994. "Los modelos de organización sociopolítica de la cultura Moche: Nuevos datos y perspectiva." In *Moche propuestas y perspectivas*, ed. Santiago Uceda and Elías Mujica, 359–87. Lima: Universidad Nacional de Trujillo e Instituto Francés de Estudios Andinos. https://doi.org/10.4000/books.ifea.2415.

Silva, Julio. 2006. *Intihuatana: Piedra para atar al sol*. Lima: Kuntur.

Spengler, Oswald. 1989. *La decadencia de Occidente, bosquejo de una morfología de la historia universal*. Vols. 1 and 2. Madrid: Ed. Espasa-Calpe.

Toro, César. 1990. *Mitos y leyendas del Perú*. Vols. 1 and 2. Lima: A.F.A. Editores Importadores S.A.

Uceda, Santiago. 2008. "Rituales funerarios de los Reyes en una maqueta chimú." In *Señores de los reinos de la luna*, ed. Krzysztof Makowski, Joanne Pillsbury, Régulo G. Franco Jordán, 314–53. Lima: Banco de Crédito.

Uceda, Santiago. 2009. "A la búsqueda de los palacios de los reyes moche." In *Informe técnico 2008, Proyecto Arqueológico Huaca de la Luna*, ed. Santiago Uceda and Ricardo Morales, 269–80. Trujillo, Peru: Facultad de Ciencias Sociales, Universidad Nacional de Trujillo.

Uceda, Santiago, and Moisés Tufinio. 2014. "Huaca del Sol como centro del poder político: El surgimiento del palacio en la Sociedad Mochica Tardía." In *Informe técnico 2013, Proyecto Arqueológico Huaca de la Luna*, ed. Santiago Uceda and Ricardo Morales, 349–70. Trujillo, Peru: Facultad de Ciencias Sociales, Universidad Nacional de Trujillo.

Urton, Gary, and Anthony Aveni. 1983. "Archeoastronomical Fieldwork on the Coast of Peru." In *Calendars in Mesoamerica and Peru: Native American Computations of Time*, ed. Antony F. Aveni and Gordon Brotherston, 221–34. Bar International Series 174. Oxford: British Archaeological Reports.

Vega-Centeno, Rafael, and Sara Lafosse. 2000. "Imagen y simbolismo en la arquitectura de Cerro Blanco, Costa Nor-Central Peruana." *Bulletin Institute Français d'Études Andines* 29 (2): 139–59.

Weismantel, Mary. 2004. "Moche Sex Pots: Reproduction and Temporality in Ancient South America." *American Anthropologist* 106 (3): 495–505.

Zavaleta Paredes, Luis Enrique, Diana Jannett Zagastizábal Ruiz, Luis Martín Gómez Acosta, Lilian Margoth Villacorta Dávila, Christian Ramiro Castillo Contreras, Luis Ángel De la Flor Fernández, Jimmy Alejandro López Gamero, Marley Pompa Calderón, and Lorena Solange Toledo León. 2015. "Cambios sociales y políticos en el moche tardío y la configuración de un conjunto administrativo en el núcleo urbano del complejo arqueológico uaca del Sol y Luna." In *Informe Técnico 2014, Proyecto*

Arqueológico Huaca de la Luna, ed. Santiago Uceda and Ricardo Morales, 265–333. Trujillo, Peru: Facultad de Ciencias Sociales, Universidad Nacional de Trujillo.

Zuidema, Tom. 1964. *The Ceque System of Cuzco: The Social Organization of the Capital of the Inca*. Leiden: E. J. Brill.

Zuidema, Tom. 1990. "Dynastic Structures in Andean Cultures." In *The Northern Dynasties: Kingship and Statecraft in Chimor*, ed. Michael Moseley and Alana Cordy-Collins, 489–505. Washington, DC: Dumbarton Oaks.

NINE

Archaeology, Temporal Complexity, and the Politics of Time

TAMARA L. BRAY

Time is fundamental to the idea of archaeology, and the way we understand time affects the way we conceive of and interpret our archaeological objects of study. Radiometric dating, chronology building, and the discipline-wide focus on culture change are all key features of archaeology as practiced today, and each is intimately entangled with specific notions about time. Yet, for the most part, time has remained a largely taken-for-granted concept within the field (though see Harding 2005; Lucas 2005; Murray 1999; Olivier 2011, for notable exceptions). The chapters presented in this volume foreground the topic of temporality, offering a sustained and critical analysis of the ways in which normative Western constructs of time have channeled our approaches to, and understandings of, the past within the framing context of Andean archaeology. As evidenced in the various contributions, taking the dimension of time as a matter of concern rather than an assumed universal suggests new ways of recalibrating not only the traditional segmentation of pre-Columbian history in the Andes—incorporating important ideas about synchronicity, persistence, and overlap—but also our own culturally specific notions of time as linear, homogeneous, and universal.

DOI: 10.5876/9781607326427.c009

In this concluding chapter, I highlight several of the key themes developed in relation to the critique of conventional Andean historiography offered in this volume. These include ideas about incorporating indigenous temporalities (e.g., adopting an emic perspective); recognizing the significance of variable temporalities (e.g., human versus nonhuman); the emphasis traditionally accorded culture change versus cultural continuity; and the politics of time. Rather than approaching time, temporality, and history as absolute and universal categories, it is important for archaeologists to take on board the insights and critiques derived not only from ethnographic and linguistic work but those emerging within modern physics and the life sciences, as well, with respect to, for instance, space-time, relativity, entanglement, and observer effect, as we go forward with the business of "reconstructing the past." Exploring the ways in which archaeological practice mediates temporal experience as well as being attuned to the possibility of divergent and multiple temporal ontologies is a key facet of the ongoing development of critical self-awareness within the discipline.

INDIGENOUS TEMPORALITIES

That time is not an absolute quality of the universe has been one of the fundamental tenets of theoretical physics for over a century. As per Einstein's special theory of relativity, time is understood to exist within—and is therefore dependent upon—space. Rather than comprising a separate dimension, time exists as part of a space-time continuum and is thus relative to a situated observer and varies from place to place. The notion of space-time is remarkably congruent with the Andean concept of *pacha*, a word that expresses the inseparability of space and time in Andean thought, as discussed by various contributors to this volume. For me, it is striking how evolving Western understandings of the nature of reality in, for instance, the realms of quantum mechanics, deep ecology, theoretical cosmology, and so on seem to resonate ever more closely with fundamental understandings of indigenous peoples around the world and their systems of knowledge.

In addition to expressing a specific ontology, the concept of pacha also mobilizes a specific politics of resistance and renewal. The present space-time of native Andean peoples is known in Aymara as *jichhapacha*, characterized as the "age of colonial oppression," which has been endured for half a millennium. With the *pachakuti*, or end of world order, brought on by the Spanish colonizers, the unity of pacha was ruptured as native peoples lost control over time, history, and (hence) their identity (Choquehuanca 2009: 75). As articulated by Aymara scholar and activist Carlos Mamani, the political project of recuperating cultural identity and control necessarily involves "reunit[ing] time and space in the unity of *pacha*" (Mamani Condori 1996: 639). The politics of time is clearly central to his analysis and agenda: "Since colonial times the wheel of our history has not moved on; time has stagnated, damaging and imprisoning us. We wish to move in pursuit of our vision, a vision which is both in the past—before 1532—but also in the future" (Mamani Condori 1996: 642). For the Aymara people, "the past is not inert or dead, and it does not remain in some previous place; [rather, it] is precisely by means of the past that the hope of a free future can be nourished, in which the past can be regenerated" (Mamani Condori 1996: 644; see also Sammells 2011). That time, however arbitrary or neutral it may seem, has a politics and a history is important to bear in mind as we proceed with our own analysis and critique of its deployment within archaeology.

An important theme of many of the chapters in this volume is the call for greater attentiveness to indigenous temporal realities and practices of time management and manipulation. In chapter 1, for instance, Swenson and Roddick note the "urgent need to consider Andean temporal practices and conceptions of history" in order to better understand the significance of observed patterns of perdurance and change in the archaeological record. This is the approach adopted by Seoane and Culquichicón in their contribution (chapter 8). Relying on linguistic evidence pertaining to Andean concepts of space-time, they highlight important divergences from Western temporal models and work from this emically oriented standpoint to interpret distinct alignments of ancient ceremonial centers on the Peruvian Coast. Chase (chapter 5) similarly attempts to work out a native temporal logic based on archaeological and ethnohistoric evidence from the Huarochirí region of central highland Peru, suggesting that local constructions of time involved the performance of idealized pasts for realization of a vital and desired present.

Attempting to reverse engineer an emic perspective on temporality, Swenson offers a detailed look in chapter 6 at the way in which distinct spatial and architectural configurations may have structured distinct temporalities on the North Coast of Peru. However, he, much like Roddick in his thoughtful chapter (3), on how archaeologists graphically construct time, also maintains a commitment to the value of an etic stance in attempting to assess the temporal ideologies of Andean cultures. Sayre also underscores the importance of an emic perspective in constructing new historical narratives for Chavin, though his concern seems to lie more with uncovering the "daily rhythms of life," for example, "the things done" or Trouillot's Historicity I, than discerning local understandings of time (chapter 2). This is an interest also shared by Swenson that again necessarily entails reliance on an etic perspective.

The most explicit critique of the unconsidered imposition of Western normative views of time on non-Western cultures and advocacy for exploring emic perspectives is presented by Wilkinson and D'Altroy (chapter 4). They make the case that for the Inka, the idea of "the past" as something dead and gone would have been a foreign notion. The evidence suggests rather a temporal ontology oriented to the present in which history was immanent in the here and now and something to be continually engaged (also Chase [chapter 5], and Seoane and Culquichicón [chapter 8], this volume). Given a "present past" rather than an absent one, the authors question the presumed universal significance of the concept of memory and the need for "re-presentation" of the past. They further highlight the fact that embedded assumptions about time can have a significant impact on how we frame and interpret the archaeological record. Whether one views history as a separate temporal domain or a component part of the here and now would fundamentally affect how one approached, for instance, decipherment of the Inka *khipu* or the social significance of an Andean *wak'a*. Ultimately, they argue for the critical importance of interpreting the material practices of other societies in light of their own temporal ontologies and logics rather than automatically imposing those of our own.

VARIABLE TEMPORALITIES

As discussed in this volume, the unilinear view of history and time typically employed in archaeology leads to the characterization of temporality as a series of nonoverlapping units. In Andean archaeology, the widely accepted chronological framework of cultural "horizons" separated by intervening "intermediate periods" is exemplary of this vision. Most of the critiques of this schema offered here and elsewhere (e.g., Cook 2004; Rice 1993; Silverman 2004, etc.) center on concerns that it oversimplifies Andean history, masks regional variability with respect to duration of artifact styles and types, and is inattentive to the social significance of interregional connections indicated. Less discussed are the politics of this framework, which, though explicitly eschewing an evolutionary paradigm when originally put forward (see Rowe 1962), nonetheless sought to "master" the cultures and histories of Andean peoples by writing them into a homogeneous, Western/archaeological framework (see Foucault 1970; Tanaka 2016; White 2002).

Another key problem with the unilinear view of time, as noted by Gavin Lucas (2015), is that it leaves us with a fairly meager number of possibilities for discussing temporal relations among archaeological (and other) entities. The "holy triad" of such temporal expressions consists basically of before (/earlier), after (/later), and synchronous (/contemporary). In reality, however, many more types of temporal relationships come into focus when we cease with the requirement of chronological segmentation and acknowledge the significance of overlap, or imbrication (Lucas 2015). For instance, in addition to the linear succession of events, phenomena, or processes, various degrees of overlap and containment (e.g., X during Y, X starts Y, or X finishes Y) also become evident and expressible (Lucas 2015: 6). Thinking about the temporality of the objects we compare in our analyses is critical not only to the development of a more nuanced and refined approach to the archaeological record but also for revising our understanding of time as relational rather than absolute.

Several chapters in this volume begin to take up the challenge of creating more sophisticated models for articulating the variable temporalities of people, places, and things. Roddick (chapter 3), for instance, centers his discussion on the way in which archaeological chronographics have channeled our vision of time as universal and unilinear. Beyond deprivileging ceramic styles as transparent measures of culture change, he suggests one method for exploring variable temporalities in the archaeological record could be to chart the rate of change of different artifact categories, styles, forms, raw materials, manufacturing techniques, and so forth, for example, different materialities, relative to one another. Such an approach would aid not only in envisioning temporal complexity but in thinking about the significance of both duration and transformation in our constructions of time, as well (see also Stahl 2012).

That Andean peoples recognized the significance of variable temporalities among the different constituents of their social worlds—and sometimes sought to align these through specific interventions—is suggested by various archaeological examples. Take, for one, the common archaeological occurrence of a person buried with a ceramic vessel purposefully drilled to create what is often referred to as a "kill hole"—a feature that effectively ends the (originally intended) useful life of the pot (figure 9.1). Human beings experience a temporality that today has an average duration of seventy years and has always involved directional movement and change along a continuum from birth to death. The often humble vessel of fired clay exhibits a wildly divergent temporal reality that can potentially endure in a steady state for thousands of years. Human intervention in the "life" of the pot suggests a clear understanding of the fact that people, things, and processes occupy different "time zones"—in other words, exist within different temporal frames—that are sometimes important to synchronize.

Spence Morrow's analysis of miniature architectural models recovered in burials at the Moche site of San José de Moro offers another example that underscores the value of incorporating a concern with variable temporalities and materialities. The maquetas found at San José de Moro were made of unfired or low-fired clay and, consequently, exhibit a very different temporality from similar such objects transformed through the process of firing. Given the Moche mastery of materials as seen across various domains of craft production, it seems likely that this was a calculated rather than expedient choice. It could thus be argued that the inherent temporalities of the substances and items included in the mortuary rituals performed at San José de Moro were of considerable importance to the people enacting them. Exploring the reasons for the choices and decisions made, as Spence Morrow has begun to do via a focus on context and scale, can only enhance our insights into Moche ontology, temporal reckonings, and systems of logic.

FIGURE 9.1. *Panzaleo jar recovered in small hemispherical burial mound containing single young adult individual excavated at site of Shanshipampa, northern highland Ecuador, in 1999; note kill hole in center of vessel body; vessel size: 14 cm tall, 17.5 cm maximum diameter, 8.65 cm rim diameter (Bray 2005).*

The monumental architectural features known as wak'as found up and down the Peruvian Coast that the Moche maquetas represent exist within a distinct, asynchronous but overlapping, temporal framework relative to the people and things with which they were (and are still) connected. That the builders and users of these enormous structures recognized the significance of the divergent temporalities they materialized is one of the principal themes of Swenson's chapter 6. In it, he details how authorities in some instances seem to have sought to synchronize human and architectural temporal realities via specific material interventions, while in others, the intent appears to have been to maintain asynchronicity and the seeming timelessness of the built environment vis-à-vis the local populace.

The sum of the archaeological evidence suggests that Andean peoples were highly cognizant of the fact that persons, places, and things existed within multiple temporal frameworks and that they sometimes sought to control or manipulate these to their own ends. For them, the multitemporal nature of the world was undoubtedly evident—in the enduring profiles of the snow-capped peaks, the daily ebb and flow of ocean tides, the cycles of the moon, the seasonal variability of rivers and streams. It also stands to reason that the inherently variable temporalities of made things, ranging from monumental architecture to clay pots, would have been equally apparent. To better incorporate the notion of temporal complexity into archaeological interpretation, we need to devise models that go beyond the human biographical experience given that the temporal properties of objects, landscapes, natural phenomena, and so on, are often radically distinct. We need models that capture the dynamic processes and transformations through which things are made, unmade, and remade in order to better theorize the temporal ontologies of past human societies, as well as our own. The chapters in this volume offer a good start in this direction.

CONTINUITY VERSUS CHANGE AND THE POLITICS OF TIME

The concepts of time and history contain within them the opposing forces of continuity and change. The fundamental tension generated as a consequence has been an issue of concern for philosophers since the days of Aristotle, who argued that time is inextricably bound up with change. There is little question that historians have generally privileged change and the chronological ordering of events over interests in duration and stability.[1] That archaeology is also "the study of culture change" has similarly been a mantra for generations of students and practitioners (see Fagan 1972; Fagan and Durrani 2016). While discernible change and movement are clearly major features of how we construe time, it is equally important for archaeologists to incorporate concerns with continuity, stability, and persistence in the study of the past—not only to further our understanding of variable temporalities but for political reasons as well (e.g., Lightfoot 2001; Oland et al. 2012; Silliman 2005; Smith and Wobst 2005).

Over the years, the ideological nature of time and its deployment as a political tool are premises that have been developed by social theorists from a variety of different backgrounds (see, for instance, Adams 1990; Bender and Wellbery 1991; Bloch 1977; Fabian 1983; Gell 1992; Thompson 1967; Zerubavel 2003). In anthropology, Johannes Fabian (1983) offered one of the earliest indictments of time's presumed innocence and was severe in his assessment of how the discipline used time to construct, separate, and subordinate its object of study, in other words, other peoples, societies, and cultures. Fabian's critique charted the ideological origins of what he referred to as political time (or Time with a capital "T"), outlined the relationship between political time and political space (or chronopolitics and geopolitics), and showed how both have been wielded as instruments of power. The way Time has structured—as well as been structured by—archaeology has also been critically discussed in recent years (e.g., Echo-Hawk 1997, 2000; Leone 1978; Lucas 2005; Shanks and Tilley 1987).

Time with a capital "T" is the familiar chronological time that underpins much of the archaeological enterprise. As defined by the *Oxford English Dictionary* (2017), chronology is "the science of computing and adjusting time or periods of time, and of recording and arranging events in the order of time, . . . and the assigning of events to their correct dates." As the historian Stefan Tanaka (2016: 162) notes, this kind of time comprises a metric that is external to the life and events that it records, arranges, and adjusts. Chronological time emphasizes linearity, segmentation, measurement, and sequence; it divorces events from the world of relations and systems of knowledge within which they were originally embedded; and it claims a privileged position vis-à-vis all other conceptions and expressions of time. But this form of temporality, like any other, has a history and a cultural fingerprint. Recognizing this, as well as the fact that chronological time does not serve all interests equally, should help open new spaces for multi-temporal approaches to the archaeological record—ones that aim to approach it from both emic and etic angles.

The way people, communities, and entire cultures construe the past depends on the way they think about time. At the core of temporal consciousness—whether historical or mythic, linear or cyclical—are narratives through which people interpret and express the nature of their lived experience (Ricoeur 1988). "For the historian [and the archaeologist], temporality is characterized by change, sequence, and coherence, and these are the qualities usually foregrounded in narratives" (Hughes 1995: 4). Change, or our awareness of time through recognition of change, is crucial to historical narratives, and so to archaeology. With other modes of temporal consciousness, as in the case of memorial and mythic forms, repetition is key to knowing, learning from, and living with "the past." Thus continuity and the unchanging nature of time are the aspects of temporality that are emphasized.

Even though the tendency is to think in terms of binaries and oppositions, as in linear versus cyclical or mythic versus historic time, numerous ethnographic and analytical studies indicate that people are perfectly capable of, and quite often do, hold multiple conceptions of time and temporality simultaneously and that the coexistence of such may well be the norm rather than the exception (e.g., Birth 2008; Farriss 1995; Howe 1981; Krech 2006; Sammells 2011; Wineberg 2001). One way we might approach seemingly disparate discourses of temporality is to view them as sources of knowledge rather than as representations of nature or reality. In other words, different engagements with and explanations of, for instance, duration and change may be seen as serving different political, cultural, or academic purposes rather than as making distinct ontological claims (see Hughes 1995).

Adopting this type of pragmatic approach to time and temporality would signal a move toward greater inclusivity on the part of archaeologists. While it doesn't necessarily entail a rejection of chronological time, it does require a leaving off from the totalizing move of slotting local histories into universalizing narratives of cultural evolution, foregoing the single-minded pursuit of culture change in our reconstructions of the past, and according greater theoretical significance to archaeological evidence for cultural continuity.

Taking seriously the issue of variable temporalities is important for the future of archaeology—not only for enriching interpretations of the archaeological record but also because an exclusive focus on chronological time does not address the interests of all potential stakeholders. Here I am thinking specifically of the Native peoples whose "pasts" we often study. While I would not argue that there is any one essentialized, indigenous temporality or singular view of the past, I would suggest that the ways in which Native people think about time does not exactly match the vision generally held by archaeologists.

This can be readily seen in the canonical writings of any number of indigenous authors and literary figures, including Vine Deloria (1994, 1995), Black Elk (Black Elk and Neihardt 2008), John Lame Deer (Lame Deer and Erdoes 2004), and Leslie Marmon Silko (1981), among others. In many of these works, Native notions of time and temporality are presented as fundamentally distinct from Western views, with "Indian time" being characterized as cyclical, mythic, or communal in nature as opposed to chronologically ordered or individually internalized. In other documents, the ubiquitous use of the phrase "since time immemorial" seen in the treaties and constitutions of North American tribes, as well as heard in oral histories, further underscores the distinctive temporal register within which Native peoples operated and operate. That this phrase retains its significance can be seen, for instance, in the various declarations of indigenous peoples contained in recent key documents pertaining to the creation of the Plurinational State of Bolivia (Ministry of Foreign Affairs 1995–2010: 125, 189, 199, etc.). For Native peoples, the notion of "since time immemorial" stands out as carrying particular cultural, spiritual, and temporal weight, while chronology and precise dates clearly do not.

Further concerns with chronological time have been articulated by indigenous intellectuals in South America. This can be seen, for example, in the discussions and documents surrounding the notion of "Vivir Bien / Buen Vivir" (e.g., "living well") developed from Native understandings of how to live in harmony with Pachamama and currently being promoted at the governmental level in both Ecuador and Bolivia (Ávila Navajas 2001; Ministry of Foreign Relations, 1995–2010; Mutuberría Lazarini and Chiroque Solano 2011; SENPLADES 2013). As explicitly stated in an open-source document prepared by the Ministry of Foreign Relations of the Plurinational State of Bolivia (2009: 9), indigenous peoples recognize as one of seven key crises presently facing the world a "crisis of time." This is described as a brutal clash between "the global time of industrial production, cyberspace, and telecommunications" and "the time of life" brought about by the collision of the "cyclical time of nature and the linear time of history and the clock."[2] In an earlier iteration of this document, the authors state that "in the face of the mechanistic and three-dimensional West, which explains history as a lineal process pre-established by Divine will, El Vivir Bien is based on a total and multidimensional vision of history that recognizes that time-space is latent and multidirectional, that we exist simultaneously in diverse universes (multiverses), and that everything is simultaneously related"[3] (Ministry of Foreign Affairs 1995–2010: 123–24). Such statements make clear that indigenous peoples of the Western Hemisphere think differently about time as well as the fact that the temporal is also political.

The benefits of making room for alternative temporalities in archaeological interpretation with respect to building collaborative relations with Native peoples can already be seen in some quarters. One recent example is that of On Your Knees Cave in Alaska, where the 10,000 year old remains of a young male were discovered in the mid-1990s (Dixon, Worl, and Swenson 2006). The positive relations that evolved out of this find between the Tlingit community, the US Forest Service, and the Denver Museum are celebrated in an hour-long video called *Kuwóot yas.éin / His Spirit Is Looking Out from the Cave* produced by Sealaska Heritage Institute and released in 2007. The story here, as told from the Native perspective, emphasizes the significance of this discovery for the scientific corroboration of Tlingit claims to the continuity of their lifeways in this region "since time immemorial." This telling in no way detracts from or hinders the archaeologist's story of the specific historic event of this individual's death, Paleo-Indian adaptations, or the coastal migration theory.

In South America, various examples of the divergent temporal concerns of indigenous peoples versus those of archaeologists are illustrated in a recent edited volume (Gnecco and Ayala 2011). In one example from the northern Andes, Colombian archaeologist Cristóbal Gnecco outlines a case involving the Nasa peoples of the southwestern region of Tierradentro and the relationship they developed with a specific archaeological site in the process of being relocated to a new territory (also Gnecco and Hernández 2008). As with other indigenous peoples of this region, the Nasa had never previously expressed interest in any archaeological remains, which Gnecco suggests was a consequence of the colonial project of "othering" (Gnecco and Hernández 2008). The incorporation of the local archaeological remains into a new narrative about Nasa ancestral roots in the newly occupied region deviates considerably from archaeological accounts of the site's significance. Nonetheless, collaborative efforts between the Nasa and local archaeologists have achieved positive social outcomes and arguably enriched understandings of local history even in the face of such divergent views of, and approaches to, the material remains (Gnecco and Hernández 2008).

To suggest how accommodating multiple timescales and alternative temporalities might also enhance archaeology from the scientific side, and to add one more case in point to those already offered in the present volume, I conclude with a brief example from Ecuador. A particular pottery type known variously as Panzaleo and Cosanga-Pillaro (see figure 9.1) is widespread throughout the northern Ecuadorian highlands (Bray 1995, 2003). Although never comprising more than 10 percent of the assemblage of any given site in the region, it is nonetheless remarkable for its ubiquity. Compositional analyses demonstrate that this ware was likely manufactured on the eastern Amazonian slopes and foothills of the Andean range and thus represented an import into the Northern Highlands (Bray 1995). Contextual data indicate that it was often a mortuary element of highland burials (Bray 1995). One of the most intriguing aspects of this pottery is its seeming lack of change across two millennia (Porras 1975: 146–48). As evidenced archaeologically, vessels from dated contexts covering a nearly 2,000-year span display little evidence of morphological, decorative, compositional, or associative change (Bray 1995).

Rather than dismissing this seemingly "radical continuity" as uninteresting from an archaeological standpoint mainly concerned with culture change, I suggest that problematizing the issue of stability vis-à-vis change in other categories of material culture and other regional trajectories has the potential to define new research horizons and offer new insights into the rhythms of social life. The persistence and retention of specific materials, styles, forms, and types do not just happen but rather require active cultural intention and intervention (Lightfoot 2001). In this regard, continuity is as much a problem to be explained as change, and even more so is the articulation between these two particular modes of temporality.

By starting from an assumption of temporal heterogeneity rather than universal uniformity, we have the potential to expand our vision of the archaeological record. Recognizing that chronology is not the only form of reckoning time opens new possibilities for exploring variable temporalities, multiple temporal scales, and different modes of temporal connection beyond sequence (e.g., iteration, recursivity, feedback, and so on). In recalibrating our temporal interests, we stand to gain new insights into neglected aspects of the past (e.g., the "silences" discussed by Trouillot [1995] and Sayre [chapter 2], this volume). By becoming more attuned to temporal complexity, it is also possible to envision a future for archaeology that would speak more to the interests of contemporary Native peoples whose histories we tend to study.

NOTES

1. An important exception to this general observation is the influential French Annales school of the mid-twentieth century, whose members emphasized long-term

processes, historical continuities, and the variable duration of social phenomena in their analyses (see Braudel 1972).

2. "La crisis del tiempo, donde el tiempo global de la producción industrial, el ciberespacio y las telecomunicaciones chocan brutalmente contra el tiempo de la vida, ocasionando una tremenda colisión de tiempos entre el tiempo cíclico de la naturaleza y el tiempo lineal de la historia, el tiempo de reloj" (Ministry of Foreign Affairs, Plurinational State of Bolivia 2009: 9).

3. "Frente a la visión mecanicista y tridimensional occidental, que explica la historia como un proceso lineal, pre-establecido (por voluntad divina), el Vivir Bien se basa en una visión total y multidimensional, que coexistimos simultáneamente en diversos universos (multiversos) y que todo está relacionado simultáneamente" (Ministry of Foreign Affairs 1995–2010: 123–24).

REFERENCES

Adams, Barbara. 1990. *Time and Social Theory*. Philadelphia: Temple University Press.

Ávila Navajas, Diego. 2001. "Presentación." In *Suma Qamaña: La comprensión indígena de la vida buena*. La Paz: Comunicación PADEP/GTZ.

Bender, John, and David Wellbery, eds. 1991. *Chronotypes: The Construction of Time*. Stanford, CA: Stanford University Press.

Birth, Kevin. 2008. "The Creation of Coevalness and Danger of Homochronism." *Journal of the Royal Anthropological Institute* 14 (1): 3–20. https://doi.org/10.1111/j.1467-9655.2007.00475.x.

Black Elk, and John Neihardt. [1932] 2008. *Black Elk Speaks: Being the Life Story of a Holy Man of the Oglala Sioux*. Albany: SUNY Press.

Bloch, Maurice. 1977. "The Past and the Present in the Present." *Man* 12 (2): 278–92. https://doi.org/10.2307/2800799.

Braudel, Fernand. 1972. *The Mediterranean and the Mediterranean World in the Age of Phillip II*. London: Collins.

Bray, Tamara. 1995. "The Panzaleo Puzzle: Non-Local Pottery in Northern Highland Ecuador." *Journal of Field Archaeology* 22 (2): 137–48.

Bray, Tamara. 2003. *Los efectos del imperio incaico en la Frontera Septentrional: Una investigación arqueológica*. Quito: Abya-Yala Press.

Bray, Tamara. 2005. "Multi-Ethnic Settlement and Interregional Exchange in Pimampiro, Ecuador." *Journal of Field Archaeology* 30 (2): 119–41. https://doi.org/10.1179/009346905791072369.

Choquehuanca, David. 2009. "Sumaq Kawsay-Buen Vivir." In *Vivir bien, mensajes y documentos sobre el Vivir Bien*, 1995–2010. Produced by El Ministerio de Relaciones Exteriores del Estado Plurinacional de Bolivia, 75–92. http://www.planificacion.gob.bo/uploads/Vivir_bien.pdf.

Cook, Anita G. 2004. "Wari Art and Society." In *Andean Archaeology*, ed. Helaine Silverman, 146–66. Oxford: Blackwell.

Deloria, Vine, Jr. 1994. *God Is Red: A Native View of Religion*. Golden, CO: North American Press.

Deloria, Vine, Jr. 1995. *Red Earth, White Lies: Native Americans and the Myth of Scientific Fact*. New York: Scribner.

Dixon, James, Rosita Worl, and Greg Swenson. 2006. "Discovery of Ancient Human Remains Sparks Partnership Documentary." http://www.colorado.edu/today/releases/2006/09/19/discovery-ancient-human-remains-sparks-partnership-documentary.

Echo-Hawk, Roger. 1997. "Forging a New Ancient History for Native America." In *Native Americans and Archaeologists*, ed. Nina Swidler, Kurt Dongoske, Roger Anyon, and Alan Downer, 88–102. Walnut Creek, CA: Altamira Press.

Echo-Hawk, Roger. 2000. "Ancient History in the New World: Integrating Oral Traditions and the Archaeological Record in Deep Time." *American Antiquity* 65 (2): 267–90. https://doi.org/10.2307/2694059.

Fabian, Johannes. 1983. *Time and the Other: How Anthropology Makes Its Object*. New York: Columbia University Press.

Fagan, Brian. 1972. *In the Beginning: An Introduction to Archaeology*. Boston: Little Brown.

Fagan, Brian, and Nadia Durrani. 2016. *In the Beginning: An Introduction to Archaeology*. 13th ed. London: Routledge.

Farriss, Nancy M. 1995. "Remembering the Future, Anticipating the Past: History, Time, and Cosmology among the Maya of Yucatan." In *Time: Histories and Ethnologies*, ed. Diane Owen Hughes and Thomas R. Trautmann, 107–38. Ann Arbor: University of Michigan Press.

Foucault, Michel. 1970. *The Order of Things: An Archaeology of the Human Sciences*. London: Tavistock.

Gell, Alfred. 1992. *The Anthropology of Time: Cultural Constructions of Temporal Maps and Images*. Oxford: Berg.

Gnecco, Cristóbal, and Patricia Ayala, eds. 2011. *Indigenous People and Archaeology in Latin America*. Walnut Creek, CA: Left Coast Press.

Gnecco, Cristóbal, and Carolina Hernández. 2008. "History and Its Discontents: Stone Statues, Native Histories and Archaeology." *Current Anthropology* 49 (3): 439–66. https://doi.org/10.1086/588497.

Harding, Jan. 2005. "Rethinking the Great Divide: Long-Term Structural History and the Temporality of Event." *Norwegian Archaeological Review* 38 (2): 88–101. https://doi.org/10.1080/00293650510032707.

Howe, Leopold E. A. 1981. "The Social Determination of Knowledge: Maurice Bloch and Balinese Time." *Man* 16 (2): 220–34. https://doi.org/10.2307/2801396.

Hughes, Diane O. 1995. "Introduction." In *Time: Histories and Ethnologies*, ed. Diane Hughes and Thomas Trautmann, 1–18. Ann Arbor: University of Michigan Press.

Krech, Shepard, III. 2006. "Bringing Linear Time Back In." *Ethnohistory (Columbus, Ohio)* 53 (3): 567–93. https://doi.org/10.1215/00141801-2006-005.

Lame Deer, John (Fire), and Richard Erdoes. 2004. *Lame Deer, Seeker of Visions*. New York: Scribner.

Leone, Mark. 1978. "Time in American Archaeology." In *Social Archaeology: Beyond Subsistence and Dating*, ed. Charles Redman, Mary Jane Berman, Edward V. Curtin, William T. Langhorne Jr., Nina M. Versaggi, and Jeffery C. Wanser, 25–36. New York: Academic Press.

Lightfoot, Kent. 2001. "Traditions as Cultural Production: Implications for Contemporary Archaeological Research." In *The Archaeology of Traditions: Agency and History before and after Columbus*, ed. Timothy Pauketat, 237–52. Gainesville: University Press of Florida.

Lucas, Gavin. 2005. *The Archaeology of Time*. London: Routledge.

Lucas, Gavin. 2015. "Archaeology and Contemporaneity." *Archaeological Dialogues* 22 (1): 1–15. https://doi.org/10.1017/S1380203815000021.

Mamani Condori, Carlos. 1996. "History and Prehistory in Bolivia: What about the Indians?" In *Contemporary Archaeology in Theory: A Reader*, ed. Robert Preucel and Ian Hodder, 632–45. Oxford: Blackwell.

Ministry of Foreign Relations. Plurinational State of Bolivia 1995–2010. *Vivir Bien*. http://www.bivica.org/upload/diplomacia-vida_tres.pdf.

Ministry of Foreign Relations. Plurinational State of Bolivia. 2009. *El vivir bien como respuesta a la crisis global*. http://www.cancilleria.gob.bo/webmre/sites/default/files/libros/03%20el%20vivir%20bien%20como%20respuesta%20a%20la%20crisis%20global.pdf.

Murray, Tim. 1999. *Time and Archaeology*. London: Routledge. https://doi.org/10.4324/9780203203019.

Mutuberría Lazarini, Valeria, and Henry Chiroque Solano. 2011. "Perspectivas y aportes desde la economía comunitaria como alternativa para el desarrollo de 'otra economía' en Latinoamérica." *La Revista del CCC 11, January/April*. http://www.centrocultural.coop/revista/articulo/211/perspectivas_y_aportes_desde_la_economia_comunitaria_como_alternativa_para_el_desarrollo_de_%93otra_economia%94_en_latinoamerica.html.

Oland, Maxine, Siobhan Hart, and Liam Frink, eds. 2012. *Decolonizing Indigenous Histories: Exploring Prehistoric/Colonial Transitions in Archaeology*. Tucson: University of Arizona Press.

Olivier, Laurent. 2011. *The Dark Abyss of Time: Archaeology and Memory*. Lanham, MD: Altamira.

Oxford English Dictionary. June 2017. *Oxford English Dictionary Online*. Oxford University Press. http://www.oed.com/view/Entry/32604?redirectedFrom=chronology. Accessed August 9, 2017.

Porras, Pedro. 1975. *Fase cosanga: Estudios científicos sobre el oriente ecuatoriana*. Quito: Centro de Publicaciones de la Pontificia Universidad Católica del Ecuador.

Rice, Don S. 1993. "The Making of Latin American Horizons: An Introduction to the Volume." In *Latin American Horizons*, ed. Don S. Rice, 1–14. Washington, DC: Dumbarton Oaks.

Ricoeur, Paul. 1988. *Time and Narrative*. Vol. 3. Chicago: University of Chicago Press.

Rowe, John H. 1962. "Stages and Periods in Archaeological Interpretation." *Southwestern Journal of Anthropology* 18 (1): 40–54. https://doi.org/10.1086/soutjanth.18.1.3629122.

Sammells, Clare. 2011. "The Aymara Year Count: Calendrical Translations in Tiwanaku, Bolivia." *Ethnology* 50 (3): 245–58.

SENPLADES. 2013. *Plan Nacional de Desarrollo / Plan Nacional para el Buen Vivir, 2013–2017*. Quito: Secretaría Nacional de Planificación y Desarrollo. http://documentos.senplades.gob.ec/Plan%20Nacional%20Buen%20Vivir%202013-2017.pdf.

Shanks, Michael, and Christopher Tilley. 1987. *Social Theory and Archaeology*. Cambridge: Cambridge University Press.

Silko, Leslie Marmon. 1981. *Storyteller*. New York: Henry Holt.

Silliman, Stephen. 2005. "Culture Contact or Colonialism? Challenges in the Archaeology of Native North America." *American Antiquity* 70 (1): 55–74. https://doi.org/10.2307/40035268.

Silverman, Helaine. 2004. "Introduction: Space and time in the Central Andes." In *Andean Archaeology*, ed. Helaine Silverman, 1–15. Oxford: Blackwell. https://doi.org/10.1016/B978-0-08-051337-9.50006-7.

Smith, Claire, and H. Martin Wobst, eds. 2005. *Indigenous Archaeologies: Decolonizing Theory and Practice*. London: Routledge.

Stahl, Ann. 2012. "When Does History Begin? Material Continuity and Change in West Africa." In *Decolonizing Indigenous Histories: Exploring Prehistoric/Colonial Transitions in Archaeology*, ed. Maxine Oland, Siobhan Hart, and Liam Frink, 158–77. Tucson: University of Arizona Press.

Tanaka, Stephan. 2016. "History without Chronology." *Public Culture* 28 (1): 161–86. https://doi.org/10.1215/08992363-3325064.

Thompson, E. P. 1967. "Time, Work-Discipline and Industrial Capitalism." *Past and Present* 38 (1): 56–97. https://doi.org/10.1093/past/38.1.56.

Trouillot, Michel-Rolph. 1995. *Silencing the Past: Power and the Production History*. Boston: Beacon.

White, Hayden. 2002. "The Westernization of World History." In *Western Historical Thinking: An Intercultural Debate*, ed. Jörn Rüsen, 111–18. New York: Berghahn.

Wineberg, Sam. 2001. *Historical Thinking and Other Unnatural Acts*. Philadelphia: Temple University Press.

Zerubavel, Eviatar. 2003. *Time Maps: Collective Memory and the Social Shape of the Past*. Chicago: University of Chicago Press. https://doi.org/10.7208/chicago/9780226924908.001.0001.

Contributors

TAMARA L. BRAY, Department of Anthropology, Wayne State University, Detroit, MI

ZACHARY J. CHASE, Department of Anthropology, Brigham Young University, Provo, UT

MARÍA JOSÉ CULQUICHICÓN-VENEGAS, Universidad Nacional de Trujillo, Trujillo, Peru

TERENCE D'ALTROY, Department of Anthropology, Columbia University, New York, NY

GILES SPENCE MORROW, Department of Anthropology, University of Toronto, Toronto, ON

ANDREW P. RODDICK, Department of Anthropology, McMaster University, Hamilton, ON

MATTHEW SAYRE, Department of Anthropology and Sociology, University of South Dakota, Vermillion, SD

FRANCISCO SEOANE, Universidad Nacional de Trujillo, Facultad de Ciencias Sociales, Programa de Arqueología, Trujillo, Peru

EDWARD SWENSON, Department of Anthropology, University of Toronto, Toronto, ON

DARRYL WILKINSON, McDonald Institute for Archaeological Research, Cambridge University, Cambridge, UK

Index